About Method

About Method

Experimenters, Snake Venom, and the History of Writing Scientifically

JUTTA SCHICKORE

THE UNIVERSITY OF CHICAGO PRESS   CHICAGO AND LONDON

The University of Chicago Press, Chicago 60637
The University of Chicago Press, Ltd., London
© 2017 by The University of Chicago
Published 2017
Paperback edition 2020
Printed in the United States of America

29 28 27 26 25 24 23 22 21 20   1 2 3 4 5

ISBN-13: 978-0-226-44998-2 (cloth)
ISBN-13: 978-0-226-75989-0 (paper)
ISBN-13: 978-0-226-45004-9 (e-book)
DOI: https://doi.org/10.7208/chicago/9780226450049.001.0001

Library of Congress Cataloging-in-Publication Data

Names: Schickore, Jutta. author.
Title: About method : experimenters, snake venom, and the history of writing
    scientifically / Jutta Schickore.
Description: Chicago ; London : The University of Chicago Press, 2017. | Includes
    bibliographical references and index.
Identifiers: LCCN 2016033873 | ISBN 9780226449982 (cloth : alk. paper) | ISBN
    9780226450049 (e-book)
Subjects: LCSH: Science—Methodology. | Science—Experiments.
Classification: LCC Q174.8 .S333 2017 | DDC 507.2/4—dc23
LC record available at https://lccn.loc.gov/2016033873

♾ This paper meets the requirements of ANSI/NISO Z39.48–1992 (Permanence of Paper).

Contents

"A matter so obscure, so difficult, and likewise so new . . ."

Toward the end of his two-volume *Treatise on the Venom of the Viper*, published in 1781, the Tuscan naturalist Felice Fontana declared: "I have made more than 6000 experiments; I have had more than 4000 animals bit; I have employed upwards of 3000 vipers and may have been deceived; some essential circumstance may have escaped me: I may have neglected some other, not thinking it necessary; my consequences may have been too general, my experiments too few in number. In a word, I may very easily have been mistaken, and it would be almost impossible that I should never have been so in a matter so difficult, so obscure, and likewise so new."[1] When I first encountered Fontana's treatise, it appeared to me as a striking document for many reasons: its peculiar theme, its proportions (more than 700 pages!), the meticulousness with which Fontana described his endeavors, his detailed ideas about proper experimental procedure (and that he carefully laid them out for the reader), his self-deprecating tone, and, of course, the sheer number of experiments that he had performed, which strike the modern reader as rather excessive.

But were Fontana's project and his approach to experimentation really so unusual? I initially thought this question would be easy to answer considering the extensive literature that has resulted from the turn to experiment in history and philosophy of science. I quickly realized that even though we have substantial bodies of literature on scientific methods, experimental practices, scientific rhetoric, and experimental reports, analyses of scientists' conceptions of proper experimental practice are few and far between. Putting Fontana's methodology of experimentation in historical perspective turned out to be much harder than I had expected.

Philosophy of science, for instance, tells us much about scientific methods. For the most part, however, philosophical investigations of scientific methods concern the construction and test of hypotheses or models and the rules and principles of reasoning. Those philosophical analyses that do deal with strategies for assessing experimental outcomes often probe one particular strategy, such as making sure that theories of instruments and theories under test are independent from one another. Even those sets of criteria for the assessment of experimental outcomes that have been provided did not offer what I was looking for. We have a good analytic grasp of epistemological criteria and strategies for the validation of experimental results,[2] but I wanted to get a sense of how these sets of criteria were interpreted by working scientists and how they developed over time.[3]

The historical literature on experimentation in the sciences has grown vast, but the rich accounts of past experimental projects and approaches that have been produced since the 1980s focus on research materials, laboratory infrastructure and instrumentation, techniques of visualization, and so forth. Scientists' methodological views are rarely discussed, however.[4] Although a great many specific issues have been studied in depth, these bodies of scholarship are largely disconnected from one another, and specific historiographical frameworks are tied to specific periods. There is substantive scholarship on each of a small group of innovators in scientific method, including Bacon, Descartes, Newton perhaps, Herschel, Whewell, Mill, and Duhem. We know quite a bit about experimental reports and the "new method" of experimentation in early modern physical science, both in England and in Italy, as well as about exploratory experimentation in eighteenth- and early nineteenth-century research on electricity. We know about experimental practices in Italian, French, and English medicine and anatomy. We also know a great deal about the rise of physiological experimentation and instrumentation in the nineteenth century and the fierce debates within the medical community (mostly during the nineteenth century) about whether "scientific medicine" would lead to more effective therapies than more traditional clinical approaches had thus far.[5] Yet we lack an overarching framework for bringing these analyses together. My motivation for putting together a long-term history of scientists' methodologies is, in part, to connect the islands of knowledge in current scholarship. My overall goal is to understand in more detail how researchers in the life sciences conceptualized their experimental practice, what rules for proper experimental procedure they endorsed, and how these notions changed from the late seventeenth century to the mid-twentieth century.

For a number of reasons, venom research is uniquely suitable for the long-term history of experimentation and the development of methodological thought in the life sciences. First, because experiments with poisonous snakes and venom were tricky, taxing, and often controversial, the experimenters felt compelled to produce quite detailed and graphic accounts of their trials (in every sense of the term). They described their own troubles, discussed and justified their own methods of experimentation, and critiqued other experimenters. Of course, significant developments in venom research and innovations in the methodology of experimentation did not necessarily happen simultaneously, but because venom researchers were generally very explicit about their methods and methodological concerns, an account of how these discussions unfolded can serve as an instructive frame of reference for future study.

I found, however, a dearth of books on venom research. A number of historical works focus on the rich cultural history of poisons and on the cultural imagery of serpents, and some historians have examined specific medical topics such as theriac or snake stones.[6] Moreover, a few herpetologists have written popular books on snakes and snake venom poisoning.[7] But no book-length study specifically traces the history of venom experiments. This, too, is surprising, for venom research has been significant for many fields in the life sciences—among others, therapeutics, pathology, physiology, bacteriology, and immunology. Over time, the aims of venom research have included—besides finding antidotes—clarifying key chemical concepts such as "fermentation," elucidating nerve functions, understanding the circulation of blood and pathological changes of blood, explaining cardiovascular functions, and illuminating immune reactions and the structure of biological building blocks. Even though the topic "venom research" might seem narrow and exotic at first, it can tell us a great deal about more general issues in the history of the life sciences: Venom research was shaped by, and in turn informed, the concepts of life, disease, and body functions that were available to venom researchers. Those concepts also influenced views about how to make experiments reliable and instructive. If, for instance, a living body is regarded as extraordinarily complex, establishing the exact effects of an intervention in the working of a living body will seem next to impossible for an investigator. As this book will show, late nineteenth-century investigators developed specific methodological strategies to address precisely this problem.

The key episodes in venom research were closely linked—another reason why this topic is so suitable for a long-term history of methodologies of experimentation. For more than 250 years, venom research was imbued with a strong sense of tradition both in terms of techniques and results and in terms of the methodology of experimentation. Indeed, as late as 1962, toxicologists praised the eighteenth-century scholar Fontana for having been the first to use "adequate controls" in his venom experiments.[8] Fontana, in turn, announced his own work as "new" but at the same time presented it as a continuation of an endeavor that had begun in the mid-seventeenth century. Venom researchers saw their works as contributions to an ongoing endeavor, engaging with and explicitly building on the work of their predecessors.

Of course, as analysts of conceptual developments in the sciences, we cannot simply take at face value the historical actors' sense of tradition, their "origin stories," and their narratives of steady progress. This book seeks to account for both continuity and change in venom researchers' endeavors to analyze the nature of venom and to find out what venom does to the human and animal body. Similarly, it accounts for both continuity and change in venom researchers' views about how experiments should be performed, how results should be checked and secured, what standards of precision should be applied, and how—and exactly what—one can learn from experiments. The tradition of venom studies that venom researchers themselves have construed and continued to evoke can be only one aspect of a multifaceted analysis. If we turn our attention to the institutional and intellectual contexts for venom research, the protocols that were applied, or the material resources the researchers had available, we find that venom researchers' projects owed much more to contemporaneous practices than those researchers' talk of tradition would lead us to expect.

My focus on snake venom research is a compromise of sorts. The topic of the history of methodologies of experimentation in the life sciences is dauntingly broad—much too broad for one book—so I must be selective. Perhaps my most controversial choice (at least to historians of early modern science) will be to dissociate throughout the book the study of snake venom from that of poisons in general. Prior to the late eighteenth century, snake venom research was an integral part of more general research into how living bodies were affected by small amounts of certain substances. My choice of focus helps limit a potentially overwhelming mass of materials so as to produce a book of tolerable length. Because venom research is situated at the intersections of so many different fields in the life sciences, methodological discussions pertaining to experiments with venom are indicative of methodological concerns in those other fields as well. My selection of examples makes it possible to produce a narrative that is thematically coherent yet cross-disciplinary in perspective. The focus on venom can thus be viewed as a strategic choice that allows the bringing together of these different subfields.

There is, of course, one considerable challenge both to the writer and to readers of this book. Many of the experiments that are described in the sources are extremely cruel and will strike readers as impermissible.[9] For the most part, however, I have decided not to place the ethical aspects of this research center stage in my analysis: neither the—admittedly sparse—comments by past experimenters nor contemporary concerns about ethics of animal experimentation.[10]

Methods Discourse

The turn to experiments that history and philosophy of science took in the 1980s was a turn away from the analysis of scientific ideas and the dynamics of theory change. This book takes another turn, shifting the attention from materials, practices, and tools back to scientific texts: their content and their organization. But this is not a shift back to the development of scientific concepts, models, and theories. Instead, the book examines specifically how questions about proper experimental procedure were dealt with in scientific texts.

Historians have long warned that the study of scientific publications alone cannot give us an adequate picture of the activities involved in scientific research:[11] Experimental reports are not a reliable source of information about what researchers really do in the laboratory. They are not transparent windows on the day-to-day research practice in which they originated. This is undoubtedly correct. Indeed, the same caveat applies to methodological thought: scientists often do not explicitly express their methodological convictions, and if they do, the expressed convictions might not capture the methodological rules and strategies that were actually applied and pursued in the project described in the report.

Arguably, however, methods-related concepts, statements, and reflections as they are presented in experimental reports are significant because they reflect the authors' understanding of the structure and organization of good experimental research. This is what makes these reports valuable resources for the history of methodology. If an approach, a methodological strategy, or a methodological concept is explicitly appealed to, one may assume that it is endorsed. It if is explicitly defended, discussed, or justified, we can at the very least state that it was of concern for the author or authors of the report. Methodological concepts, statements, and reflections are important indicators of scientists' views about proper experimental procedure in this sense.

Sociologists of science often read scientists' writings as a means with which to persuade relevant audiences (within the scientific community and beyond) of the value of the claims that are made in those writings.[12] Publications are treated as artifacts of "literary technologies" through which the outcomes of events and activities in the laboratory are articulated and brought into publishable form.[13] A number of scholars, most notably Steven Shapin, Simon Schaffer, Larry Holmes, and Paula Findlen, have argued that if we pay attention to the social organization of the community of experimenters and its place in society, we will become aware that several features of experimental reports are attempts to appeal to others within and outside the community. These features include the reference to witnesses (Shapin and Schaffer), the amount of procedural detail that is related in an experimental report and the proportion of narrative and argument (Holmes), and the narrative style of the writing (Findlen).[14] Scientists' methodological views may also become part and parcel of literary strategies to situate one's work within a community of, say, Newtonians, Baconians, or Popperians.[15]

Persuasion and positioning are certainly important goals of text production (as of any form of communication, scientific or otherwise).[16] But even if there is general agreement on this point, other questions regarding the content, organization, and historical development of methodological views are still wide open: What methodological concepts and strategies were referred to, described, and discussed in different periods? What challenges were perceived as being the most severe? How was "proper [experimental] procedure" established in experimental reports?

General recourse to social factors does not fully explain the content, organization, and history of methodological discussions concerning experimentation. This book shows, among other things, how the development of methodological views about experimentation resonates with the history of broader ideas about life, body, and body functions. As we will see in more detail, these notions affected the content of scientists' methodological conceptions. Even the very practice of experimenting might generate novel methodological thought.

Moreover, experimental reports are shaped by certain historically changing conventions of textual organization. In his studies of the development of scientific writing, Charles Bazerman once warned that we should not overestimate the freedom that experimenter–authors have to bend rules and take advantage of local resources as they see fit for their purposes. Bazerman thus urged that we must recognize the role of "accepted genre and style" in defining current work.[17] Heeding Bazerman's admonition about the formative and constraining role of available literature for methods discourse, the following chapters consider books on scientific methodology, manuals on writing scientifically, and emerging editorial guidelines, along with the experimental reports themselves. Even though we cannot assume that scientists actually followed these guidelines, these writings on scientific writing are telling, indicating the kinds of constraints and rules that confronted scientists at a given time.

In the twentieth and twenty-first centuries, experimental reports are highly standardized and typically follow the same scheme—sometimes called the IMRD (introduction, methods, results, discussion) scheme—and the methods sections are typically completely devoid of details and of justifications of techniques and procedures. Sociologist Karin Knorr-Cetina described the methods sections in these reports as highly decontextualized: "full of brand names of instruments, lists of materials, and descriptions of procedures tied together by nothing but sequence"; methods become "a catalogue of sequential manipulations stripped of both context and rationale."[18] It is still not clear, however, when and how these methods sections came into being. Of course, even early modern researchers told their readers—often at great length—how they had proceeded. In their writings, accounts of methods were integral parts of the narrations of experiments. But as late as 1950, the IMRD scheme was not always strictly and consistently applied.

Prior to the mid-twentieth century, accounts of methods were often rich in detail; descriptions of procedures were usually tied together by narratives of experimental procedures, and the rationale for using specific techniques and approaches was discussed as well. For my analysis, I thus prefer the broader term *methods discourse* to *methods sections*. I use the term *discourse* in a nontechnical sense to refer to all kinds of methods-related statements in scientific writing about experimentation, including explicit commitments to experimentalism, descriptions of protocols, explanations of methodological concepts, and justifications of strategies of experimentation.[19]

Precisely because scientists are often silent about methodological views and commitments, the long-term perspective is especially important for the history of methods discourse. If we find that a particular procedure or methodological conception is described and justified in a textual source, we can conclude that the procedure or conception is novel, unusual, or perhaps contested; hence, the methods-related statements and discussions that are provided in scientific writings become all the more significant.[20] Because extended coverage of methodological issues indicates what is not yet, or what is no longer, taken for granted, putting methods accounts and methodologies in historical context and perspective is crucial. Studying the history of experimental reports with an eye to innovations in their structure, the emergence of new concepts, and changes in how arguments were framed, evidence presented, methods described, and methodological approaches justified and defended can give the analyst a sense of what later reports tacitly presuppose.

Methods discourse is multifaceted. It comprises the description of actual, localized research techniques and experimental designs, more general experimental strategies and criteria to assess proper procedure, and broader, even more general commitments, such as the commitment to experimentation as the principal means of knowledge acquisition. This book examines methods discourse on all three of these levels, albeit with an emphasis on "middle-term" methodological concepts and strategies.[21] Are experimental procedures and techniques described in detail, or are they reduced to a list of brand names and acronyms? What methodological strategies and criteria are mentioned? Are they discussed—or even defended? Are methods-related topics presented in separate sections, or are they integrated in an overall narrative? Questions like these guide my analysis of experimental reports.

Actors' Terms and Reconstructions

This book is, in a sense, an extended commentary on Fontana's statement quoted at the beginning of this introduction—an attempt to situate his rather astounding methodological pronouncement in historical perspective. The attempt to make sense of a striking methodological statement by an eighteenth-century scholar has resulted in a book that traces the entwined histories of venom research, of scientists' methodological views about experimentation, and of the organization of methods discourse in experimental reports. The geographical and chronological span of this account is vast. My story ranges over continental Europe, Britain, and the Americas and spans three centuries. Obviously, since the early modern period, the practices and tools of biomedical experimentation have as profoundly changed as have the conceptions of life, disease, poison, and the body. Also, the institutional contexts of research in different countries were sometimes radically different even within the same time period. Many more episodes could have been included in this book. Each episode I selected for inclusion could be probed more deeply and could be presented in much richer detail.

What insights can be expected of such a selective long-term history? Will it not simply show that with the advancement of experimentation over the centuries, the methodology of experimentation progressed as well—that methods and methodologies became more sophisticated, more rigorous, and more complex? In a sense, this book does tell a story about progress: scientists are "getting better" at experimenting, and their methodologies are becoming more complex and more sophisticated over time. But, more important, this book discusses in what sense specifically experimentation and the methodology of experimentation advanced, as well as what factors contributed to the development of experimental approaches and methodologies. There is a surprising twist to the story of progress, because, as we will see, methodological advancement really means developing strategies for managing and perhaps overcoming ever new difficulties of biological and medical experimentation.

My own analytic approach sits somewhere in between the history and the philosophy of science. I offer a history, but historians of science might find my analysis mainly driven by current philosophical issues. After all, I analyze the historical development of methodological concepts and strategies such as variation, replication, independent confirmation, and so forth, which are usually discussed in philosophy papers.

Philosophers of science, conversely, might find my approach mainly historical, both because I do not subject the methodological arguments and concepts of past scientists to formal analysis and because I do not measure the distance between past methodological notions and ideal epistemic conditions of successful confirmation and test. Moreover, several of the figures who populate my account were not luminaries, leading figures, or innovators of methodologies but merely representatives of their time. Their writings exemplify "normal methods discourse," as it were. In contrast to my approach, philosophers' analyses of experimental reasoning strategies often consider cases—generic situations, plausible scenarios that might or might not be inspired by actual scientific episodes and that can be used as conceptual tools to think through philosophical issues.

Changing the features of such scenarios demonstrates the epistemic significance of certain conditions, showing that such conditions must obtain for an argument to have epistemic force. Turning to actual scientific episodes can then show the philosopher whether, and by how much, an actual argument falls short of the epistemic ideal.

Because my analysis of past methods discourse draws on discussions in current philosophy of science, I make occasional reference to "ideal" epistemic situations. This is not to measure past scientists' methodologies against standards that were not theirs, nor to suggest that their methodologies were somehow lacking. To me, current philosophical terminology and philosophical reconstructions are helpful for the purpose of characterizing past scientists' concepts and strategies. But in my account, philosophically informed reconstructions are complemented by an analysis of the changing meaning and use of actors' terms and conceptions.

Generally speaking, the historical study of methodologies of experimentation requires a combination of reconstructions of experimental designs and strategies in current analytic terms and sensitivity to actors' terms. If we try to ground our histories of experimentation just in reconstructions of experimental designs and strategies based on interpretations of experimental reports, experimental practices now and then would often appear quite similar. It would be easy to find "forerunners" of, say, John Stuart Mill's "method of difference" or modern control experiments in seventeenth-century experimental practice. Only if we bring our reconstructions of experimental designs together with a careful study of the experimenters' own conceptualizations of their experimental practices do we get a sense of their own ideas about what aspects of their practices were indicative of proper procedure—and do we realize, for instance, that the method of difference did not play such a key role in seventeenth-century experimentation as it did in the mid-nineteenth century. Both actors' terms and interpretative reconstructions of experimental designs and methodologies in present-day terminology are important, and together they tell us something about the epistemic force that past scientists attached to specific strategies of experimentation—and thus something about the historical development of methods discourse.

CHAPTER ONE

Argument, Narrative, and Methods Discourse

One very common and very general programmatic commitment can be found in almost all accounts of experiments from the early modern period: the commitment to experimentalism or to the "new experimental method." Of course, systematic manipulations of natural objects to gain knowledge of nature were performed long before the seventeenth century—in alchemy and in medicine, for example.[1] But in the early seventeenth century, when new instruments such as the telescope and the microscope, along with the extension of trade and travel, had revealed new and strange facts, natural and experimental philosophers—most notably Francis Bacon—took it upon themselves to order and systematize the new knowledge that had been gained through experiments and observations, as well as the methods through which such knowledge was gained. In the second half of the seventeenth century, scholars began to pursue experiments on a larger scale in the context of new organizations, including the Florentine Accademia del Cimento, the Royal Society of London, and, a little later, the Paris Académie des Sciences.

The early modern scholars presented their experimental projects as "new"—as a "new organon" for the investigation of nature, a new challenge to the authorities, a new concern with matters of fact. Those investigators who were attached to the budding societies and organizations committed themselves expressly and in print to the new experimentalism. For instance the authors of the *Saggi di naturali esperienze fatte nell' Accademia del Cimento* of 1667, the Cimento's sole and collaborative official publication, declared that the institution's "only task" was "to make experiments and to tell about them."[2]

The "New Experimental Method" in Tuscany,
London, and Paris

The "new experimental method" had many supporters at the Tuscan court, among the Royal Society, and at the early French Académie des Sciences and other related French institutions. Founded in 1657 to invigorate Tuscan science and to develop and promote research in the spirit of Galileo, the Cimento Academy provided an institutional setting for some of the most notable Italian natural and experimental philosophers. Its aristocratic patrons encouraged experimental research, often requesting that specific topics be examined, and provided resources to aid these investigations. Although the academicians complied with these requests, they often took Galileo as their authority and model for matters of experimentation.

The men of the Royal Society were committed to Francis Bacon, of course. Bacon's emphasis on the importance of experiment, method of induction, and warnings against the idols of the mind became key reference points of the New Philosophy of the Royal Society, which stipulated the importance of performing experiments—and performing them oneself rather than relying entirely on the opinions of authorities, an ideal encapsulated in the succinct motto of the Royal Society: *Nullius in verba.*

Like the members of the Tuscan court, the French sovereign expected by promoting science to magnify the ruling monarch and acquire benefits for the French people. Like London's natural and experimental philosophers, the *sçavants* at the French Academy initially endorsed a Baconian framework for their pursuits: they warned of prejudices, highlighted the practical utility of the advancement of knowledge, and encouraged the collective use of instruments in public demonstrations and experiments. Jean Chapelain, one of the organizers of the academy, evoked the Baconian spirit of the enterprise, writing that one of the academy's principal goals was "to banish all prejudices from science, basing everything on experiments, to find in them something certain, to dismiss all chimeras and to open an easy path to truth for those who will continue this practice."[3]

All this is well known. It is, however, less appreciated (at least outside the community of early modernists) that the Italian, French, and English investigators held varying views about what exactly the commitment to experimentalism involved. Early modern scholars in different institutional contexts expressed and explicated their commitments in subtly different ways; they made their appeals to different leading experimentalists, and their opinions about the merit and value of older philosophical approaches to the study of nature diverged quite substantially.

Methodological Statements

The early members of the nascent Royal Society had one clear favorite: Francis Bacon. The concrete philosophical commitments of the academicians at the Cimento, however, were rather diverse and encompassed scholastic as well as mechanist notions.[4] The Proem of the *Saggi di naturali esperienze* illustrates how the academicians combined a commitment to experience and, ultimately, to Truth (capital T), with a commitment to the aristocratic authorities. The academicians humbly dedicated their work to the most serene Highness Ferdinando II, "who contributed so much with the power of his most happy favors to such new and stupendous discoveries and to the opening of an untrodden path to the more exact investigation of truth."[5]

In the *Saggi di naturali esperienze*, the expressed commitment to experimentalism is combined with (and subordinate to) deductive, geometrical reasoning. According to the authors of the *Saggi*, the best means for the investigation of truth was, in fact, geometry; and only when geometry abandoned the natural philosophers should they turn to experiment: "As one may take a heap of loose and unset jewels and seek to put them back one after another into their setting, so experiment, fitting effects to causes and causes to effects—though it may not succeed at the first throw, like Geometry—performs enough so that by trial and error it sometimes succeeds in hitting the target." Experiments can give us knowledge about causes and effects—but the academicians were by no means overconfident in their evaluation of the knowledge that could be gained by experiment. In the end, experiments might help in getting out of Plato's cave, yet experimenters must proceed "with great caution lest too much faith in experiment should deceive us, since before it shows us manifest truth, after lifting the first veils of more evident falsehood, it always makes visible certain misleading appearances that seem to be true."[6]

The Italian academicians signaled to their patrons that they were developing and promoting research in the spirit of Galileo. But in Tuscany, the call for experimentation did not include a clean break with older scholarly traditions. Even the motto of the Cimento Academy—*provando e riprovando* (checking and checking again)—suggests a rather intricate relation to past scholarship. The expression is taken from Dante's *Divina Commedia*, and the original context in Dante's canto is the discussion of a procedure for examining a celestial phenomenon. Some historians have thus suggested that the motto was chosen to highlight a break with the past and the rise of a new kind of experimental science in which matters of fact take prevalence over speculation. The Proem of the *Saggi di naturali esperienze* is very explicit about the necessity of reviewing older experiments, reconsidering the claims of ancient authorities, and "verify[ing] the value of their assertions by wiser and more exact experiments."[7] Historians of Tuscan science have described this concern with critiquing previous authorities as part of a "civil conversation" but also as the beginning of what Jay Tribby has called a "codification of impatience" with ancient authorities.[8]

At the same time, the choice of a line from Dante clearly signals an appreciation for Tuscan literary heritage.[9] The reference itself, however, is ambiguous. In the relevant passage (*Paradiso*, canto II), Beatrice responds to Dante's query about the nature of the moon spots by showing him that both his offered explanations are wrong. One explanation she refutes by reasoning, the other (that moon spots are caused by the reflection of light from the moon's uneven surface) by what appears to be an empirical demonstration involving three mirrors. At first glance it seems plausible to assume that this is an endorsement of experimental tests for theoretical hypotheses, but in the original context, Beatrice merely asks Dante to consider an arrangement of mirrors: no experiment is performed. Moreover, Beatrice finally offers a metaphysical explanation of the moon spots that does not rely on experience.

Dante's line as it appears in the *Saggi di naturali esperienze* involves a subtle and creative reinterpretation of the original text. What is more, the academicians gave the expression a new twist: the writers of the *Saggi* called for performing experiments again and again. This reference to toil and perseverance is emphasized in the Proem of the work but is not implied in the original canto.

In the Royal Society, by contrast, experimentalism was more forcefully defended than in the Italian context, and the society's members distanced themselves more explicitly from past scholarship. True to the Baconian spirit, Royal Society members advocated experimentalism as the preferred alternative to reasoning from first principles. The members of the Royal Society—represented by their spokesman, Thomas Sprat—were quite iconoclastic, declaring that the new experimental method might well make the old system of natural philosophy "fall to the ground." But,

Sprat demanded, "What can we lose, but only some few *definitions*, and idle *questions*, and empty *disputations*? Of which I may say as one did of *Metaphors*, *Poterimus vivere sine illis*. Perhaps there will be no more use of Twenty, or Thirty obscure Terms, such as *Matter*, and *Form*, *Privation*, *Entelichia*, and the like. But to supply their want, and [*sic*] infinit varity of *Inventions*, *Motions*, and *Operations*, will succeed in the place or words."[10] Sprat's statement is much more uncompromising than the cautious note in the *Saggi di naturali esperienze*. Still, the commitment to experimentalism is very general: it is programmatic and might appear quite provocative, as Sprat's pronouncements illustrate, but it does not offer much guidance beyond the notion that experiments are an important means of obtaining knowledge about nature. To guide the experimenters' actions, this general commitment needs to be supplemented with more concrete instructions about how to be a good experimentalist: How should experiments be performed? What is the proper procedure? What are the pitfalls, and how might they be avoided?[11]

On the one hand, specific projects require specific arrangements and designs. To determine the spring of the air, one must know how to build an air pump, empty it, seal it effectively, and so forth. To perform blood transfusions, one must know how to constrain an animal, collect blood from it (including how much and with what instrument), and inject that blood into another.[12] On the other hand, some views about proper procedure are both more specific than a general commitment to experimentation and at the same time more general than the concrete procedures and arrangements for a particular experimental endeavor. Consider, for instance, Galileo's famous statement that he had repeated "a full hundred times" his experiments with the inclined plane as described in the *Dialogues Concerning Two New Sciences*.[13] This statement is more concrete than a general commitment to experimentation. It tells us not only that an experiment was performed but also that the experiment was repeated many times—perhaps not exactly "a full hundred times" but definitely more than once or twice. At the same time, the statement is not bound to the specific context of experiments on falling bodies. Other experiments might also be repeated a hundred times.

These types of statements are methodological statements: They do not inform the reader of specific techniques and procedures that are particular to a concrete investigative task. Instead, they tell the reader something about how experimental findings should be obtained and secured. Methodological statements thus tell us something about what experimentalism means for an individual experimenter or for a group of experimenters.

Such statements encapsulate what counts as proper procedure. They are a "contact zone" between localized research techniques and approaches and broader commitments such as programmatic agendas. They transcend individual and localized experimental contexts. Notions about proper procedure often have a longer life than the methods, instruments, and techniques for tackling specific experimental questions, yet their content might change even as the general commitment to experimentalism prevails. This book examines statements such as these, and their history.

Boyle on Methods of Experimentation

Sometimes methodological statements play only a minor role in accounts of experiments; at other times, they are very conspicuous. Take Robert Boyle, for example: He was perhaps the most prominent "Baconian" experimentalist at the time of the formation of the Royal Society.[14] He conducted plenty of trials, many of them for the Royal Society, and wrote numerous accounts of his experiments. Boyle repeatedly emphasized that only experimentation, not familiarity with the beliefs of authorities, was the road to truth, adding that it was important to broaden the empirical foundation of knowledge by collecting more experimental results and that knowledge about nature could be attained through careful and critical assessment of the outcomes of observations and experiments.

At the same time, Boyle was very much concerned with the messiness of actual experimental practice and supplemented the Baconian conception of the new scientific method with some practical guidelines for fact-gathering. Bacon had a different goal. His commitment to experiment and observation as it is advanced in the *Novum Organum* was a means to an end—namely, to eschew the old Aristotelian and medieval scholastic philosophy, according to which certain knowledge about nature was derived from first principles. Bacon's *Novum Organum* is first of all a book on methods of reasoning from experiments. The critical part—Book I—discards traditional logic. Book II, the constructive part, offers a systematic account of the "New Scientific Method."

The critical part of Bacon's project aimed at uncovering potential impediments to the advancement of the new science. The famous typology of "idols of the mind," the obstacles to epistemic progress, is a part of this project. The typology comprises four different sources of false ideas, including falsehoods arising through general psychological tendencies, individual prejudices and expectations, the vagueness of language, and previously taught but erroneous worldviews or doctrines (such as the Aristotelian system and traditional logic). This typology of sources of falsehoods presents a systematic overview of the things that might obstruct the generation of human knowledge in all contexts—and especially, of course, when knowledge is generated using the "old" method of natural philosophy.

The typology of idols of the mind focuses almost entirely on those cognitive factors that present obstacles to scientific advancement because they lead to the formation of false ideas. The book outlines and defends a systematic method of reasoning from facts to "natures." Bacon's typology of idols of the mind (and indeed the entire project of the *Novum Organum*) is a very optimistic account of the scientific method. The expressed aim is to become acquainted with sources of error in order to avoid or remove false beliefs: "forewarned is forearmed" appears to be the general motto. The typology as Bacon drew it up reminds the reader that falsehoods could have diverse origins: They could originate in the general way in which all humans think and judge, in the human makeup (i.e., in human sense organs), in personal idiosyncrasies, or in the subtleties of language. Yet they all spring from the activity of the human mind, and they can all be identified and eradicated.

Book II, the constructive part, outlines the proper procedures for the systematic interpretation of nature—that is, for the ordering of facts in tables of presence, absence, and degrees and for proper induction from them. In it, Bacon described the methods for processing results of systematic experiments and observations to discover the "form natures" of things. The challenges of collecting those facts that are to be included in the tables—*viz.*, the concrete challenges of observation and experimental research—are occasionally mentioned, but they are not systematically discussed.[15]

Boyle turned his attention from the activities of the mind to the particulars of experimentation. He was much less confident than Bacon that the challenges of fact-gathering could be completely overcome. Boyle had little to say about error as such—he took for granted that an error was a false belief about nature. Such a false belief could come into being if one did not conduct experiments. The point is that if one did conduct experiments, it was by no means guaranteed that truth could be attained. Experiments could be unsuccessful or false, and indeed Boyle pointed to specific cases in which "learned Writers" had formed false beliefs on the basis of "contingent Experiments."[16] Bacon's project was optimistic; Boyle's was apologetic. Sometimes experiments do not succeed and thus might lead the experimenter into error. Overall, however, we are better off making experiments than not making experiments.

The difference between a more broadly "Baconian" commitment to experimentalism and the explicit concern with various practicalities of experimentation is most clearly visible in some of Boyle's early writings, such as the "Two Essays, Concerning the Unsuccessfulness of Experiments" of 1661, written a few years before the formation of the Royal Society. The "Two Essays" were geared toward the active experimenter, and particularly toward those experimenters dealing with chemicals. Together with an essay on niter and one on the qualities of fluidity and firmness, they form a series: "Certain Physiological Essays" (1661).[17] The series begins with a "Proemial Essay," and it is in this introductory text that Boyle invoked Baconian ideas, especially the view that experiments and observations help uncover false opinions. In the "Two Essays," however, he emphasized failed, "unsucceeding" experiments, including experiments that fail to produce a certain (previously obtained) effect, experiments whose outcomes are tainted by unrecognized interfering factors, and the like.

Boyle expressed the hope that an experimenter who was aware of potential troubles ahead would be better prepared to deal with them, which readily calls to mind Bacon's motivation for putting together a list of idols. But Boyle's concrete project was quite different. He attempted to identify the factors responsible for failure (in that an experiment did not produce the expected effect), which could in turn lead to false beliefs in the misinformed or ignorant experimenter. In Bacon's typology of idols, the impediments to scientific progress are cognitive (the idols are, after all, idols of the mind). For Boyle, too, falsehood-producing factors might be cognitive—he referred to the investigator's fancy and imagination, for example. But more often than not, other factors made the experiment unsuccessful: bad instruments, the "unskilfulness of the Tryers of the Experiments," and especially "the particular or mistaken properties of the Materials imploy'd about them."[18] What Boyle had in mind was, above all, the impurity of the chemical substances that were used for experiments. If experimenters unwittingly used chemicals that were contaminated or even intentionally adulterated, their experiments would not succeed.

The second of the "Two Essays" casts an even wider net and discusses more generally "the Contingencies to which Experiments are obnoxious upon the account of Circumstances, which either are constantly unobvious, or at least are scarce discernable till the Tryal be past."[19] In each field, experimental practice was prone to circumstances or contingencies. For Boyle, these circumstances, like accidents for Aristotle, "can hardly be discours'd of in an accurate Method, (which their nature will scarce admit of)."[20] A systematic treatment of contingencies being impossible, he resigned himself to examples. The discussion of circumstances and contingencies ranged over anatomy, therapeutics, and, again, chemistry, as well as over gardening and glass-blowing—which, for Boyle, were experimental contexts, too. Gardeners and anatomists must reckon with the variability of plants and dissected bodies, physicians must deal with variations in their patients' constitutions, and everyone must be aware of "some wantonness or other deviation of Nature"—or, as the first essay put it "the effects of an unfriendliness in Nature or Fortune to your particular attempts, as proceed but from a secret contingency incident to some experiments, by whomsoever they be tryed."[21]

In short, experiments were unsuccessful for all sorts of reasons. And yet there were certain measures one could take to avert total failure. Boyle insisted that the experimenter "try those Experiments very carefully, and more than once, upon which you mean to build considerable Superstructures either theoretical or practical, and to think it unsafe to rely too much upon single Experiments, especially when you have to deal in Minerals."[22] A cautious experimenter who performed experimental trials repeatedly had good reason to be confident. Most experiments did succeed, and the gain from successful experiments was so great that it was worth the risk of failing.[23] Thus experimenters should not feel discouraged; after all, neither would a merchant quit after having lost one vessel.

I have already characterized Galileo's statement about repeating experiments with the inclined plane as a "methodological" statement, and we have just seen that Boyle, too, referred to repetitions. Indeed, in his essay on "unsucceeding experiments," he referred to repetitions as a methodological imperative, pointing out the danger of relying overmuch on single experiments while he recommended—indeed, demanded—that an experiment be performed more than once.

What should we make of such an imperative? Galileo's commitment to repetitions has been read as an "Aristotelian" formula through which Galileo assured his reader of the uniformity and hence the reliability of the experiments described. The formula evoked the Aristotelian notion of experience as "knowing how things usually behave." This technique of assuring readers of the authenticity of an experience and of experiential conviction was frequently used by scholastic, mostly Jesuit writers in the early seventeenth century.[24] Even if this explains Galileo's methodological statement, however, Boyle's reference to repetitions was not formulaic, and the idea that the members of the Royal Society were Aristotelians after all is perhaps a little far-fetched. It appears more likely that Boyle's concern with repetitions really did spring from his and his fellows' encounter with all kinds of contingencies in experimentation.

Boyle's early essays on the "unsuccessfulness" of experiments are methodological essays: They are concerned with the practicalities of experimentation and with proper procedure, and in this sense they are more specific than the programmatic commitment to experimentalism—yet at the same time they transcend the context of particular experimental projects. The methodological essays deal with problems that all experimenters face—nature's contingencies, impure materials, and inept workers (or their own limited skills)—as well as with quite general measures for addressing them: repetitions, cautious distrust, and carefulness.

In his essays on airs, air pumps, bodies, and chemicals, too, Boyle had much to say about proper procedure—both about the project at hand and about experimental trials more generally. Steven Shapin and Simon Schaffer's classic analysis of Boyle's so-called literary technologies draws attention to significant features of Boyle's writings, including his prolix descriptions of successful and failed experiments, his many references to witnesses, and his display of "gentlemanly" virtues such as modesty and self-deprecation.[25]

These methods-related statements are quite diverse—there are descriptions of instruments and procedures specific to particular projects, more general references to proper procedure (the employment of witnesses), and equally general references to the persona of the experimenter (modesty and self-deprecation). Shapin and Schaffer's main point was that proper experimentation in the context of the Royal Society required, above all, properly virtuous experimenters.

Here I shift my attention to the statements concerning proper procedure, examining them in greater detail. To do so, I take a look at Boyle's account of another set of experiments more relevant to the main theme of this book: his experiments with viper poison. For Boyle, these experiments were just one small part of his work on poisons, which in itself constituted only a fraction of his entire experimental work. Nevertheless, the account is an instructive example of how Boyle incorporated methodological notions in his (early) writings on experimentation—and is also a prelude to the controversy that unfolded across the English Channel only a few years later.

Boyle's report on experiments with poisons comes in two versions—a very early (and unpublished) longer manuscript and an abbreviated account. The latter forms part of a general essay on the usefulness of experimental natural philosophy to the mind (in the study of nature) and to the body and fortune (in the improvements of medical treatment, gardening, husbandry, and the like). Boyle's early essays combined Helmontian chemical philosophy with a commitment to the "new" experimental philosophy.

When it was published, the short section on poisons became part of a general apology for animal experiments. Just like other experiments, animal experiments could be used as a tool to correct the false beliefs of the ancients as well as the false beliefs of physicians who were not experimentalists and would merely trust authorities.[26] In the context of this essay, Boyle did not dwell on contingencies, impure substances, or any other causes for failure. On the contrary, he stressed that active intervention—including the dissection of an animal to observe the changes that the intervention had produced—was much more informative than passive observation and thus could arbitrate pathological disputes.[27]

Experiments with poisons were of special interest for pathology. Poisoning formed a specific class of disease—namely, diseases brought about by an element external to the body, not by an "internal distemper" of the patient. Animal experiments could reproduce those diseases, because the disease-producing element could be introduced at will into the animal body. To illustrate this point, Boyle envisioned a series of experiments in which poison was administered in various ways: "not only by giving Beasts poisons at the mouth, but also by making external applications of them especially in those parts where the Vessels that convey Blood more approach the surface of the Body, and also by dexterously wounding determinate Veins with Instruments dipt in Poysons (especially moist or liquid ones) that being carried by the circulated Blood to the Heart and the Head." The outcomes of the trials, Boyle hoped, could establish "whether their strength be that way more uninfringed, and their operation more speedy (or otherwise differing) then [sic] if they were taken at the mouth."[28]

Considering that Boyle's experimental writings on airs, pumps, and chemical substances were full of circumstantial descriptions and have been treated as paradigmatic prolix reports, it is striking how few details Boyle provided in his descriptions of the experiments with poison. The one experiment that Boyle did perform himself was significant not so much for the manner in which it was performed but because the results he obtained did not agree with the beliefs of older authorities (including Aristotle). This experiment examined the question of whether certain body parts of vipers were poisonous, as Aristotle had assumed. Boyle reported matter-of-factly that he had fed to a dog various parts of a viper that traditional sources held to be poisonous: the head, the gall, and the tail. He had fed another viper's head and gall to a different dog; both dogs survived.

He also offered information about other people's findings, again without going into too much detail. An "inquisitive friend" of Boyle's assured him that he had set out to perform a similar trial, having given a dog "a dozen heads and galls of vipers, without finding them to produce in him any mischievous symptom." Moreover, "the old man . . . that makes viper wine, dos it (as himself tels me) by leaving the whole vipers, if they be not very great, perhaps for some months, without taking out the gall, or separating any other part form them in the wine, till it have dissolved of them as much as it can."[29] Based on these reports, Boyle inferred that established views about the poisonous parts of the viper must be wrong. He suggested an alternative interpretation of the findings as they presented themselves to him. Because the dogs had survived their meals, no poison could have been contained in them.

And yet vipers brought about death. Accordingly, Boyle concluded that viper venom could not be a material substance. Instead he assumed "that the venom of vipers consists chiefly in the rage and fury, wherewith they bite, and not in any part of their body, which hath at all times a mortal property."[30] He compared this process to the bite of a mad dog, whose teeth were not poisonous until the dog went mad.

Boyle drew this assumption from Andrea Bacci's treatise *De venenis et antidotes prolegomena* of 1586 but did not develop this view, nor did he try to find further experimental evidence for it. Indeed, the notion that rage and fury were involved in causing disease was a key element in the theory of disease proposed by Belgian chemist Jean Baptiste van Helmont. According to van Helmont, diseases were caused by an indwelling vital principle, the Archeus—which, if incensed, whether by a troubling idea or by harmful material substances, produced disturbance and corruption in the body.[31]

For Boyle, the experiments on vipers were only one project among many others, certainly not as essential to his experimental philosophy as were his experiments with airs and pumps. Elsewhere, however, snake venom research was both a topic of considerable medical interest and a topic of fierce controversy, as we will see. Both the Aristotelian question of whether body parts of snakes were poisonous and the Helmontian view of disease became driving forces for the controversy that unfolded between the Tuscan naturalist Francesco Redi and his French critic, the apothecary Moyse Charas.

As these two experimenters criticized and attacked each other, they were compelled to make their experimental methods and procedures explicit.

Layers of Methods Discourse

Experiments need to be communicated if they are to be of consequence for the scientific community. Even though many early modern experiments were performed in public—at meetings of the Royal Society, at the Tuscan Court, at the Paris Academy—it was important to let faraway scholars know of their results. Reports of experiments were put on paper and circulated. The early modern scholars used quite different textual genres for that purpose: They produced book-length treatises and collections of accounts of experiments such as the *Saggi*. They wrote letters addressed to other scholars as well as to their patrons, which were often widely circulated, printed, and published in novel scientific journals such as the *Transactions of the Royal Society* and the *Journal des Sçavants*. These two publications also carried reviews and reports of various experimental projects and publications.

Such accounts of experiments announce what experiments were performed and what was found and often present their reader with additional discussions of the significance and implications of the findings. Boyle, in fact, turned this distinction between work performed and results obtained into a guideline for writing reports. In the preface to *New Experiments Physico-Mechanical touching the Spring of the Air*, he distinguished between "narratives of experiments" (detailing what had been done) and "discourses made upon experiments" (interpreting what the findings meant). This distinction had normative implications for Boyle, who insisted that a strict distinction between the elements of narrative and the elements of discourse be maintained in the account of an experiment—a goal that Boyle himself did not always reach in his own writings but that encapsulated the Royal Society's preoccupation with "matters of fact."[32]

To capture the dual function of relating what had happened in an experiment and conveying the significance of the events, historian Larry Holmes proposed distinguishing between "narrative" and "argument" in scientific writings about experiments.[33] The narrative element comprises descriptions of what experiments were performed, of experimental set-ups and procedures, of specific circumstances, and of the results obtained. The argument deploys evidence from observations and experiments to advance and support a claim to knowledge.

Holmes stated: "This duality of literary form reflects the dual function of the research paper as both a summary of the author's current *findings* and an account of investigative work that she or he has carried out during a previous period of time."[34] Drawing on an analysis of research papers published by eighteenth-century chemists, Holmes showed that different research papers might vary in the proportion of narrative and argument (or "story" and "presentation of findings and their significance"), not least depending on the social contexts in which the work was performed. In the tightly knit community of full-time researchers at the French Academy, authors of experimental reports did not have to establish authority, and they did not have to establish whether the experiment had actually been performed and how.[35]

Narrative elements include narrations of concrete events, vivid descriptions of scenarios, and records of the feelings and emotions of the experimenters. Narratives often (but by no means always) say something about the consecutive steps of the investigation. Elements of argument, by contrast, comprise statements of findings as well as of those findings' significance and of the implications and consequences that might follow from them. The argument's structure and presentation tell modern readers something about the mode of reasoning that the authors endorsed, but not necessarily the order in which the steps of the research were actually performed. The argument might be laid out in a deductive fashion, with general statements expounded first and findings then presented to support those ideas. The argument might also be laid out inductively, with several matters of fact presented first and more general conclusions derived from them. For Bacon, it was expressly a part of the new method that facts be collected first, then ordered in tables, then reasoned about. But we will see that some of his contemporaries chose modes of presentation that did not reflect the Baconian schema exactly.

The categories of narrative and argument are helpful analytic tools with which to compare accounts of experiments across contexts and, as we will see hereafter, across historical periods. At the same time, Boyle's own accounts of experiments show that some relevant distinctive features and elements of writings on experimentation do not neatly fall into either of the categories "argument" and "narrative." His writings contain, as we have seen, very general endorsements of experimentalism. These endorsements are neither narratives of work carried out nor discussions of the significance of particular findings. Those more specific statements invoking methodological criteria and ideas—statements about repetitions, witnesses, and the care taken with the investigative project—also escape the distinction, as do the passages vindicating those methodological ideas, such as references to contingencies, variability, and the "wantonness" of nature.

Methodological discussions and imperatives in experimental writings are more than just particulars of specific events; they express, reinforce, and sometimes challenge dominant methodological ideas about good experimental practice. Such statements provide additional support for the overall argument expounded in the experimental report, because they tell their readers something about how the experiments were done and by what means the experimental outcomes were secured. They assure readers of the validity and significance of experimental outcomes.

Of course, it would be a mistake to take accounts of methodological strategies at face value, reading them as faithful representations of strategies that the researchers actually followed. Nevertheless, descriptions and discussions of methodological issues are valuable resources for the history of methodology, because they reflect the experimenter–authors' understanding of the structure and organization of good experimental research. My reading of the early modern texts about experiments suggests a threefold distinction of textual elements: Accounts of experiments combine elements of narrative and of argument as well as methods-related elements. The methods-related parts are also complex, with several layers distinguishable. First are the descriptions of the design of an experiment or observation. I am thinking not so much of the very specific recipes for performing individual experiments in local contexts. Rather, I am thinking of procedural methods or series of steps such as the steps involved in performing a color indicator test or a test of spontaneous generation. Hereafter I will use the term *protocol* to refer to this procedural information. It is apt, because we commonly associate *protocol* with an explicit, binding procedure. Such a protocol prescribes certain procedures for experimental investigations— just like diplomatic protocol prescribes certain procedures for visits of foreign state officials. The term is also reminiscent of *experimental protocol*, a term that is used in modern science to refer to the written procedure for carrying out an experiment.

Second are scientists' (or experimental philosophers') conceptualizations of procedures to secure empirical results, such as repetitions or precautionary measures for the identification of impurities. I call the conceptualizations of such procedures "methodological views" or "methodological statements" (or "strategies"). Boyle's example shows that these statements can be formulated as imperatives about what ought (or ought not) to be done.

Third, there are broader commitments to experimentation as the main road to knowledge about nature. Finally, there are different perspectives on methods and methodologies: explicit reflections on, justifications of, and defenses of protocols, of methodological views, and even of broader commitments—such as the commitment to experimentalism.

My umbrella term for all this is *methods discourse*. I use the term *discourse* in a wide, nontechnical (and, I should add, non-Foucauldian) sense. Unlike Boyle's "discourse made upon experiment," which refers to the interpretation of experimental findings, I intend my "methods discourse" to cover all layers and aspects of methods-related accounts—namely, explicit commitments to experimentalism, protocols, and everything in between: methodological statements, explanations of methodological concepts, methodological imperatives, and justifications of strategies of experimentation. Methods discourse covers both authoritative expressions and justifications of commitments to and straightforward descriptions of procedures.

The distinction among narrative, argument, and methods discourse and the distinction among the different layers of methods discourse are not clear-cut. Reflections on methodological views, for instance, might at times invoke broader commitments to experience, impartiality, or the like. Naming of witnesses or reporting of repeated trials can be regarded as part of the narration of the work done while telling the reader how experimental outcomes were secured—and thus both are part of methods discourse. Which distinction we make depends on our specific goals as analysts: Do we want to trace the history of certain experimental designs and practices— say, the role of tasting in chemical experiments? Or are we interested in the historical development of broader programmatic commitments, such as Baconianism or—yet broader—experimentalism? Nevertheless, the distinctions are valuable analytic tools for understanding methodologies of experimentation and their history. Distinguishing between narrative and argument helps prepare the ground for analysis of the patterns of reasoning by which experimenters interpreted their findings. Distinguishing among the different layers of methods-related accounts allows for a fine-grained analysis of historical developments, offering an analytic tool with which to capture continuities and discontinuities in scientists' conceptualizations of proper experimental procedures both across local contexts and across historical periods. For instance, all the early modern experimenters were committed to some form of experimentalism, even if a locally specific one. Moreover, in one form or another, commitments to experimentalism continued to be made at least until the late nineteenth century. Protocols, however—experimental designs and procedures—are often bound to specific contexts and goals. Methodological views, methodological reflections, and broader commitments are typically transcontextual, spanning longer periods and broader spaces.

We already have some indication of what factors might constitute the layers of methods discourse and what forces might drive changes in methodological concepts and reflections. We know, especially from early modern historians, that the organization of the scientific community and its status in society affects the methodological reflections that are incorporated in scientific texts because the experimenter–authors seek to appeal to those who will judge the worth and value of their work. Thus, for example, someone soliciting funding from courtiers might well seek approval of his findings by appealing to aristocratic witnesses.

In addition, Boyle's example suggests that methodological statements reflect, at least to an extent, the very experience of trying to manipulate variable bodies, elusive airs, and recalcitrant mechanical devices in an orderly fashion. His urgings that experiments be repeated can be understood as a result of the experimental philosophers' trials, conducted to find recognizable patterns and structures in animal guts and mixtures of primitive and simple elements.

The following chapters examine past reports of experiments with an eye to exposing and reconstructing the layers of methods discourse they incorporate. In a sense, these chapters examine what today we would call the methods section of a scientific paper, tracing its history and probing its structure.

CHAPTER TWO

Many, Many Experiments

Historians of Italian science have argued that the work carried out at the Accademia del Cimento was primarily a manifestation of courtly life, shaped by the directives of the Medici.[1] The Proem of the *Saggi di naturali esperienze* in fact expresses various general commitments of the members of the Cimento Academy: commitments to their patrons, commitments to experience, and commitments to the authorities of the "new science." The reports of experiments that were carried out in the context of the academy reflect these larger commitments, but the methods discourse is often significantly more elaborate than the general views expounded in the Proem of the *Saggi di naturali esperienze*. Francesco Redi's extended studies of viper venom illustrate how more concrete methodological views and conceptions of experimentation were integrated in experimental reports to establish proper procedure.[2]

Snakes were in many respects significant to savants in early modern Tuscany.[3] The scholars were well versed in a long and rich tradition of snake symbolism, and they were specifically interested in the role that snakes played in therapeutics. The meat of vipers was an essential ingredient of theriac, an ancient remedy for snakebite and various other health troubles.[4] Many investigations of snakes and snake venom were undertaken for the straightforward practical reasons of finding an antidote for calamities produced by snake bites and of exploiting the assumed medicinal function of viper meat. Experiments on snake venom were conducted within the larger contexts of discussions about body functions and the nature of disease; about blood, its role in the body, and its circulation; about nerve function and the theory of animal spirits; about chemical and mechanical philosophy; about iatrochemistry; and about the analogy between the actions of poisons and the actions of specific medicines. All these issues are reflected in the debates about the nature and working of the venom, possible treatments, and antidotes.

Refuting Old Fables through Many Experiments

Redi's extended study of viper venom began in the early 1660s. He mostly relied on animal experiments, but he also used some evidence from dissections of snakebite victims as well as from observations of the effects of snake venom on humans. Redi presented the report of his observations and experiments, the *Observations on Vipers* (1664), in a letter to the secretary of the Cimento Academy, Count Lorenzo Magalotti. The substantial letter comprises discussions about the nature and status of experimental evidence; a survey of a range of sources pertaining to snake bites and snake venom, beginning in antiquity; and a novel interpretation of the nature of snake venom together with reports of supporting experiments that Redi and his colleagues had performed. A second letter was published in 1670 as a response to criticism leveled by the renowned French apothecary Moyse Charas (although, as we will see later, it was not addressed to Charas).[5] The second letter, as well as the dispute about the nature and operation of viper venom that unfolded between Redi and Charas, will be considered in the next chapter.

Redi commenced his work at the request of the Grand Duke of Tuscany, Ferdinando II, who urged Redi to investigate the nature of the venom, including where it was stored and how it worked. For his experiments, Redi could make use of the numerous vipers that had been sent to the court from Naples for the production of theriac. The letter contained Redi's answer to the request. A substance existed that was responsible for the adverse effects of the bite—namely, the "yellow liquor" that was discharged from the viper's teeth. Taken by mouth, this substance was innocent. Put in wounds, it was fatal to humans and animals. The letter also provided an opportunity for Redi to outline some concrete ideas about how to perform experiments successfully.

Redi's letters have been read as a direct response to aristocratic interests and expectations at the Tuscan court.[6] Historians of early modern Italian science have suggested that the expectations of the members of the Tuscan court shaped the narratives of the experimenters' projects. The narratives had to present subjects worthy of the status of the patrons and had to make the study of nature "lavish, costly and entertaining."[7] Redi's account of the experiments with vipers was certainly engaging and thrilling, full of curious and gory details. Nothing could be farther from the dry, standardized, and schematic protocols we find in modern scientific journals. For instance, to establish whether the bile of the viper was a poison, as traditional sources had it, Redi deferred to the testimony of Jacopo Sozzi, the viper catcher. He gave a vivid description of Jacopo's performance. The man listened to the learned dispute about the effect of viper bile and was "just able to contain himself so as not to laugh" and then, "grinning, he took a viper's bile and diluting it with half a glass of fresh water, tossed it off with unflinching face; he gave to understand how mistaken the above-mentioned authors were."[8] Jacopo also drank half a glass of wine mixed with "all the liquor" and "all the foam and saliva that this excited, irritated, pressured, beaten serpent could shoot forth"; he drank this concoction "as if it had been so much pearly julep."[9]

Such stories are more than vignettes for the amusement and thrill of patrons, however. Redi's letter draws on these episodes to advocate experimentalism and to stake a claim, and the descriptions of the trials in his letter also indicate that the experiments were performed reliably. Paying attention to the different layers of methods discourse helps show how Redi's letter does all these things to good effect.

The authors of the *Saggi di naturali esperienze* attempted to strike a balance between commitments to the "new" experimental method and commitments to various authorities, patrons, and traditions old and new. Redi's letter employed the same technique. It begins with ideas and formulations taken from the *Saggi*. Like the *Saggi*, the letter is framed with a plea for experience as the ultimate arbiter of truth. It opens with Redi's affirmation that he found himself ever more "firm in my attention of not trusting the phenomena of nature if I do not see them with my own eyes." Like the authors of the *Saggi*, Redi stressed that experimental evidence outweighed the word of authorities, and he was also quite explicit about the need to check ancient authorities' views against experience. He assured his readers that he "loved" the ancient philosophers—Thales, Anaxagoras, Plato, Aristotle, Democritus, Epicurus, and all the other "princes of philosophy"— but insisted that not all they had written was true. Time and again Redi emphasized the importance of experiment, criticizing those who blindly followed the ancients, those who "put their hands over their eyes," and the inveterate fool who refused to look through Galileo's telescope so that he could continue to repudiate Galileo's findings.[10]

As a leading member of the Accademia della Crusca, Redi was at the same time appreciative of the Tuscan literary and cultural heritage. In his letter to Magalotti, he paraphrased Dante,[11] and he did so to claim that one should not follow authorities blindly. He noted that many writers, ancient and modern, were "like parrots, they write and read and they believe the most solemn untruths coming from the most credulous and inexperienced of common scribblers, those with the finest quality of sickening brilliance."[12]

In the mid-seventeenth century, numerous wondrous tales surrounding snakes and snake bites existed, such as the notion that if a serpent was put in the middle of a ring made from the foliage of betony, it would beat itself to death with its tail—or the notion that it was fatal to drink the wine from a bottle in which a viper had drowned. Previous and contemporaneous authorities had suggested that it was the viper's bile that caused harm, that certain body parts (such as the head) were poisonous, or that the harmful effects of the bite were caused by something immaterial.

Redi reviewed some of the older tales and beliefs about snakes, such as, for instance, in his discussion of the likely causes of Cleopatra's death, Jay Tribby's prime example of Redi's conversation with ancient authorities.[13] Redi, insisting that ancient beliefs must be critiqued in light of new evidence, offered accounts of his own experiments in response to the ancient fables. According to Tribby, the episode with the viper catcher who swallows bile in a public demonstration is exemplary for the confrontation between a "tradition of Great Texts by Great Authors"—the received opinions informing a group of onlookers—and the "bold, textually unmediated gesture" of the experimenter.[14]

It is correct that the spectators immediately questioned these findings, but we must look carefully at what exactly was at stake. The onlookers were skeptical because they questioned the experimental procedure. They likely did refer to a "great text" by a "great author," but that text includes accounts of experimental trials: Galen's treatise *On Theriac to Piso*. This work contains the description of a drug test strikingly similar to the experimental arrangement that Redi's letter presents. In the relevant passage of Galen's treatise, two trials are described. First, "poisonous beasts" are placed among wild cocks that had received theriac, and "those who have not drunk theriac die immediately, but those who have drunk it are strong and stay alive after being bitten." In the second trial, a purgative drug is given to someone who had been given theriac. The person is not purged, so evidently the theriac could protect effectively.[15]

Redi reported in his letter that his critics suspected trickery: The viper catcher might have taken theriac or some other potion that acted as an antidote. Even though the viper catcher denied the suspicion, the others requested that his testimony be backed up with another test. And it is this issue that the letter then addresses.

Redi told readers how he had established the reliability of procedure. Because the witnesses who were present were not entirely convinced by the demonstration that viper bile was harmless, the group performed animal experiments, feeding the bile to two large pigeons, a dog, and two cocks. Again, the letter offers a vivid description: A peacock and a turkey were fed two gallbladders, "and I had a cat eat four entrails without the gallbladder being removed; I can tell you that it greedily licked its lips."[16] All animals survived the meal without any ill effects.

There is more to these vignettes than just a thrilling story and a bold gesture. They also point toward concerns about whether experimental procedures are reliable and informative. These concerns are more concrete than the general commitment to experimentalism that we find in the *Saggi* and more concrete than the opening and closing statements in Redi's letter, which demand that experimenters see for themselves. In Redi's letters, the notion of repeating one's own experiments is quite conspicuous. Early on, Redi stated that he could not trust the phenomena "if they are not confirmed by iterated and reiterated experience."[17]

It has been noted that this declaration corresponds to the motto of the Cimento Academy, *provando e riprovando*.[18] This motto is ambiguous, however, because it does not specify what it is that needs to be checked and checked again—the beliefs of the authorities? One's peers' experiments? One's own experiments? Redi's letter stressed the repetitions of Redi's own trials.[19] He noted that many experiments had been done, often multiple times. The old fables were refuted "by experiments made many many times."[20] Redi had "observed [. . .] very often" that the yellow liquor spurted into the wound when the viper struck and killed.[21] The effects of bile and of the yellow fluid were tested on a number of different animals. And Redi had demonstrated the fatal effects of venom taken from dead vipers and put in wounds in more than a hundred experiments.

Redi put much emphasis on the issue of multiple experimental trials, but it is not entirely obvious how these statements should be read. Some historians have argued that we need to take them with a grain of salt, for they, too, are reflections of the social context in which the experiments were performed. Paula Findlen, for example, has suggested reading the references to many repetitions as a political gambit to display the power of the Tuscan court. She has argued that the emphasis on many repetitions in the academicians' writings underlined the wealth of resources the academy (and, by implication, the Tuscan court) had at its disposal.[22]

There is a famous passage from *Experiments on the Generation of Insects* in which Redi stated that he had opened 20,000 oak galls to check whether they had spiders and had found none. In this passage, Redi turned against an opponent, Pietro Andrea Mattiuoli, who had claimed that oak galls produced spiders.[23] The reference to 20,000 experiments—experiences—could plausibly be interpreted as a strategic move to impress and discourage his opponent. For Findlen, this abundance of objects was linked to truth and accuracy—Redi was obliged to settle the issues that the Grand Duke set before him "not by the powers of his own wit but by mustering the material and eloquential resources to his advantage."[24]

Another possible reading of references to multiple trials would be that Redi simply borrowed the formula from an acknowledged authority on experimentation. We already saw that Galileo used the reference to "a hundred repetitions" to assure readers of the uniformity, and hence the reliability, of his experiments with falling bodies.[25] Redi frequently used literary models: he rephrased in his letter some passages from the *Saggi di naturali esperienze*, and he also paraphrased Dante. He might perhaps have echoed Galileo's statement in other passages, thus investing his account with authority borrowed from the eminent scholar.

The most plausible interpretation of references to multiple trials is, however, that Redi invoked repetitions for different purposes in different contexts. There is the reference to 20,000 supporting observations, which would surely impress most scholars and discourage them from opposing Redi. There are references to "many many" experiments when Redi rejected the ancients and their fables about snakes. A more oblique case is the reference to "more than one hundred experiments in various animals" when Redi reported that the venom taken from dead vipers was fatal, too, even that taken from vipers "dead for two or three days."[26] As in the experiments with oak galls, the experiments with venom from dead vipers were, in fact, a refutation of the position held by a number of Redi's precursors and contemporaries (including van Helmont and Boyle)—that the mental or emotional state of the snake had something to do with the fatal effects. Dead snakes will not be angry.

Moreover, it is likely that Redi saw practical value in repetitions of experimental trials. At least sometimes, we should assume that the references to multiple trials described a genuine methodological strategy that he had actually followed. The literal interpretation of these references suggests itself if we consider that Redi offered explicit justifications for his methodological strategy of multiple trials.

The key experiments—the experiments with bile and the introduction of the yellow liquor in wounds—were performed on different kinds of animals and were repeated several times. In this context, Redi advanced two justifications for repetitions of trials. He reminded his reader that "as you know well, there are many things which behave as food for some kinds of animals, that in another species produce the effects of a venom, or other unusual or annoying mishaps."[27] What was poison for one might be food for another.[28] Performing experiments on different kinds of animals helped reveal these differences.

More important were repetitions of experiments on animals of the same kind. Redi expressed his astonishment about Marco Aurelio Severino,[29] "most versed in knowledge of the viper and greatly experienced," who had been convinced by "only two experiments" that the yellow liquor was not lethal when it was put into wounds.[30] He insisted that "often times it happens that these genuine causes, for some unknown or unseen hindrance, cannot produce their effects."[31] Because of this possibility, it was crucial to repeat experiments many times so as to erase the effects of any unseen hindrance. We are reminded of Boyle's concern with the "wantonness" of nature. Redi bolstered this statement with another quote from Dante, but he also described several instances in which he had experienced how his own efforts were thwarted by impediments—some known, some unknown. He had been present when "sheep, dogs, and cocks having been furiously bitten by vipers in the country at the height of the afternoon did not die";[32] he had seen a cockerel die after having been bitten by a viper whose teeth had been cut off and whose venom had been squirted out; and of the ducks and pigeons whose wounds had been smeared with venom, one had survived. Only by performing experiments repeatedly could one establish which findings were reliable.

The letter draws together narratives of experiments and methods discourse to produce an effective and forceful argument. Redi turned against traditional beliefs about snake venom and against older authorities, advancing an alternative. The letter tells us how he and his fellow experimenters had performed the trials; references to witnesses and repetitions showed that they had followed proper procedure, all while reinforcing the ideal of good experimental practice. Several experiments are described, testing the effects of bile and of the yellow liquor, and together these descriptions gently lead the reader away from the ancient beliefs about snakes: First the reader learns that viper bile was not poisonous. All experiments performed at the Cimento Academy suggested that bile was harmless when swallowed. Experiments in which bile was introduced into wounds of chicks, pigeons, a rabbit, a lamb, and a hare showed that the bile did not do any harm even when it came into contact with the blood of the victim. Then the reader learns what happens if the yellow fluid from the viper's teeth is taken by mouth: again, nothing. Finally, the reader learns that the yellow fluid has fatal effects when it comes into contact with blood. As a whole, the chain of experiments thus demonstrated that ancient beliefs about vipers and poisons were erroneous.

Getting the experiments and observations right, however, does not necessarily mean to be in a position to explain the findings. To those who were familiar with earlier debates, it must have been obvious that Redi's narrative did contain an oblique reference to and an implicit refutation of the Helmontian theory of disease, but Redi did not expressly turn against van Helmont. Even after many experiments, Redi did not provide what Boyle would have called a "discourse upon experiment": he did not attempt to explain how viper venom caused death. This was in keeping with the commitment to abstinence from theory that is expressed in the preface of the *Saggi di naturali esperienze*. Redi offered diverse possible explanations of the working of venom without committing himself to one. Venom might work by some unknown occult power, by cooling and freezing or heating and drying the heart, by dispersing and destroying its spirits, by depriving it of feeling—or perhaps by impeding the movement of the heart itself, making the blood congeal, or by making the blood coagulate in the veins.[33] Redi declared that he found himself unable to establish which of these explanations was the correct one (if any); and he did not suggest any further experiments that might help decide between the various possibilities.

Parisian Challenge

Five years after Redi communicated the results of his research to Magalotti, he was challenged by a Frenchman. In 1669, Moyse Charas published a book entitled *Nouvelles expériences sur la vipère* (in English translation *New Experiments* [!] *Upon Vipers*, 1670),[34] in which he cast doubt on Redi's investigations and offered an explanation of the working of viper venom.[35] Charas did not have any qualms with Redi's claim that vipers possessed glands that produced yellow liquor—in fact he reported having observed them himself and having shown them to "knowing Physitians"—but he believed that the yellow liquor was harmless.[36] He thought that the viper's anger made its bite fatal and that theriac was an effective antidote.[37]

In 1670, Charas had already performed a public preparation of theriac and had published a treatise on the panacea, which detailed the process of the preparation. The title page of his 1669 book identifies him as apothecary to the king's brother. In 1672, he earned a leading position as *sous-démonstrateur de chimie* at the Jardin du Roi (Jardin Royal des Plantes) in Paris.[38]

Charas's book begins with a description of the anatomy of vipers and of dissections of healthy snakes. It ends with a survey of remedies made with ingredients taken from vipers that were useful to therapeutics. In this part, no tests or experiments are described but only the preparations of the medicines, the manner in which they should be taken, and the effects they would have. For the exchanges between Charas and Redi, the middle part of the book is the most significant. It is entitled "Experiments Upon Vipérs" and comprises about 100 pages. The text describes experiments with animals—vipers biting dogs and small birds. Charas described in much greater detail than Redi had the alterations of the inner parts that could be found in dissections of the animals. Based on these experiments, Charas advanced an explanation of the nature of viper venom and explained why theriac worked against snakebite. Like Redi, Charas conducted experiments in the presence—and with the aid—of other scientific gentlemen.

Redi's letter primarily targeted the beliefs of the ancients; indeed, even though he ostentatiously abstained from interpretations, the letter was organized in such a way that the reader could not but dismiss the tales and errors of bygone times. Charas's book targeted Redi—or, rather, a "famous man," a "person very intelligent" whose views were "contrary" to his own.[39] It is much more explicitly an argument for (and against) certain positions, and although it does contain many descriptions of experiments, the accounts of the activities are far less entertaining and amusing. There are no grinning viper catchers, no potions, no vipers drowning in wine bottles—just reports of bites, tormented animals, dissections, and lesions.

Like Redi and many other early modern experimenters, Charas committed himself to experimentalism and distanced himself from the scholarly tradition of reasoning from first principles. His commitment to experimentation is much less explicit than the programmatic statements of the Italian academicians, being mostly conveyed through the structure of his book and the way in which the middle part of his book is organized. The book brings to mind Boyle's strict distinction between "narratives" and "discourses upon experiments": Charas described a number of experiments that had been done at his house over a period of a couple years—mostly experiments with dogs and small birds that were bitten in various body parts by vipers. In particular, he described, with more attention to detail and in less colorful terms than Redi, how and where the animals were bitten and the lesions that could be found in dissections of these animals.

The dissections of dogs, pigeons, and pullets that Charas and his colleagues performed showed various degrees of alterations of the blood and coagulations in various places—for instance, around the heart and in the gut. According to Charas, the experiments and dissections clearly showed that the venom affected the blood, causing it to coagulate. He then offered interpretations of his findings, stressing that the experiments he had related inadvertently forced certain conclusions upon him—or, as he put it, "do insensibly oblige us to deliver our thoughts concerning the Venom of Vipers."[40] More experiments were then described to support these thoughts concerning viper venom, and the part ended with some general reflections on the mechanical operation of antidotes.

With respect to the methodological issues that Charas discussed, the contrast with Redi's letter is also quite striking. We have seen that Redi highlighted one methodological strategy—the frequent repetition of one's own experiments—and that he explicitly justified this strategy by reminding his readers of the variations among animal species and of "unseen hindrance," the accidents and contingencies that might impede the performance of experiments. Charas obviously had Redi in mind when he performed his experiments, but he did not claim to have reproduced Redi's, or indeed anyone else's, experiments. Instead, he advanced his own views about the matter, supported by a description of the evidence he himself had obtained. Like Redi, he emphasized that he had performed many experiments, but he did not address the issue of impediments. Rather, he emphasized the uniformity of experimental outcomes across a number of trials. Precisely because his findings had disagreed with Redi's, he had been particularly careful and exact, and he had confirmed his views "by a very great number of Experiments, which have always been found alike, in the truth, we here assert, and of which we shall make evident and irrefragable proof."[41]

Moreover, although he did point out that he had performed a "very great" number of experiments, what he evidently meant was that he and his group had carried out numerous different experiments (as detailed in the book: each trial has its own subsection, complete with subtitle: "The Biting of a Dog in his Ear," "Another Biting upon a Dog," and so forth). He stated that he had performed "divers" or "many" experiments, and when on occasion he specified the number of repetitions he had performed, the numbers were comparatively small (three times, six times, certainly not "one-hundred times").

Unlike Redi, Charas did not shy from speculating about the mechanisms of the working of venom. These speculations informed the narrative throughout. Even his initial descriptions of bitings and dissections are permeated by references to "irritated" and "angered" snakes and "weakened spirits." He noted, for instance, that the corruption of the blood disturbed its circulation, which "hinders the communication of the Spirits through the whole body, depriving the noble parts of them, as well as of the pure bloud, which was wont to bedew them, and destroying them indirectly, by causing this privation of Spirits and of the good liquor, whence depends their subsistence."[42] Having described these experiments and dissections, he reasoned that the ultimate cause of the fatal effects of the viper bite lay in the imagination of the viper. As he put it: "The imagination of the Viper being irritated by the *idea* of revenge which she had fram'd to herself, gives a certain motion to the spirits." Charas assumed that the spirits were being pushed out through the teeth of the viper. They communicated the anger to the victim.[43]

Only in his second work on viper venom did Charas explicitly refer to van Helmont, but in his earlier work he evidently had the Helmontian notion of disease in mind when he treated the animal spirits as moved by some kind of vital principle that was, in turn, aroused by an irritant. The Helmontian conception had great explanatory power: It explained the difference between the bite of an angered viper and that of a viper whose jaws were forced open. The application of force made the viper retain its spirits; the freedom of action was required for releasing the spirits.[44] Moreover, Charas was able to explain why the bite of a viper was less dangerous after the snake had been made to bite on a slice of bread. This was because the bread crumbs that stuck in its teeth hampered the movement of the spirits. In addition, this conception of the nature and working of the venom could also explain the power of certain remedies—in particular, volatile salt. The salt parts acted mechanically: they hooked themselves onto the vexed spirits and drove them out of the body.[45]

The experiments that are described after the thoughts he had felt obliged to deliver to the reader all supported this explanation. The experiments not only confirmed that the yellow fluid was harmless when ingested but also showed that the yellow liquor was harmless when introduced in a wound and that the bite of an angered snake was fatal even when no yellow liquor was involved. What is notable about these experiments is that Charas derived the main support for the position from comparisons. These comparative trials were designed specifically to establish the causes of observed effects—a task that must come after the careful observation of effects, as Charas pointed out. Echoing Bacon, he said that the plan for his book was to describe the search for causes only after various experimental effects had been described, he was not entirely consistent in the execution of his plan. The experimental part begins with descriptions of dissections of various birds after they had been bitten—they all showed the same lesions. But in one section, entitled "EXPERIMENTS of the Biting of Vipers, made upon Pigeons and Pullets," Charas reported a more informative trial, as if he wanted to give the impatient reader a glimpse of what was coming later. That trial involved a comparison and conclusions drawn from it and should thus have been discussed in the last interpretive section of this part: "it will not be amiss," he wrote (although, by his own standards, it was!) "here to relate the different success in two Pigeons, we caused to be bitten equally and in the same place by an angered Viper. One of them we made to swallow the weight of about half a crown of Theriack, a moment before it was bitten, giving nothing to the other."[46] The former survived, the latter died in half an hour. The surviving bird was bitten again after the first ordeal and died from the second bite.

The point of this comparison was, of course, to demonstrate that the "different success" of the two trials was due to the presence and absence of the presumed cause for the recovery: theriac. Charas added an interpretation to his finding—strictly speaking, also out of place in this particular part of the book—in which he explained the different success as a result of the battle between the joint forces of animal spirits and theriac and the evil spirits of the viper.[47]

That same comparative strategy is used to great effect in the second, "demonstrative" section of the experimental part of the book. Charas devoted an entire chapter to the question of whether the yellow liquor was fatal, marshaling evidence from comparative experiments designed to answer that question. Some experiments examined the effect of the yellow liquor alone and found it ineffective; others examined the effect of the bite of angered vipers without yellow liquor involved and found it lethal.

A pigeon wounded under the wing and in the leg received some yellow liquor "drawn from the gums of two enraged Vipers," and both wounds were immediately covered to prevent the venom from flowing out.[48] The pigeon was not affected, and the wounds healed. A cat, several pullets, and more pigeons were treated in the same way, "always with the like success, and without any offence to the Animals."[49] The same trial had been repeated on a dog—three times, even twice in one day. Charas stressed that the dog had been wounded toward the bottom of the ear so that it could not lick the wound, and "no mischief at all" had followed.

Charas described experiments specifically designed to show the effect of the snake's anger. He referred to comparisons of "the biting of the Viper angered" and those of "a Viper, which was made to bite by holding its jaws, and by passing its great teeth into the body of some animal" (a procedure that we may assume would not anger the snake). He found "a quite manifest difference" between the effects of the bite of an angry viper and that of a viper that was not angered. In the second case, the wound healed and no "sinister accident" occurred.[50]

A pigeon died from a bite even though the viper had first been made to bite several times on a slice of bread (so that no yellow liquor could have been left in the teeth). The pigeon died slowly, but that, Charas explained, was because the bread crumbs had stopped the pores of the teeth and thus impeded the exit of the vexed spirits. The main point was that the death had occurred "without any mixture of the juyce which had been altogether emptied."[51] In all these experiments, comparisons of trials are described to demonstrate that certain factors are responsible for certain observed effects.

Redi emphasized multiple trials; in Charas's work, the emphasis is on comparisons. It would be mistaken to assume that comparative trials were more advanced or sophisticated experimental strategies than "mere" repetitions, however—repetitions of trials and comparative trials have different purposes. Repetitions have practical relevance and are employed to deal with practical challenges. They are performed to stabilize an experimental situation and to make the findings more secure. A comparative trial, by contrast, is the precondition for a process of reasoning from experiment.

Charas's reasoning immediately reminds us of Bacon's new scientific method and his conception of privileged instances, specifically the fourteenth of those: the "decisive instance" or instance of the crossroads. But we need to consider this carefully. For Bacon, the instances of the crossroads are crucial instances in that they show which of two possible causes (or forms) is the true cause of the nature that is being investigated. Bacon characterized these decisive instances as very illuminating, being invested with great authority.[52] Notably, however, Bacon's point was about the function of these instances. Certain observations or experiments were called "crucial" not because they had a specific identifiable structure but rather because they could demonstrate a certain contested position or help decide between two alternative hypotheses. The various experiments that

Bacon discussed to illustrate the fourteenth of his privileged instances for inductive reasoning fill this role, as does Newton's famous *experimentum crucis* from the *Opticks*.

On the other hand, a number of examples of comparative experimental trials precede Charas's work as well as Bacon's; indeed, I have already drawn attention to Galen's experiments with cocks and theriac. Another example, and one that might well have been known to Charas, is Bernard Gordon's investigation of theriac as an antidote to poison, which includes a comparison of this type.[53] Gordon, too, examined the efficacy of theriac, proceeding just as Galen had. He advised that the experimenter "take two pheasants, cut off their crests, apply a poison to the wounds (or administer it orally) and wait until they begin to stagger. Then put theriac on the crest wound and in the drink of one of them: if this one lives and the other dies, the theriac is good."[54] Considering that there were prominent precedents of comparative experimentation, it is not surprising that neither Redi nor Charas commented much on the practice itself. For them, performing many experiments—numerous, and repeatedly—was the strategy worthy of note.

Reading Charas's treatise with Redi's letter in mind helps us see that an overall commitment to experimentalism can be realized in quite different ways. Both Redi and Charas were experimentalists, both called for many experiments, and both demanded that interpretations and thoughts be based strictly on their outcomes—but they had different understandings of what it meant to carry out many experiments. The two investigators highlighted different strategies for establishing proper procedure. Only Redi expressed concern about the "wantonness" of nature; presumably this had something to do with his repeating his own experiments much more than Charas repeated his. In the second round of their dispute, the discrepancies between the two approaches would become more apparent. The dispute was also an occasion for more explicit reflections on exactly how the experimentalist imperative should be put into practice as well as on how the practical challenges of experimenting with living beings could be addressed.

CHAPTER THREE

Trying Again

Redi was certainly not impressed by Charas's objections. A few months after the publication of the *Nouvelles expériences*, he responded to the challenge with another letter. He avoided any direct confrontation with Charas, however, by addressing his response to Alessandro Moro (Alexander More, of Scottish origin, pastor in Paris) and the Parisian physician Abate Pierre Michon Bourdelot. This was an interesting choice: Bourdelot was a known supporter of scientific work and a facilitator of learned exchanges among French and foreign scholars.[1]

Whereas Redi's first letter had contained many vivid and entertaining descriptions—of the grinning viper catcher, the reactions of witnesses and spectators, and so forth—the second letter shifted the emphasis from narratives of experimental trials even more emphatically to methods discourse. If Redi had aimed in his first letter to entertain his patrons with stories of amazing experiments, the second epistle was more obviously and explicitly an effort to demonstrate that his, not the French experiments, could reveal the true nature of viper venom.

Even though Redi's conception of poisons clearly differed from Charas's, he continued to be silent about interpretations and speculations based on his findings. The controversy between the two scholars was strictly limited to the establishment of the cause of death from snakebite: Was the cause a material substance, the yellow fluid, or something immaterial, conveyed by the viper's anger? The debate between the two scholars brought methodological issues to the fore, and it highlighted the challenges involved in the practical application of strategies of experimentation. Compared to Redi's first letter, very little was said about the merits of experimentalism, and a great deal was said about methodological issues and details of experimental procedures.

Repetitions and Decisive Trials

Redi began his second letter with a review of available interpretations of the cause of death from snakebite. He quoted extensively from Charas's description of the French experiments and briefly recounted his own main findings. He also drew attention to the French group's having designed experiments specifically to decide between the two competing views of the nature of venom. Its members had directly compared the effects of the yellow fluid as it was instilled in wounds with the effects of the bite of an angered viper. The outcome of the French experiments had been contrary to his own findings. Who was correct, and what was the explanation for the conflicting outcomes?

Simply trying to follow the right protocols did not guarantee the success of an experiment, Redi cautioned: success depended on the manner in which an experiment was carried out, in particular on the care and precision of the experimenter. In his second letter, he offered more specific descriptions of his experimental procedures. One crucial factor for the success of the experiments he had performed was how the victim was wounded and how the venom was inserted into it. The operation had to be "highly accurate"; if the wound was too narrow, the venom would not enter, yet if it was too large, the bleeding would flush out the venom. He also noted that the potency of the venom varied with the size of the animal and with the area where it was bitten, especially with the number of veins and arteries in the place of the bite.[2] Redi's letter thus raised suspicions that the French group might not have proceeded with quite enough care. The bulk of the letter consists of reports of new experiments that had been made possible by a fresh supply of vipers from Naples. These new experiments were narrowly focused on examining the effects of the yellow liquor on wounds. Redi took great care to explain what measures he had taken to make his experimental findings more secure.

He did take the occasion to reiterate his general commitment to experimentalism, but the main point, again, was the crucial importance of performing many repetitions of experimental trials. The uniformity of experimental outcomes was the main concern here—the indication that a set of trials was reliable. Redi stressed, "I do not put much faith in matters not made clear to me by experiment although I do not rashly depreciate them as false, rather, wishing them to be true, I test them by experiment, not only one, or a few more for peace of mind, I wish to see many many, and I mistrust myself and always doubt if I could be deceived, as has often happened to me when I trusted a single, hastily done experiment."[3] Throughout the letter, Redi kept insisting that experiments be repeated many many times, and as before, he frequently noted that he had "tried again and again."[4] The numbers of trials he mentioned are much greater— twelve vipers, eight doves, fifteen vipers, ten large pigeons, and so forth— than the numbers mentioned by Charas.

In the parts of the letter in which Charas's views are most forcefully refuted, numbers are particularly abundant. The heads of twelve vipers were cut off, their venom collected and put in the wounds of eight doves; many other vipers were killed, their venom used to wound ten pigeons; the experiment was repeated with eight viper's heads—and so on and so forth.[5]

The experiment that supported his view particularly well and that seemed to oppose Charas's views most directly—the experiment with venom collected from dead snakes—involved numerous animals and many repetitions. Redi experimented with venom collected from twelve vipers that were "quite dead."[6] Moreover, he used old, dried venom that had been collected from the heads of 250 vipers and that had then been neglected for a month. The venom—whether liquid or dry—was invariably fatal.

Charas, on the other hand, had tried to establish this finding based on only one experiment, as Redi noted. On one occasion, one single viper had bitten and killed five pigeons, and Charas had inferred that one single viper could kill in succession as many animals as it happened to bite. Redi did not voice any doubts about the event itself, but he insinuated that it was a one-off: "I would have wished that those gentlemen, as confirmation, would have continued to have many other pigeons and many other different animals of various sizes be bitten by the same viper which had killed those five doves, to see if verily that angry venom was endowed with infinite power."[7] He, of course, had properly sought confirmation for his results—his findings were confirmed by experiments on sets of ten cockerels, five ducks and three pigeons, again on twelve pigeons, then on five rabbits and three lambs. The animals bitten last always survived, even though the viper "showed marked signs of vexation, anger, and fury" throughout the entire set of trials.[8]

Charas had also reported that an animal bitten by a viper had been cured by eating the head of another viper. It was only at this point in the discussion that the replication of experiments—or, in Redi's terms, the "imitation" of another investigator's work—became an issue. Redi asked why some of his experiments did not yield the same results as Charas's even though the experimental design was similar. Like Charas, he had fed viper heads to two dogs—one head half cooked, the other crushed in a mortar, per Charas's description. Both dogs were subsequently bitten in the ear, and both survived—which, on the face of it, suggested that Charas was right and that viper meat could cure snakebite victims. Yet "a doubt arose" in Redi, as he put it, and he repeated the experiment without feeding the dogs viper meat. Both dogs, despite having been bitten in the ear, survived.

The letter continues with an account of similar experiments with a number of different animals. Redi fed viper heads to a cockerel, a capon, two dogs, eight more cockerels, two kittens, two hares, and six doves. All animals died from snakebite, but they were bitten in the thigh or the tongue, not in the ear. Like Charas, Redi varied the preparation of the viper meat: he fed some heads raw; of others he prepared "a delicious stew of parboiled viper heads" or "broth of this meat"[9]—but no animal escaped death, even though in some the place of the bite had also been smeared with viper blood as an additional antidote.

Besides repetitions and variations of procedures, the letter also mentions comparative experiments. In light of the disagreement with Charas about the cause of death from snakebite, performing such decisive trials was the key task. To establish that the yellow liquor was the real cause of the animal's death, the experimenters had to make sure that just the yellow liquor entered the wound of an animal, as well as that the wound alone was not fatal. Moreover, they had to show that the bite of the angered viper did not do any harm if no yellow liquor was involved.

In his letter, Redi relied on comparisons to demonstrate that his experimental procedure showed what it was supposed to show—namely, that it really was the yellow liquor that caused death, not the wound that the bite of the viper inflicted on the unfortunate animals. Redi reported that he had punctured the thighs of ten pigeons with broom straws sharpened like arrows and dipped in the yellow liquor collected from many dead vipers. The poisoned straws were left in the wounds. All pigeons died after two or three hours. Because one could argue that the pigeons had died of the wound alone, he also wounded four other pigeons with straws that had not been dipped in the fluid, and these birds survived.[10]

Redi challenged the French group to make new observations. Notably, he did not say that he would convert to the French view if the new experiments yielded results that were contrary to his own findings. The only concession he made was that "we could join in saying that we have discovered a truth hidden until now, that is, that the venom of the French vipers consists of an imaginary idea of anger aimed at vengeance, and that of Italian vipers has its seat in that yellow liquor so many times mentioned by me."[11] Redi made the same kind of argumentative move in other contexts.[12] It was probably meant to be ironic—but, as we will see, the critics and commentators (including Charas and Bourdelot) understood Redi as being indeed prepared to believe that French viper venom worked in strange ways.

Redi gave the comparison between the trials with the two broom straws a prominent place in his narrative, and it thus functioned as a "decisive instance" in the argument. But the emphasis of methods discourse was elsewhere. In the text, Redi was especially careful to describe the repetitions he had performed and to point out how many trials he had carried out. He ended with the wish that the French gentlemen "would have continued to have many other pigeons and many other different animals of various sizes be bitten by the same viper which had killed those five doves, to see if verily that angry venom was endowed with infinite power, as I have sought to do in clarification."[13] Judged by the weight that Redi put on repetitions, Redi was more concerned with the practical challenges of stabilizing experiments than with the conclusions one could draw from decisive instances. *That* one could draw these inferences from successful comparative trials did not seem to be in doubt.

Charas's Response

Perhaps not surprisingly, Charas also remained firm in his position. In 1671, he responded with a new treatise, the *Continuation of the New Experiments concerning Vipers*.[14] He was displeased because Redi had communicated his new findings to various people at Paris and elsewhere but had not even honored him with a copy. Nevertheless, Charas was ready to respond.

Like Redi's second letter, Charas's second communication is more narrowly focused on the core of the dispute: the real cause of death from snakebite. Once more, Charas set out to demonstrate the innocence of the yellow liquor and the fatal force of the vexed spirits. He had two experimental tasks complementary to Redi's: First, he needed to devise experiments in such a way that only the enraged spirits but no yellow liquor could have acted upon the animals. Second, he needed to perform experiments to show that the yellow liquor did not have any ill effects if applied to wounds. (Both investigators agreed that the yellow liquor did not do any harm if taken by mouth.)

The investigators agreed on the basic principles that should be followed in experimental research: Claims to knowledge had to be based on matters of fact. Experimental trials had to be carefully executed. A sufficient number of trials had to be performed for the outcomes to be reliable, and comparisons were informative. But we have already begun to see that within these general commitments and principles there was much room for disagreement. A close reading of Charas's second communication shows that Redi and Charas had in fact quite different views about how exactly these principles should be put into practice, as well as about the relative significance of these principles. The disagreements begin with the two scholars' ideas about what social environments were conducive to experimental research.

Historians of early modern science have illustrated how the aristocratic context of the Tuscan community shaped the selection of research topics, the design of experiments, and the communication of findings. The French group noticed, too, and its members were nettled. Charas complained that Redi's experiments had been made "by order and at the expence of so great a Prince, who is as curious as he is Munificent" and who had provided Redi with inexhaustible resources. Charas, however, had done "all from my self, at my own charges, and with a meer desire to discern truth from falsehood."[15] Charas insinuated that his own work was more disinterested and honorable than the work of his adversary: it had been carried out at considerable personal cost and in a public space, whereas Redi's experiments had not been made in public—and Redi was, after all, a paid dependent of the court.

Charas also lamented that it was easy enough for Redi to request numerous repetitions, he having an abundance of resources at his disposal. The wealth of the Cimento Academy enabled Redi to set methodological standards so high that it was impossible for others to meet them. In view of his own lack of funds, Charas did not think he was "obliged to multiply expences when the truth was found sufficiently clear'd up, and all the by-standers acknowledged that I had made sufficient tryals for every experiment."[16]

Charas directly engaged with some of the critique Redi had leveled against his way of experimenting. His comments tell us that even seemingly

straightforward methodological statements lend themselves to different interpretations. The methodological imperative that "many many" experiments ought to be performed is a perfect example. I noted earlier that Redi
had accused the French group of not having performed enough trials. But
how many trials were sufficient?

Charas declared that he could beat Redi at his own game. He indig-
nantly pointed out that Redi "hath contented himself with having made
some of the yellow liquor to be swallowed by one only Man, one only Duck,
one only Kid, thence to know and to assure himself of its innocence when
swallow'd, without making a greater number of Experiments."[17] The reader
could see that Charas had done even more than Redi, and Redi really did
not have any reason to blame Charas for not performing enough trials.

To the modern reader, it is obvious that Charas's interpretation of the
notion of "sufficient trials" differed from Redi's. In one experiment that
Redi had criticized, a viper had been made to bite five pigeons. All had
died. When Redi requested "more trials" he surely meant that many experiments should be performed multiple times. Charas took Redi to refer
to that specific experiment. Charas stated that more trials were not necessary because Redi himself had found that the yellow liquor was exhausted
after the first or second bite. If this was the case, then it was indeed remarkable that the fifth pigeon still died. Charas thus pointed out that he
had "done more than enough to maintain my Reflexion, and that I was
obliged from that time to seek the venom elsewhere than in the yellow liquor, in regard it did not intervene, *according to him*, in the three last bitings."[18] Charas reiterated this argument later in the text, quoting several
examples taken from Redi's letter. Redi had reported cases in which one
and the same viper had bitten and killed more than one or two animals.
According to Charas, Redi should have seen that this evidence contradicted his original claim (that there was no yellow liquor left after a couple
of bites), and Charas stated that he could not "but be amazed, that all
these Experiments have not been able to change his opinion or at least to
suspend it."[19]

Charas also presented new experiments designed to demonstrate that
the yellow liquor was not responsible for the lethal effects of the bite and
that the rage of the viper killed the animal. Like Redi, he proceeded by
comparing the effects of yellow fluid on wounds with the effects of a bite
by an angry viper. For the latter experiment, Charas used a "lusty" viper whose gums had been carefully cleared of yellow liquor. It was made
to bite five pigeons and two pullets, having been angered and irritated

with pincers at every bite. At the same time, six pigeons and pullets were wounded in different places and yellow liquor inserted into the wounds. After a couple of hours, all those animals that had been bitten by the angry viper were dead. The other animals, however, merely showed lividness around the wounds. Two days later, these animals were "very well," and their wounds had healed.[20] This episode demonstrated that the yellow liquor was innocuous, in direct opposition to Redi's view.

Charas also obtained results that seemed to contradict his own ideas, as he readily admitted. No doubt a fascination with new medical techniques played a role when some of the bystanders proposed an experiment in which the yellow liquor was administered by way of transfusion "that hath lately been made in divers parts of *Europe*."[21] Charas related that he had three experts, "[o]ne Physitian and two Chirurgions" perform the transfusion. The pigeon died. But Charas was undeterred by the unexpected outcome. He had an explanation for it, stating that the bird had died from the loss of blood, not from the yellow liquor. The experiment was then repeated with extra care. This operation succeeded; a few days later the pigeon was seen to be very well, just like most all of the other animals that had received the yellow liquor without a transfusion. "Yet true it is," Charas admitted, that another one of them had died. Again, he could explain why: The animal had died near a very hot furnace that Charas was using to distil spirits and oil of tobacco. It was obvious that the animal had in fact died from the piercing vapors or from excessive heat, not from the yellow liquor.[22]

This set of trials show perhaps most clearly that the general commitment to experimentalism allowed for very different interpretations and could be operationalized in different ways. What does it mean to base one's beliefs on experimental evidence? At what point should one give up one's beliefs? How many repetitions were required? In what way might accidents affect experimental outcomes? We saw in the last chapter that Redi justified his call for multiple reruns of experimental trials with the reference to accidents and contingent circumstances. Charas, too, assumed that accidents and contingencies might interfere with experimental outcomes. But he used the reference to contingencies to explain away certain findings. A few of the experimental outcomes seemed to speak directly against Charas's position, and he resolved the conflict by invoking contingencies. He speculated that peculiar circumstances might have produced certain unexpected results, such as heavy blood flow, excessive heat, and deadly fumes. In this way he could fit into his theory those experimental results that undermined his position.

To the modern reader, it might be striking that Charas did not provide any additional support for the validity of these explanations and speculations, which to us seem ad hoc. Apparently, the question of the legitimacy of auxiliary hypotheses was not a question for Charas. Evidently his approach was still within the scope of experimentalism, even though he offered very imaginative interpretations of the evidence from observations and experiments in light of his theory. Redi, too, criticized Charas's experiments and practices, but he did not comment on what appear, to us, to be rather daring ad hoc hypotheses and a refusal to give up one's beliefs in view of conflicting evidence.

In fact, some members of Charas's group were unconvinced by Charas's view—Charas ascribed this to the considerable reputation Redi enjoyed in Paris. As he put it, they were "yet inclined to doubt, whether the yellow Juyce were always innocent" but they refused to undertake experiments to clear up the issue.[23] Apparently these critics were more concerned about idols of the mind—so to convince these skeptics, Charas took an additional step. He replicated—imitated, that is—his adversary's experiments and failed to obtain the results that the Italians had obtained. Like Redi, Charas gave yellow fluid to pigeons and two kittens. He made a point of "engaging a person whom I knew most favourable to Signor *Redi*, to make himself the incision, and intromit the yellow liquor as he should think best." Apparently Redi's warnings of bleeding wounds were lost on Charas, who reported that this gentleman severed the animals so much that he, Charas, "could not hold to tell him smiling, he did more than Signor *Redi* himself said in his Letters he had done."[24] Even though one of the cats was very poorly, none of the animals died. And all pigeons and pullets wounded with freshly cut heads of vipers survived, even though Charas had proceeded by "thrusting the teeth into them as deep as I could."[25] Finally, three young physicians whom Charas had sufficiently persuaded of the truth of his claims tried the experiments on themselves: they wounded their fingers with a dead viper "having yet all her yellow liquor."[26] None of them experienced any inconvenience beyond what they might have suffered from the prick of a needle. As the evidence of these experiments clearly showed that Redi was wrong, said Charas, he was unable to comprehend why Redi's animals had died after the application of yellow liquor to wounds.[27]

The experiments on the therapeutic effects of viper meat offered another occasion for the experimenters to negotiate proper procedure. Redi replicated Charas's experiments and had come to deny that swallowing viper heads was an effective treatment of snakebite. Charas cast doubt on Redi's conclusions. After all, Redi's first results had agreed with the evidence obtained by Charas. Dogs that had swallowed vipers' heads after having been bitten had survived the bite. Only the variation of the original experiments had suggested that the cure was not effective.

Redi had simply diverged too far from the original design: Charas had used medium-sized dogs and Redi large ones, on which no certain judgment could be made. Also, the fatality of the bite varied according to the degree of anger of the viper and the place and depth of the bite. Small animals like rabbits and pullets did not have the strength to resist the enraged spirits, and a bite in the tongue was always fatal because of the great number of nerves, arteries, and veins in the tongue). Moreover, Charas could refer to a case history in his favor. A young man who had been present at Charas's experiments was bitten by a viper while trying to hold it. Charas reported that the young man had so much faith in the truth contained in Charas's books that he immediately decided to eat the head and neck of the perpetrator. Charas applauded this and served the broiled head, neck, heart, and liver of the viper. To be on the safe side, he made the young man swallow some volatile salt of vipers, too. He was pleased to report that on the following day the man had recovered nicely, even though his wound had been much deeper than that of the foreign gentleman in a similar case that Charas had related in his first treatise. None of the "grievous accidents" that had befallen that gentleman had happened to the young man. Charas concluded that he had demonstrated the innocence of the yellow liquor.[28]

He then offered a number of additional arguments against Redi that did not draw on experimental evidence. The consistency of the yellow liquor was such that it could not circulate as swiftly as would have been necessary for the venom to produce its effects. Moreover, if the yellow liquor were really effective, it would also act when swallowed, as the stomach was connected to the blood through the thoracic duct.[29] Nor did he go for Redi's apparent compromise. He insisted that the nature of the yellow liquor must be similar in France and Italy. On the one hand, the appearance of the yellow liquor of the French vipers was very similar to what Redi had described for Italy. On the other hand, Charas's vipers had come from diverse parts of France—some provinces, in fact, very close to northern Italy—and he had not experienced any differences between the potency of vipers from these various areas.

Charas now explicitly invoked van Helmont, paraphrasing extensively from van Helmont's work *De Tumulo Pestis*, which advances the notion of archeal diseases. Charas argued that his notion of the action of enraged spirits was in perfect agreement with van Helmont's view that the idea and imagination of terror that could be transmitted from animal to animal and from animal to man. Charas drew attention to the case of the mad dog, which was known to have "the power of communicating his evil to all sorts of animals"—why, then, could one not say the same thing about a viper?[30]

What can we make of Charas's attempt to refute Redi's results? In hindsight, it would be easy to dismiss Charas's efforts as faulty, because he presumably did not take enough care to clear the fangs of vipers from venom and made much too deep incisions when he inserted yellow liquor into wounds. Or we could say that the episode demonstrates once again how much the Tuscan court empowered its experimenters—as Charas himself lamented, Redi's findings counted for his peers because Redi was such a reputable figure. My point, rather, is that the dispute tells us much about how the general commitment to experimentation was fleshed out. It tells us what kinds of methodological concerns were deemed most important— repetitions in Redi's case, reasoning from decisive instances in Charas's case. We see how flexible and fluid were the methodological statements employed by early modern experimentalists. Even when experimenters agreed about, say, the importance of repetitions, they could still disagree about how many repetitions were necessary and why, as well as about what exactly it meant to repeat an experiment. Even when they agreed that beliefs should be based on experiments, they could still have different ideas about how experimental evidence should be used in arguments. Moreover, we see how the dispute brought local differences in procedure to the fore.

Publicity

One way in which seventeenth-century scholars kept abreast with the "new experimental philosophy" at home and abroad was through reviews and commentaries in the budding scholarly journals. Redi and Charas were both well regarded scholars whose letters and treatises attracted enough attention to be reviewed. Reviews of their works and commentaries on the dispute appeared in France and also in England. Reading reviews is instructive for historians of methods discourse, because reviews tell us something about the perceived value of empirical findings as well as of the perceived quality of the experimental work that was being reviewed. The first review of Redi's endeavors, which was published in the French *Journal*

des Sçavans of 1666 and in English translation in the *Philosophical Transactions*, not only reports Redi's findings but also comments on his commitment to experimentation and on his methodology.[31] The commentary
is quite positive. It emphasizes that Redi had "performed his undertaking
with much exactness" and explicitly mentions that Redi had carried out
"many reiterated experiments."[32] The author offers a summary of Redi's
main insights concerning the nature of the venom, the manner in which it
kills, the chemical properties of the salt of vipers, the alleged fatality of the
spittle of fasting persons, and the antipathy of vipers to certain things. The
reviewer concludes that all those odd properties that had previously been
ascribed to vipers had been "refuted by so many experiments made by
this *Italian* Philosopher, that it seems to [the French reviewer], there is no
place left for doubting, after so authentick a testimony."[33] Evidently Redi's
audience took notice of Redi's concern with sound experimental practice.

Charas's publication was also quickly reviewed in the *Philosophical
Transactions*. That review remained largely descriptive of Charas's investigations. The reviewer did note that Charas had based his view on many
experiments and that his experiments had been witnessed by "many Physitians and others." His presentation of Charas's interpretation of the evidence sounds rather cautious. In an apparent effort to stay neutral in his
assessment, the reviewer simply noted that Redi believed that the cause of
death was the yellow liquor, but Charas would "have it to be in their vexed
and enraged spirits."[34]

In 1672, the *Philosophical Transactions* printed a letter by Thomas
Platt, sent from Florence. This publication is an excellent tool for gaug-
ing how widespread, and how established, experimentalist commitments
were in the larger community of early modern investigators, as well as how
their commitments were interpreted and what roles references to concrete
methodological strategies played in early modern intellectual discourse
more generally. Platt reported that there was disagreement among him
and his friends in Florence about whether venom was material, as Redi
had claimed, or involved animal spirits. The gentlemen met—in private,
at the house of Signor Magalotti—to clear up the issue. Platt took as a
starting point the work of Marin Cureau de la Chambre,[35] whose general
theory of poison suggested that the angered spirits gathered the venom
from the blood and conveyed it to the teeth of the viper. Platt introduced
Charas simply as a "new ally of *M. dela Chambre's* opinion."[36] Among the
people present were—besides Magalotti—Dr. Francini, a fervent advocate of Redi's views. The group decided to perform their own experiments

to settle the question. Francini conveniently had available a box of viper heads, freshly cut off. Two pigeons were purchased from the market to make sure that they did not have any antidote in them. They were wounded with a viper head that had been cut off "about 7 or 8 a clock the same morning."[37] Platt carefully described how the wound was made by thrusting the protruding teeth twice into the pigeon's breast. Both pigeons died.

The next morning saw a repetition of the experiment with six pigeons and a cock. Platt noted that the audience had grown larger (it now included Vincenzo Viviani[38] and others). We may assume that the new spectators came out of curiosity. For Platt, who named them, they must have been welcome witnesses who could testify to the events. The group followed Redi's experimental procedures and first established whether bite wounds alone might cause death.

One can gather from Platt's description that the group did not take Redi's descriptions of procedures as protocols that they needed to follow very closely. A certain degree of looseness in the performance of the experiment was obviously acceptable; as "the first thing that Dr. Francini did, was, to thrust several thorns of Rose shrubs into the breast of one of those Pigeons, to manifest, that such accidents, as might befall those that should be wounded by the Teeth of the dead Vipers, were not meerly caused by the wound. And whereas one of the company began to make some nice reflexions, and to take some of the heads to measure the just proportions of their teeth to see what difference there might be betwixt them and the thorns, this made the Doctor lose patience; and soon taking a pin, which was none of the least, he gave to the first Pigeon, that he could lay hands on, a very deep wound in the breast, which no sooner was got free but began to leap and frisk about the room, as if it had not been concern'd in the least."[39]

After these preliminary trials, the experimenters began to work "in good earnest"—that is, they concentrated on testing the doctrine of the vexed spirits. Although Platt reported the events in vivid detail, he never lost track of the argument. Two pigeons died having been wounded by dead viper heads, and the doctrine lost plausibility. Yet another series of comparative trials was made, this time with stalks of a broom. Two were wetted with yellow juice, each thrust into a pigeon's breast and left there. One bird was wounded with a third stalk, which had not been treated with yellow juice. Only the third pigeon survived and "continues to this very day in Signor Magalotti's Pigeon-houle as brisk and as fat as ever." While

they were at it, the experimenters decided to try whether the swallowing of a viper head was indeed a remedy against viper bites. According
to Platt, Dr. Francini "smiled at that phancy" but nevertheless set about
performing one trial, again employing a comparative strategy. A cock was
bitten by live vipers and a pigeon wounded by a dead viper's tooth, both
after having swallowed a viper head. Both died; the cock after fifteen minutes, the pigeon in less than four, so the viper head could not have done
much good.[40]

The news of the experiments performed at Signor Magalotti's attracted
much attention, and Platt reported that many others had been eager to see
them. A few days later, the group thus gathered again in Magalotti's garden,
joined by various luminaries, including two English gentlemen, Thomas
Frederick and John Groscall, as well as Lorenzo Bellini, the mathematician
Pietro Salvetti, and Abbot Strozzi. Various animals, including pigeons, pullets, and a young bitch, were bitten, and all but one capon died (the dog was
in fact killed by a servant who could not bear to see it suffer). Finally, the
pigeon that had been wounded with the untreated stalk was brought out
and found to be thriving.

It is not clear whether the spectators came because they had an interest in the nature of venom or whether they just came along for the thrill.
Platt was certainly occupied with the scientific question at hand; writing
to the Royal Society, he must also have had an interest in establishing that
he and his group had followed proper procedure. It is thus revealing to
read Platt's affirmation that he had related only what he could "confi-
dently affirm to have been an eye-witness of." Seeing for oneself was
clearly important, as was the convenient presence of many well-known
witnesses. For Platt, the main point of this event was that Redi's position
had been "confirmed by the Testimonies of so many persons, that are the
more able to be judges of them, because their understandings are such,
that 'tis not possible to impose upon them."[41] Platt noted that the group's
experiments had really only confirmed what Redi had already shown; he
would leave any interpretations to his English colleagues.

Platt did not make any comments on what constituted proper procedure. In light of the exchange between Redi and Charas, however, it is
noteworthy that comparative experimentation carried so much weight in
this account. Moreover, a few of Platt's candid descriptions of peculiar
circumstances (of the doctor's impatience and the maid's compassion, for
instance) suggest that strictly following a protocol was not a top priority
with this particular group. Finally, Platt's letter also suggests that at least

some of the repetitions of trials were performed to satisfy the curiosity of the public, not on methodological grounds (although gaining further support for Redi's view was probably a welcome side effect).

Platt also refrained from commenting on the findings. Other commentators were less reserved and did include reflections on the outcomes of the experiments. Like Platt, these reviews did not comment on experimental strategies and procedures, even though they occasionally took sides. The review in the *Philosophical Transactions* of Charas's second essay, for instance, faithfully reports the new experimental evidence and the challenges Charas had posed to Redi. The review did not have anything to say about experimentation or experimentalism, but the author did have an axe to grind. He evidently was an advocate of corpuscularism, and this was the reason why he disagreed with Charas. The spirits must be material, not immaterial. Surely what Charas must really have had in mind was that the spirits were "invisible" not "immaterial"—"for, 'tis beyond all doubt, that those irritated Spirits are corpuscles."[42]

In contrast to this reviewer, the recipient of Redi's second communication, Bourdelot, acted as an intermediary. His brief treatise, entitled *Recherches et observations sur les vipères*, is a very interesting document for a number of reasons. Bourdelot opined that although Redi's letter was a useful contribution to the discussion, the dispute between him and Charas had not been resolved. He offered an interpretation of the available evidence that reconciled Charas's and Redi's views. Yet this approach was out of sync with Redi's and Charas's endeavors, not least because Bourdelot based his argument on what he considered common knowledge as well as on anecdotal evidence.

Bourdelot took seriously Redi's suggestion that the discrepancy in the outcomes of the French and Italian experiments might result from differences between French and Italian vipers. Bourdelot noted that in France, the liquor flowing from the viper's teeth was indeed not yellow. Moreover, he argued, it was well known that lues venerea was maligner in hotter climates and that animals were more venomous in Africa than elsewhere. He also put forward another explanation that integrated Redi's and Charas's positions: perhaps the choleric breath of the viper invigorated the liquor. He supported this view with various examples of ill effects that had been caused by bad breath: the gardener whose rotten teeth tainted the graft he had held in his mouth before he inoculated a tree, the breath of wind that could corrupt meat (especially in thunder and lightning), the butcher having eaten onions or garlic whose blow makes beef and mutton livid, and the bite of redheads—which, as was commonly known, was venomous. Considering these facts, it seemed likely that the rage of the viper amplified the effects of the venom. These pieces of information could be regarded as "experience" or empirical facts, but they were not established through experiments in the sense that Redi, Charas, and many of their critics had in mind.

Bourdelot's contribution to the debate was an exception, however. Most commentators at least related their positions to Redi's and Charas's experiments even if they did not perform any experiments themselves. In 1683, the English physician Edward Tyson published a long paper on the anatomy of the rattlesnake that also contained a brief assessment of the dispute between Redi and Charas. Tyson outlined the two contrasting opinions and pointed out that both investigators had made experiments to support their claims, but he remained silent on the design and process of the experiments themselves. He did offer reflections on the experimental findings, invoking analogies and teleological considerations. He reasoned that Redi must be right because the form of the teeth was so obviously designed for the purpose of conveying the yellow juice into a wound and because there were analogies between viper poisoning and the unpleasant effects of the nettle. Nettles would create pains and swellings, but Tyson was quite confident that they were not produced by angered nettle spirits.[43]

The reviewers' comments show, first, that a decade after the controversy between Redi and Charas had taken place, the dispute had not entirely been settled. Although most of the commentators sided with Redi and took the yellow liquor to be the cause of disease and death from snakebite, Charas's views had not been completely discarded. Second, the reviews tell us that the commentators on the dispute largely confined their assessments to the competing explanations of the nature of snake poisoning. In the exchanges between Redi and Charas, questions regarding proper procedure had come to the fore, but the commentators did not address these in detail; moreover, if they did comment on procedures—as Platt did—their concerns differed from those of the two antagonists.

The late seventeenth-century debates about snake venom show how instructive it is to pay attention to the layering of methods discourse. Tracing these different layers allows for a more finely grained comparison of experimental projects across different contexts, and we begin to see the contours of historical developments in the experimenters' conceptualizations of proper procedure.

With a few exceptions (such as Boyle's methodological essay and the reviews of the controversy between Redi and Charas), all the early modern texts I have considered contained quite detailed and often vivid narratives of concrete experiments. We read about grinning viper catchers, smiling experts, frisky pigeons, lusty vipers, writhing dogs, and gushing wounds. Witnesses were often identified by name. Local opportunities and resources—the Florentine viper catcher, the French enthusiasm for the new blood transfusion technique—shaped the concrete experimental procedures. Instruments and techniques were described, but most of the time rather sketchily.

The commitment to experimentalism transcended different local contexts and historical periods. In the debates about poisons and vipers, all experimenter–authors expressed some kind of general commitment to experimentation. Both Redi and Charas appealed to their experiments as an empirical basis and confirmation of their ideas about the nature of viper venom and its working in the human and animal body. In this general sense, both researchers presented themselves as experimentalists who were keen to establish matters of fact through experimentation. But this commitment could take on different forms, as the comparison between the *Saggi di naturali esperienze* and Sprat's history of the Royal Society illustrates. Bourdelot's attitude to anecdotal evidence for the effects of poison shows that at the time not all scholars subscribed as fully as their English colleagues did to an experimentalism that was entirely experiment-based or that required "seeing for oneself." The Royal Society demanded *Nullius in verba*. Redi required that anecdotal evidence had to be tested by experiment. Bourdelot paid close attention to empirical facts ("experiences"), but for him, this could mean anecdotal evidence, too, if the evidence came from a reputable source. As we will see in later chapters, the commitment to experimentalism remained part and parcel of accounts of experiments throughout the eighteenth and nineteenth centuries. The reasons why the writers expressed their commitments in their writings, however, changed quite dramatically over time.

The experimenter–authors' commitments to experimentalism did not commit them to one specific textual genre for the reporting of experiments. Some experimenter–authors made a point of avoiding speculations or interpretations of their experiments (like Platt), others put their thoughts in separate sections (like Charas), and still others presented the sequence of experiments in such a way that narration and methods discourse together formed an argument (like Redi). Redi's compelling argument-as-story, Charas's tightly organized sequence of experimental evidence–thoughts–experimental confirmations–reflections, and Platt's news coverage of local events show that there were several possible formats for the experimental reports.

Besides the very general and very common commitments to the new experimentalism, there are also more concrete conceptions of proper experimental procedure, such as the requirement to do experiments more than once and in the presence of witnesses, to exercise caution, and to restrain one's imagination. Historians have drawn attention to the key role of witnesses and witnessing in early modern experimentation, and snake venom research is no exception. References to witnesses (the more illustrious and knowledgeable the better) were deployed by virtually everyone. Besides witnessing, there were other strategies that were used to establish proper procedure. Even though these strategies were also accepted by many, they were interpreted in different ways by different experimenters. For instance, what exactly it meant to be careful and to pay attention to the circumstances of experimentation differed from context to context. Even the seemingly straightforward requirement of performing many experimental trials lent itself to quite different interpretations.

Opinions were also divided regarding the relative importance of those strategies. This becomes very clear when we compare Redi's and Charas's methodological views—and especially when we read the reviews of the dispute. No experimenter–author motivated or reflected on repetitions of experimental trials in quite the same detailed way as Redi and Boyle. Both investigators highlighted the strategy of "many many" repetitions, and they told their readers why it was so important to repeat one's experiments several times.

The first reviewers highlighted multiple trials as a mark of proper procedure (no doubt inspired by Redi); for Platt, the testimony of well-known witnesses was crucial. Notably, for both Redi and Charas, *comparative* trials became more and more important as the controversy unfolded. Comparative trials became the strategy of choice in the debates about snake bites as other investigators made similar experiments and modeled the work of Redi and Charas. The direct confrontation between the two opposing conceptions of the causes of death from snakebite stimulated the application of this strategy.

Nevertheless, we should not take this shift of emphasis from repetitions to comparative experimentation as an indication that comparisons were the novel feature in methods discourse. Comparative experimentation can be found well before the seventeenth century. In the late seventeenth century, the practice of comparison was well established. It is thus not surprising that the early modern experimenters used the comparative approach in their endeavors. For the same reason, it is not surprising that they did not specifically highlight, conceptualize, or defend any of their efforts to compare experimental procedures. We see how illuminating it is to put methods discourse in historical perspective and how crucial it is to ask whether and to what extent experimenter–authors provide justifications for methodological conceptions.

CHAPTER FOUR

Newtonian Poison

Richard Mead's work on poisons is rarely mentioned in the newer historiography of venom research. Eighteenth-century investigators of viper venom, however—notably Felice Fontana—took the chemical and microscopical analyses Mead performed seriously enough to refer to and to critique them. Mead's work spans almost the entire first half of the eighteenth century. He published the first edition of his *Mechanical Account of Poisons*, which includes a study of viper venom, in 1702, very early in his professional life. Judged by the number of editions of the work, the book was a great success. Numerous editions and revisions, some of which were even pirated, appeared in the first half of the eighteenth century. In 1745, toward the end of Mead's very distinguished career, a thoroughly revised (third) edition came out in which Mead presented a completely different account of the working of poisons, and in 1747, a fourth edition appeared (a corrected version of the third edition).

Mead was highly valued as a practicing physician and leading member of professional organizations in England. But neither his contemporaries nor indeed historians of science and medicine have greeted his written works with equal enthusiasm. Especially the latter have found nothing original in his writings and have suggested that through his books Mead mainly sought to carve out a good position for himself in the medical community. Historian William Coleman was perhaps the most impatient with Mead's work. To him, Mead seemed not much more than Newton's mouthpiece—"whose every gesture was to preserve vital and holy the example of this 'Chief Philosopher'"—and a rather incompetent one at that.[1] According to Coleman, Mead's essay on poisons exemplifies "both the obscurity and contradictions of assumed Newtonian influences in biology," because Mead never managed to live up to the ideals to which he expressly aspired.[2]

There is something to this characterization of Mead and his work. The mechanical account of poisons is above all an exercise in "new medicine." The English physician devoted only a small part of his book to viper venom; he did not conduct many experiments himself. Nevertheless, the work is instructive for my purposes. Mead's book—more precisely, the sequence of its many editions—showcases the state and dynamics of medical thought in the first half of the eighteenth century and its effect on research into poisons. The evolution of Mead's views reflects the general move away from mechanistic theories in medicine, the influence on medical writing (especially in England) of Newton's *Principa* and *Opticks*, and a redirection of attention from the blood to the nerves as key functional parts of the healthy and the diseased body. Precisely because Mead was so conscious of what others required of a respectable scholar as well as of a good piece of medical writing, his text and the revisions he made to it over the years tell us something about conventions for writing, expectations for medical scholarship, and the different ways in which methods discourse could be integrated in a treatise about experimental research. In particular, the book further illuminates the broad scope of experimentalist commitments in the early modern period. As we will see, Mead was a steadfast experimentalist who performed very few experiments and had little to say about proper procedures.

An Eclectic Collection

At first glance, Mead's professional environment was very different from the Italian context of Redi. Mead had just begun practicing medicine in Stepney, his home town (now a part of Greater London) and was trying to make a name for himself when he published his book on poisons.[3] The Italian scholar was a well-respected figure who worked with a group of investigators at a scientific academy and who had ample resources at his disposal thanks to the financial means of the Florentine court. However, Mead's work was not as remote from the late seventeenth-century Italian scene as it might seem. Mead was well acquainted with the Italian mechanistic tradition in medicine and anatomy through both his studies and his travels. He had studied the theory and practice of medicine with the avowed Newtonian Archibald Pitcairne[4] in Leiden, where the Italian tradition of mechanical anatomy and physiology was highly valued. Also, after taking his degree, Mead gained first-hand knowledge of Italian medicine, traveling extensively in Italy and even earning a medical degree at Padua. Moreover, he was well aware of the recent history of research on viper venom. In 1702, the dispute between Charas and Redi had still not been completely resolved, and Mead took it upon himself to clarify the issue under discussion. Ultimately, he sought to devise a general mechanism for the working of all poisons.[5]

Mead's 1702 book is a collection of five essays on poisonous things: vipers, tarantulas and mad dogs, poisonous minerals, opium, and bad airs. In the preface, Mead noted that his original plan had been to extend this work also to include contagious diseases, but, alas! "the humour of Scibbling would not hold out."[6] Only in his later work on the plague did Mead bring together his account of poisons and his ideas about the nature of disease contagion. In fact, as I will show, the transformation of Mead's theory of poisons parallels the transformation of his views on the nature of the plague in the early 1720s.

Mead's book tells us how sweeping the early modern notion of experimentalism was. Mead turned against those who would argue without assistance of experiments and insisted that progress could only be made based on matters of fact. Nevertheless, for himself, experimentalism meant, above all, a commitment to selected authorities in mechanical philosophy. Mead admired Newton; he also praised the works of the mechanical philosopher Pitcairne and of Pitcairne's close friend Lorenzo Bellini.

His main programmatic point was to try to carry the mechanical account of poisons as far as possible while at the same time grounding it in matters of fact. His expressed aim was to "Discover the Footsteps of Mechanism" in the working of poisons and to "Unravel the Springs of the several Motions" on which effects depended. He reflected that if it was possible to bring such "abstruse" phenomena under "the known Laws of Motion," it was even more plausible that other appearances in the animal body were also due to "such Causes as are within the Reach of Geometrical Reasoning." Knowing the cause of a disease was the first step toward its removal.[7]

This is experimentalism with a Newtonian flavor. Its distinctive feature is the considerable optimism about the beneficial role of mathematics for medicine. In his essay on Newton, Newtonianism, and the roles of experiment in seventeenth- and eighteenth-century scientific endeavors, Friedrich Steinle has identified three kinds of Newtonianism. Steinle distinguishes between experiments designed against the background of mathematically formulated theory (the Newtonianism of the *Principia*), experiments to collect a broad set of data (the Newtonianism of the *Opticks*), and experiments as heuristic tools to devise hypotheses about causes (the Newtonianism of the *Queries*).[8] Mead was a representative of the first kind of Newtonianism, the mathematization of experimental evidence. "Mathematics," for Mead, meant "Demonstration and Truth," and Mead laid out quite an ambitious project for medicine, expressing the hope that argumentative rigor could help transform medical practice into a science. Mathematics also served as demarcation between the professional and the quack: true physicians knew numbers, math, and geometry.

Mead's own use of mathematics remained largely promissory. His work did not quite live up to his Newtonian aspirations, as he himself modestly admitted and as many historians have been keen to point out. Mead's book is a rather eclectic mixture of ideas associated with mechanistic anatomy, medicine, and chemistry. Each of the essays in the book describes symptoms of poisoning and considers remedies, and each includes a more or less detailed explanation of the working of the respective poison. There are some common and recurrent features in these explanations: the poisons are generally described as acting on the blood and obstructing its circulation, the action is regarded as a process of fermentation, and Mead assumed that the properties of the particles contained in the poisons damaged the internal fabric of the body mechanically (through pricking and tearing).

The essay on viper venom opens the collection. Mead aligned himself with Redi's view that the yellow liquor caused the symptoms following the bite. He did note that symptoms of poisoning might be more or less severe depending on the circumstances, such as the climate, the season, the size of the snake, and the snake's greater or lesser rage; but usually the nature of the symptoms would be the same in every bite.

The essay offers assorted insights from microscopy, chemical testing, and animal experimentation to support the idea that viper venom works by mechanical operation. Mead announced that his account was based on matters of fact, but he did not hide from his readers that those facts had mostly been provided by others. He had not seen everything with his own eyes. He treated other peoples' findings as trustworthy even though it was "far more easie to Spin out a false Notion into precarious Reasonings, than to make faithful experiments."[9] As commentators then and now have noted, Mead himself was pretty daring in his own reasoning as he pieced together hypotheses and explanations from various sources. What is remarkable is not Mead's warning about precarious reasonings and the spinning out of false notions but rather that Mead told his readers very little about how to make faithful experiments. Unlike Redi's and Charas's extensive empirical investigations, Mead's own practical work appears to have been confined to some microscopical observation and chemical testing and a few experiments. Mead told his readers what he had found, but not much about how exactly he had proceeded. We learn that he managed to extract drops of liquid from a viper's fangs by letting it bite on something solid. He then put the drops under the microscope. What he saw under the microscope very much suggested a mechanical operation of the venom. He found at first "a Parcel of small Salts nimbly floating in the Liquor, but in a very short time the Appearance was changed, and these saline Particles were now shot out as it were into *Crystals* of an incredible Tenuity and Sharpness, with something like Knots here and there, from which they seemed to proceed, so that the whole Texture did in a manner represent a *Spider's Webb*, tho' infinitely Finer, and more Minute; and yet, withal so rigid were these pellucid *Spicula*, or *Darts*, that they remained unaltered upon my Glass for several Months."[10] Quite possibly, particles like these could damage the internal fabric of the body through pricking and tearing.

To determine the chemical nature of the fluid, Mead made use of the by then well-established indicator test. Despite the difficulty and hazard involved in experimenting with such a small quantity of such a dangerous substance as viper venom, he had "plainly seen that it does, as an *Acid*, turn the Blue Tincture of *Heliotropium* to a Red Colour." Mixing the venom with syrup of violets did not give quite as clear results, but "it did really seem to induce in this a *Reddish Hue*." In any case, Mead was certain it did "not at all change it to a *Greenish* Colour, as it would have done if any ways *Alcalions*."[11]

When Mead commented on the controversy between Redi and Charas, he did mention experimental procedures, but his comments remained fairly general and conjectural. He sided with Redi. The notion that the force through which the venom was thrust into the blood determined the venom's efficacy explained why some experiments with viper venom did not succeed in killing the animal and also why there was a discrepancy between Charas's and Redi's experiments. Charas might simply not have been forceful enough in administering the venom; hence the venom was not active enough to affect the blood in a fatal way. Mead suggested that prejudices might have played a role in shaping the experimental outcome: Charas's timidity might have been due to his theoretical commitments, "in as much as there is oftentimes a great deal of difference in the Event of Experiments, when made with Purpose, and a Design that they should

Succeed, and when Timorously and Cautiously managed, lest they should unluckily overthrow a darling *Hypothesis.*"[12]

Mead did not attempt to repeat any of the experiments himself; instead he relied on the results of three animal experiments his friend Dr. Areskine had made in Paris, expressly "in order to put this Matter out of all doubt."[13] There was nothing innovative in the design or performance of these experiments. In the first series, six pigeons were bitten by a live viper one after another. The first three animals died, the fourth recovered, and the last two did not show any sign of great inconvenience. In the second experiment, two pigeons died, having been wounded by the head of a dead viper. In the third experiment, two pigeons received previously collected yellow liquor through wounds. Both died within two hours. All this agreed very nicely with Redi's claims that the yellow liquor was poisonous when introduced into the blood through wounds and that it was used up during the first two or three bites. Mead reported that according to his friend Areskine, another French gentleman, du Verney at the Académie, had made "not only These, but also several other Experiments of the same Nature" with similar success.[14] Mead merely stated that the proofs were so convincing that no further experiments were necessary—besides, he said, the viper's mouth was obviously designed for the purpose of discharging venom.

The essay ends with a brief discussion of therapeutics, and only in this context did Mead make one brief comment that was related to concrete methods and procedures. Mead considered possible remedies for snakebite, such as sucking the wound and the application of snake stones, as well as the merits of viper meat as a remedy for certain illnesses. In both cases, "faithful experimentation" was presupposed. Mead's aim was to show that assumptions about the physical properties of spiky salts had explanatory power. Sucking bite wounds was obviously effective because it prevented the salts of the venom from harming the blood. Because snake stones had a spongy and porous texture, they could, to an extent, also extract the salts from the wound, but they did not do so very effectively. This explained the mixed results of trials with snake stones as remedies (so, again, faithful experimentation was presupposed). Only when Mead turned to an assessment of a certain specific that viper catchers were using—apparently with great success—did he refer to an actual experiment. He found that the specific was *Axungia viperiana*—the soft fat of the viper's body—which, rubbed into the bite, immediately cured the person bitten. To examine its effects, he had a young dog bitten in the nose and treated the wound with *Axungia*. The dog recovered and was well the following day, so the treatment must have been effective.

In light of the great emphasis that Redi had placed on "many many" repetitions it is noteworthy that Mead initially seemed to have considered one single trial sufficiently demonstrative. His methodological interests did not extend to the particulars of proper experimental procedures. Apparently only because some of the gentlemen who witnessed the trial were not quite convinced of its demonstrative powers, Mead saw the need to continue. His critics did not take issue with the experiment itself; they objected that other explanations could not be ruled out. The recovery might have been due to the dog's spittle rather than the viper fat. Mead had the dog bitten again, this time in the tongue, and no *Axungia* was given. The dog died within four or five hours, so its spittle could not have done much good.[15] Mead added that he had successfully repeated this trial (just once).

Mead's own main contribution to the debate was an explanation of the way in which the venom worked—mechanically, of course. Mead was able to explain the interaction between the spiky crystals and the blood as fermentation. The explanation is rather convoluted, combining reports of experiments Mead had garnered from the *Philosophical Transactions* with an assortment of ideas about fermentation, acid–alkali mixtures, and the nature of blood. To put together this explanation, Mead helped himself liberally to ideas presented in the main works of Boyle, Pitcairne, Bellini, Bernoulli, and others.

The notion of and interest in fermentation had a long tradition in chemistry, reaching back to Paracelsus. In the late seventeenth century, it had become a label for a wide range of natural processes, including the formation of winds, the germination, flowering, and decay of plants, and the heat of the blood.[16] Toward the end of the seventeenth century, different conceptions of fermentation took shape, and the notion of fermentation also became part of the chemistry of acids and alkalis.[17]

In an article published in the *Philosophical Transactions*, Edward Tyson had already described the fermentation produced by the mixing of coagulated human blood with cobra venom. Mead referred to Boyle's and Pitcairne's claims that there was nothing of acid in human blood and that arterial blood was an alkali. When mixed with venom, arterial fluids thus exhibited the characteristic phenomena of acid–alkali mixtures "according to the known Principles of Chymistry"—so Mead assumed.[18] In the first step of the argument, Mead used the concept of fermentation in a general sense to refer to changes of fluids. Following Bellini's theory of disease [*De Stimulis*], Mead explained that the spikes of the pungent salts of the venom, when mixed with the blood, would change the cohesion of the globules, thereby affecting the degree of fluidity and impulse of the parts so that the "very Nature [of the blood] will be changed, or in the common way of speaking, it will be truly and really *Fermented*."[19]

Mead then relied on Bellini [*De Fermentis*] and especially to Newton's *Principia* to explain in general terms the nature of fluids: In all fluids, there is not only contact between the parts but also cohesion or—as Newton had demonstrated—attraction. If the cohesion of the parts was changed, "an Alteration of the Nature of the Fluid" occurred—that is, a process of fermentation "as the Chymists express it."[20] Mead then advocated a version of the mechanical interpretation of fermentation as a change in the physical properties of parts that Bernoulli had expounded in his 1690 book *Dissertation on Effervescence and Fermentation*. In it, Bernoulli spoke out against an occult quality as the cause of these processes. He explained these processes in mechanical terms as the result of an action of a sharp body on another body, through which the body acted on was destroyed and the air contained in it was expelled.[21] Mead borrowed from Bernoulli to explain the process of fermentation in mechanical terms as an outcome of the mechanical properties of the salt of the venom. The blood fermented because the acute salts pricked the globules of blood, thereby destroying the walls of the globules. This freed the active substance imprisoned in the globule, and the fluid thus discharged served as a vehicle to disperse the venom. When the "acute Salts" mingled with blood globules, which contained a very subtle fluid, they changed the cohesion of the blood corpuscles. They "do prick those Globules, or *Vesiculae*, and so let out their imprisoned active Substance, which expanding it self in every way, must necessarily be the Instrument of this speedy Alteration and Change."[22]

Like Charas, Mead let the explanatory power of his account speak for itself. This happy mixture of ideas could explain many features of the effects of snake bites. Above all, his approach could account for one puzzling factor—namely, the speed with which a small portion of venom could swiftly affect the entire body. The spikes of the salts pierced the blood globules, and the active substance that was thus freed served as a vehicle to disperse the salts all over the body. At the same time, the fermentation of the blood disturbed the circulation, interrupted the secretion of the spirits, and obstructed the flow of the bile (thus producing jaundice). The changes of the blood then affected the fluid of the nerves, which caused nervous complaints like convulsions and sickness.

Mead also claimed that the force through which the venom was inserted into the blood mattered to the efficacy of the venom. This means that experiments did not completely mimic the process of viper poisoning, because the bite of the viper was more forceful than the experimental procedure of first making a wound and then instilling the venom. He did not draw any further conclusions regarding the epistemic force of experiments but simply presented his theory as a guideline in the search for an antidote: its nature had to be such that it could overcome the venomous ferment. Later in the text, Mead noted that chemical theory could explain why poison taken by mouth was ineffective: the action of the stomach turned acid substances into alkaline ones, thereby breaking their spikes and thus rendering them harmless.

Early modern scholars expressed different views about how to combine empirical data and theory: Redi put forward an argument about causes but refrained from discussing possible explanations for the working of venom. Charas advanced an explanation and showed how much he could explain with it. Mead took the same approach, and he also had some explicit comments on the issue. On the one hand, he was not altogether against hypotheses. He found hypotheses acceptable if they could explain the phenomena. He noted, for instance, that his mechanical concept of venom could account for many features of the effects of snake bites—indeed, (so Mead claimed) better than other theories did. "From such an *Hypothesis* as this," he noted, "(and, it may be, not very easily from any other) we may account for many of the surprising Phaenomena in the Fermentations of Liquors; and as precarious as it seems, its Simplicity, and Plainness, and Agreement with the forementioned Doctrine [Bernoulli's mechanical theory of disease], will, I believe, recommend it before any other to those who are not unacquainted with *Geometrical* Reasonings."[23] Newton, of course, had argued that hypotheses or conjectures had no place in natural philosophy proper, even though they might be entertained for heuristic purposes.

Historians of science have chided Mead for seeing "no incongruity between full liberty in casting hypotheses on a most *ad hoc* basis and claiming that his hypotheses enjoy the protection of mathematical vigor. His remarkable statement to this effect betrays the logical infelicities of prominent elements in a so-called Newtonianism."[24] Anita Guerrini even maintains that the ambitious young doctor referred to Newton solely to gain social capital. She doubts that Mead ever read the *Principia*.[25] Although it is correct that despite his avowed Newtonianism Mead made no attempt at meeting Newton's concrete methodological requirements for meticulous inductive reasoning, and that his arguments do not appear very rigorous to the modern reader, his explicit emphasis on these matters surely helped make Newtonianism a plausible option as a programmatic commitment and as a style of reasoning from experiments.

Newtonian Bodies

Mead published the first edition of the *Mechanical Account of Poisons* at the very beginning of his career. In the years following its publication, he quickly became an influential member of the medical community in England. He was elected to the Royal Society in 1703 and to the College of Physicians in 1709, after Oxford had recognized his M.D. from Padua. Having moved to London, he became one of the governors of St Thomas's Hospital, and he was also instrumental in the founding of Guy's Hospital. Mead was personally acquainted with many luminaries of his time, including George II and Newton, whom he attended during the latter's last days of illness.

In the mid-1740s, when the third and fourth editions of his book on poisons were published in quick succession, Mead could look back on a long and distinguished career as physician and a somewhat less distinguished but certainly prolific career as medical author. Perhaps his most popular work was a treatise on the plague, first published in 1720 (just after the arrival of the plague in Marseilles, in 1719). The book went through six editions during that year, and several more followed; it was translated into Latin, Dutch, and Italian. Besides the book on the plague, Mead also published treatises on the sun's and moon's influences on the animal body, on diseases mentioned in the Bible, and on smallpox and measles.[26] But it was his discourse on the plague that proved momentous for the development of his work on poisons. Only then did his book on poison become more than just a collection of facts and hypotheses about poisonous things: it now became an account of the operation of poison that was informed by Mead's own thoughts on disease. Still, the product remained eclectic; Mead did not fully reconceptualize the work. Instead he inserted new sections, which sit somewhat awkwardly with the remnants of the original edition. Newton still loomed large, but Mead now relied on Newton's matter theory—the matter theory from the *Opticks*—rather than on the methodological rules from the *Principia*. Newton's theory of ether now did the explanatory work that the mechanical philosophies had done in the 1702 edition. This new orientation had some consequences for the *Mechanical Account*.

The most remarkable feature of the two works is how intricately they are intertwined. Both books developed from tools intended to promote Newtonianism and mechanical philosophy (and Mead) into actual contributions to medical scholarship. Both, however, drew largely on empirical evidence and theoretical concepts provided by people other than Mead.

Mead thoroughly revised the discourse on the plague in the early 1720s; the resulting eighth edition appeared in 1722. In the 1740s, after retiring from his professional duties, he took the time to revise some of his earlier works, including (yet again) the discourse on the plague and the account of poisons. The ninth edition of the former, which contained additional changes, was published in 1744; a revised edition of the latter appeared shortly afterward. The treatise on the plague presented the affliction as a contagious disease caused by a poison, the plague contagion.[27] One of the major innovations in Mead's account of the plague appeared already in the revised eighth edition of 1722. It was a reference to Newton's matter theory as drawn from the *Opticks*. This work became the new point of orientation for Mead's thought on disease. Toward the end of the chapter on the origin and nature of poisons, Mead briefly considered the cause of the disease. Until the eighth edition, the plague is assumed to work mechanically, like the viper venom, just as Mead had described it in the 1702 collection of essays on poisons: the plague contagion affected the blood and body fluids by giving them "corrosive Qualities" that in turn produced inflammation and gangrene.[28] Referring to Queries 18–24 of Newton's *Opticks*, in which Newton had outlined his ether theory, Mead also surmised that this process could be subsumed to the laws of interaction between particles and that the operation of the plague contagion somehow depended on a "*subtile* and *elastic Spirit* diffused throughout the Universe."[29] But he added that Newton had not explained in detail the operations of this spirit, so the exact nature of these operations remained unclear.

In the ninth edition, published in 1744, Mead developed the brief reference to Newton and extended it to a more comprehensive theory of the working of the plague contagion. The section moved to the next chapter, titled "On the cause of disease." Mead now assumed that the contagion affected the nerves first. But he still relied on Newton to explain this process. Mead drew attention to two phenomena that a theory of poisons had to explain. One was the nature of the symptoms, which were for the most part nervous complaints: "rigours, tremblings, heart-sickness, vomitings, giddiness, and heaviness of the head, an universal languor and inquitude, the pulse low and unequal."[30] The other was the considerable speed with which the poison acted.

Renouncing his own previous position, he now stated that the blood could not become corrupted quickly enough for the symptoms to be caused by it. Drawing on Newton's *Opticks*, Mead argued that the nature and suddenness of the effects of the disease "must be owing to the action of some corpuscles of great force insinuated into, and changing the properties of, another subtile and active fluid in the body; and such an one, no doubt, is the nervous liquor."[31] Mead again compared this fluid with Newton's ubiquitous subtle and elastic fluid. Taking a cue from Newton's Query 24, he assumed that the nerve fluid, the animal spirits, somehow incorporated much of this elastic fluid, which thus possessed great energy and was very susceptible to alteration. In keeping with some elements of his earlier account, he used chemical concepts to explain that the mixing of the contagion and the nervous fluid could be regarded as the fermentation that occurred between two "chymical spirits" when they were put together.[32]

The new edition of Mead's essays on poisons appeared shortly after the ninth edition of the discourse on the plague. The new version of the work drew heavily on Mead's discourse on the plague. It still contained the old preface, but it was framed by a six-page advertisement, in which Mead explained some of the changes he had made, and by a new introduction. He made light of his change of opinion: "*Dies diem docet.* I think truth never comes so well recommended as from one who owns his error."[33] Perhaps the most significant change was that the new introduction now offered the general account of the working of poisonous substances that the first edition had lacked. Main passages of it are cribbed nearly verbatim from the 1744 discourse on the plague. Mead explicitly referred to the discourse to reconfirm that all he said about animal, vegetable, and mineral poisons applied to disease contagion and vice versa. Poisons did not affect primarily the blood; they harmed the fluid of the nerves first. Mead still suggested that one of the effects of poison on the animal body was an alteration of the blood, but he now proposed that this was a secondary effect. The vitiated nervous fluid would cause irregular circulation, interrupted secretions, and stagnation of the blood, which would then produce additional ill effects in the body of the victim. Only the flow, not the nature of the blood, would be changed.

In the first edition of the *Mechanical Account*, Mead relied on matters of fact as they were provided by others. The fourth edition is quite remarkable for how Mead combined "old" evidence with the new theoretical perspective and some new facts. As before, he relied mostly on the reports of others. His contribution was limited to providing an account that made sense of the evidence available.

The main reasons for adopting a new account of the working of poison were the same as the ones Mead gave in the discourse on the plague: the rapidity with which the poison worked and that the symptoms of poisoning were all nervous complaints. Mead added two pieces of pertinent empirical information, again provided by others. One stemmed from an article published in the *Philosophical Transactions* of 1727–28 whose author reported that the bite of a rattlesnake had killed a dog in less than fifteen seconds. The other came from James Keill's 1718 measurements of the velocity of bloodflow. These measurements indicated that the blood circulated much too slowly to account for the speed with which the poison took effect.

Like in the discourse on the plague, Mead assumed the presence of animal spirits or nervous fluid in the nerves. Once more referring to Queries 23–24 from Newton's *Opticks* as well as to a letter from Newton to Boyle, written in 1678 and published in the 1740s, Mead repeated his statement that animal spirits were something substantial—a thin, volatile liquor of great force and elasticity and a part of the universal elastic matter. The nervous fluid—quick, active, forceful, and capable of receiving alterations from other bodies—was the medium through which poisons worked in the animal body. Due to the speed and great activity of the nervous fluid, poisons could immediately affect and disturb the entire animal economy, thus producing the violent spasms, pains, palpitations, and other symptoms of poisoning. Mead retained the notion of fermentation to explain the working of poisons. The actual interaction between the poison and the (nervous) fluid is likened to fermentation of chemical liquors, whereby fermentation is now explained as "indeed no more than the attraction and repulsion of the particles of different bodies, when they come together."[34] This account applied to all cases of poisoning, including the bite of a mad dog, the intake of opium, and the inhalation of bad air. In the essay on vipers, Mead merely referred the reader to the general explanation presented in the introduction.

The first edition of the mechanical account of poisons marshaled empirical data specifically to support Mead's initial theory of the working of the venom. In the later edition, the passages on microscopical observations of spiky crystals are left unchanged. Nothing follows from them. The chemical–mechanical discussion of salts and fermentation is omitted from the new edition. The great activity and force of the nervous fluid and its potential for being affected by other forces is now crucial. The reports on chemical tests with the venom differ significantly from the earlier account: Mead reported new tests performed on a mixture of human blood and viper venom. The first edition of the essay, by contrast, describes results obtained from indicator tests with undiluted venom, which showed that the venom was an acid. The fourth edition reports pairs of tests: Each test was performed twice, once on the mixture of blood and venom and once on plain blood. Both liquids looked the same after the venom had

been added to one. When the tests with heliotropium and syrup of vio-
let were performed, no alterations were visible, and no fermentation or
change of color occurred; the mixture thus did not exhibit the characteristics of acid–alkali mixtures that Mead had observed in his first book. In all
experiments, the mixture behaved just like plain blood. Neither spirit of
niter nor salt, tartar, or lemon juice produced any fermentation or change
of color in the fluids.[35] Mead did not explicitly tell his readers what to
make of these experiments, but one may assume that the tests were meant
to support his claim that the venom did not affect the blood. If it did not,
then the ill effects could well be caused by a change of the nervous fluid.

The revised edition mentions again the animal experiments Areskine
and du Verney had performed in Paris. In addition, it reports new experiments that Mead and his colleagues had carried out "with a view to the controversy between Signor *Redi* in *Italy*, and Monsieur *Charas* in *France*."[36]
More than fifty years after that controversy had taken place, it was still
significant enough for Mead to engage with it. The experiments seem to
have been similar to those earlier ones performed by Redi himself. Mead
still did not discuss any methodological issues concerning experimentation,
and he left to his readers the interpretation of the evidence.

The first series of experiments, in which "several animals, dogs, cats
and pigeons" were bitten by an enraged viper and "generally died, some
in a longer, others in a shorter space of time," seem to have been performed to demonstrate the nature of the symptoms of poisoning. Mead
reported that it was constantly observed that all animals were affected by
"sickness, faintings, convulsions, *etc.*"[37] These are best explained as nervous afflictions.

One pigeon was wounded in the breast by the head of a viper three
hours after the head had been cut off. The bird died as if from the bite
of the live snake, and one may conclude that Mead took this as evidence
to support the claim that the yellow liquor was the poisonous part of the
snake and that the anger of the viper was not essential for the poisoning.
Finally, Mead and his colleagues had a sharp hollow steel needle made,
"in shape not unlike to the Viper's tooth," into which they put a drop of
venom and wounded the nose of a young dog. This animal recovered. But
a pigeon wounded in the breast with the same needle "suffered as from
the bite, and died in about eight hours."[38] Confronted with the discrepant
outcomes of the experiment, Mead merely inserted his comment from
the first edition that "there is a great deal of difference in the success of
the same experiments, when faithfully and judiciously made, and when

they are cautiously and timorously managed, lest they should happen to overthrow a darling hypothesis."[39]

Having thus warned the reader once again of the danger of prejudice in experimentation, Mead reported the last experiment he and his colleagues had performed: they had tasted the spiky salts. None of the men suffered any ill effects from the tasting, but all agreed that it tasted "very sharp and fiery, as if the tongue had been struck with something scalding or burning."[40] As before, Mead's concern was to offer an explanation that fit the facts that he had assembled. The saline spikes were broken up in the mouth and in their passage through the digestive tract so that they could not do any harm. Together with the chemical investigation of the mixture of poison and blood, these experiments and assumptions supported—or at any rate did not contradict—Mead's notion that the venom acted on the nervous fluid.

Writing Mechanically

Mead's mid-eighteenth-century biographer (who, by the way, did not refer to the link between Mead's works on poisons and on the plague) approved of Mead's conversion, noting, "In his younger days he imagined he was able to account mechanically for the effects of several poisons, by their mixture with blood; but when he was improved by age and experience, he was fully convinced that there is, in all living creatures, a vehicle infinitely more subtil, an ethereal and invisible liquor, over which poisons have a real tho' inexplicable power. Such is the progress of science."[41] This author saw in Mead's conversion chiefly a laudable move away from the old mechanical philosophy.

Indeed, the different editions of Mead's treatise echoed the developments of medicine and matter theory in the eighteenth century. The transformation of Mead's views between 1702 and 1745 was obviously driven not by any new experiments Mead had performed but rather by the encounter with Newton's *Opticks*, by Mead's wish to promote Newtonian thought, and by the redirection of medical attention from the blood to the nerves.

Chapters 2 and 3 have shown that both Redi and Charas were expressly concerned with the methodological aspects of experimentation and that methods discourse was shaped, in part, by practical problems of experimentation. Mead expressed a commitment to "matters of fact" and to experimentation as the final arbiter of truth, but apart from that, we find only an occasional warning against the dangers of prejudice—nothing comparable to the detailed methodological discussions one finds in Redi's and Boyle's texts or even in Charas's work. What little discussion of methods and methodologies is in Mead's treatise does not seem to be informed by any practical challenges he might have encountered in his research. Mead advocated "Newtonian" experimentalism, but his commitment to the "mathematical ideal" from the preface of the first edition—which is included in the text of the third and fourth—appears to be even more perfunctory in the later editions. The arguments and explanations offered in both versions of the text are not organized tightly and do not refer to numbers, math, and geometry as one would expect from someone who subscribed to a mathematical ideal, nor are the experiments designed against the background of a mathematical theory.

Much of the evidence of the 1702 text—notably the results from animal experiments and the microscopical observations—appears again in the later edition but in support of a different theoretical explanation of the modus operandi of the venom. The microscopical observations of the spikes of the salt reappear unchanged in the third edition, and there is no indication that Mead had repeated them. They no longer play a key explanatory role. They might have been used to explain the pungent taste of viper venom, but this was not how Mead used them; the tasting of the venom did not play any part in the argument of the later edition. The evidence from the chemical tests in the 1745 essay flatly contradicts the outcomes of the indicator tests in the earlier essay. But Mead did not address this point; he just remained silent about the chemical nature of the venom.

For the history of methods discourse, Mead's *Mechanical Account of Poisons* serves as a proof of concept. It helps us read Mead's *Mechanical Account* as an experimentalist treatise even though it contains almost no discussion of research techniques, experimental strategies, and criteria for proper procedure. Distinguishing among different layers of methods discourse helps capture the continuities and discontinuities between conceptualizations of proper procedure across historical periods.

CHAPTER FIVE

Experiment as the Only Guide

Mead's writings on poisons did not have any long-lasting effect on venom research. Felice Fontana's substantial experimental project on viper venom, by contrast, was pivotal. In the eyes of toxicologists as well as of historians of science, Fontana's work is a milestone. His *Treatise on the Venom of the Viper; on the American Poisons; and on the Cherry Laurel, and some other Vegetable Poisons* appears in almost all bibliographies related to venom research, both old and recent. Throughout the nineteenth and early twentieth centuries, his research served as a starting point for investigations of snake venom, and his experimental methods were praised as exemplary. Until the twentieth century, toxicologists cited his writings, and especially his methodological views, as a model for their own investigations. He was often characterized as innovator in matters methodological. As late as 1962, American herpetologist Laurence Klauber described Fontana as "the first of the great experimentalists to use adequate controls."[1] Fontana's work is a treasure trove for the history of methods discourse, because he was very explicit about methodologies of experimentation. At the same time, his methodological ideas are not easy to interpret and situate—although he was an innovator for many, his writings also seem to bear striking resemblance to early modern texts. This chapter and the next will discuss the content and organization of Fontana's methods discourse. Close reading is crucial; only if we pay attention to the exact arrangement of Fontana's account can we assess whether and in what respects the methodology of experimentation he expounded really was innovative, unique, and pathbreaking, and how much he owed to tradition.

Like Redi, Fontana was naturalist to the Grand Duke of Tuscany (the reform-minded Pietro Leopoldo), and—also like Redi—he had ample resources to pursue his investigations. Fontana was the director of the Florentine Museum of Natural History. In this capacity he traveled to major European cities, visited instrument makers, bought instruments for the collection, and conversed with the luminaries of his time. He was in charge of the famous collection of anatomical wax models housed by the natural history museum in Florence. His works, excerpted and reviewed in major scientific journals such as the *Philosophical Transactions*, were quickly translated into English.[2]

Fontana's extensive treatise was published in two volumes in French in 1781, then in English translation six years later. It comprises a collection of texts on various topics—including the microscopic structure of nervous tissue and the regeneration of nerves—which were written at different times. According to the title, the treatise covered several poisons—viper venom, "American poisons" (especially curare or "*Ticuna*," the arrow poison), and cherry laurel—but the book really is a treatise specifically on viper venom. Only a small portion of the text deals with those other substances.[3]

The organization of the work is rather confusing.[4] There are two books (one from the 1760s, one from the 1780s) and two volumes, but part of the second book is included in volume I. The first volume contains a translation of Fontana's early essay on experiments with snake venom, published in Italian in 1767. The Italian text refers to prior debates between Redi and his adversaries as well as to Mead. Considering that Mead's *Mechanical Account of Poisons* did not contain much original research, its service as one of the main points of orientation for Fontana's own investigations is worthy of note. In the introduction to the Italian text, Fontana acknowledged the achievements of both Redi and Mead and announced that he mainly wished to clear up some errors. These errors were not, as one might expect, theoretical errors but rather were incorrect experimental findings (although Fontana had something new to say about disease as well).

Fontana's second (French-language) treatise on viper venom—the bulk of which is printed in volume I of the *Treatise on the Venom of the Viper*—is above all a response to Balthazar-Georges Sage's publication on the chemical nature of viper venom and on volatile alkali as an antidote, a work that was at odds with Fontana's view. Fontana's book presents experiments made after the completion of the Italian treatise. These experiments refuted not only Sage's position but also Fontana's own previous findings, and Fontana took the occasion to comment at length on experimental practices and methods. The book even contains a methodological essay, entitled "On the Source of Many Errours," which deals with the problems and pitfalls of physiological experimentation.[5] The second volume of the work concludes the discussion of the new experiments on vipers. It also contains the additional material dealing with the other poisons as well as microanatomical observations (including a discussion specifically on the problems of microscopical observation) and a supplement featuring descriptions of some more experiments and scattered comments on vipers and viper venom.[6]

How Theories Fall and Vanish before Experiment

Fontana's Italian treatise on venom from 1767[7] confirmed Redi's (and Mead's) view that the yellow liquor flowing from the viper's protruding teeth was the substance that contained the poison. The text begins with new observations of the teeth of the viper and the "vesicles" containing the venom, intending to clear up some anatomical facts that Redi had gotten wrong. The main part of the Italian treatise addresses the two issues that Mead had discussed—namely, the chemical nature of the yellow liquor and the mechanism of its working. Fontana showed that both Mead's observations and experiments as well as his conclusions were wrong.

The series of editions of Mead's *Mechanical Account* mirrored the development of eighteenth-century medical thought rather than the development of Mead's own insights into the nature and working of poisons. The symptoms of snakebite were first explained as fermentations affecting the blood, then—when the focus of disease theory shifted from the blood to the nerves—as a nervous affliction. Fontana explained those symptoms in light of the latest development in medicine, the rise of *Anatomia animata* and its associated conceptions of irritability and sensibility.[8] Not only did he discuss and refute Mead's claim that viper venom was a salt, but he also advanced the view that the fluid affected the "irritable" tissues. Unlike Mead, however, he linked his theory to his own experimental findings.

Putting together the first edition of his *Mechanical Account*, Mead borrowed extensively from other scholars' writings. Fontana, too, was inspired by other people's ideas—theoretical as well as methodological—but he forged his own approach. Fontana's new explanation of the working of viper venom is framed by a passage drawn from Redi's first letter on vipers. In it, Redi had presented a range of possible explanations for the working of the venom, although without committing himself to one.

Fontana proposed that venom killed by destroying a fundamental principle of life: the irritability of the muscles. This explanation was, of course, not among those put forward by Redi; it could not have been. Fontana drew on the physiological and pharmacological experiments that were ongoing at the time, Albrecht von Haller's investigations of the irritability and sensibility of tissues being the main source of inspiration.[9]

The work opens with a commitment to experimentalism. We have already seen that such a commitment could take quite different forms—it could be combined with a warning against deceptive appearances, as in the *Saggi*; it could be combined with a turn against the old systems of philosophy, as in Bacon's *Novum Organum*; or it could mean a broad commitment to "matters of fact," as in Mead's case. Fontana's opening statement reads: "It is agreed at present that there is no other guide in a search into natural truths, than a knowledge of facts; it is only on facts that the philosopher can hope either to establish a reasonable system, or to form a sound judgement of those already established. Observation is alone capable of dissipating the mists that envelop the hidden causes of the phenomena of nature."[10] Throughout the book, Fontana reiterated that experimentation was the road to truth, that he had performed experiments to establish this or that, that the most brilliant philosophical systems were worth nothing unless they were founded on good experiments, and so on. Fontana's experimentalism committed him to variations of experimental designs and to tracing all the circumstances of an experimental trial.

In chapter 4, we saw that Mead's experimentalism had a Newtonian ring to it and that, just like experimentalism more broadly, "Newtonian" experimentalism, too, comes in different flavors. Mead called for an experimentalism based on numbers, mathematics, and geometry. Haller expounded another form of Newtonian experimentalism in his treatise on irritability and sensibility. Because Haller did not want to engage in the discussion of the delicate issue of whether animals had a soul, he avoided the matter by a Newtonian move, stating that he did not wish to speculate about why some parts of the human body were irritable and others were not. The source of these properties, he said, "lies concealed beyond the reach of the knife and microscope, beyond which I do not chuse to hazard many conjectures, as I have no desire of teaching what I am ignorant of myself."[11] For Haller, being a Newtonian meant not devising any speculative hypotheses. Fontana advocated the Newtonianism of the *Opticks* and the Newtonianism of the *Queries*. Like Mead, Fontana made a reference to the "creative genius of Newton." Unlike Mead and Haller, Fontana pointed out that such genius was needed to derive the "causes of the laws" that regulated observations.[12] As we will see later, he also followed Newton in his view that knowledge of the nature of causes was outside the realm of science.

Fontana spent more than a hundred pages putting his findings before his readers. He described numerous experiments on the nature and action of venom, and he commented extensively on experimental methods. As in the early modern writings, methods discourse in Fontana's work goes well beyond a general commitment to experimentalism. Along the way, and as an integral part of the account of his experimental report, he offered detailed comments on proper procedure. After decades of intense observations and experiments, experimenters had become painfully aware that disagreements among observers were rather frequent—observations might be disproved by others or even by one's own continued efforts. And disagreements were hard to resolve. It was therefore necessary to do more than just add further observations to the existing jumble of facts; it was necessary "to discriminate nicely, to compare the experiments of my predecessors with my own, to trace and develope [sic] all the circumstances of them, and in short, to discover what may have occasioned so great a variety in the opinions of these observers, and in their manner of seeing. Such is the true motive that has induced me to give an account of the experiments which follow."[13]

True to his word, Fontana described not only a host of experimental trials but also all sorts of conditions, circumstances, and particulars that surrounded his experiments. The result is a bulky, somewhat tedious report showing Fontana's readers how he had arrived at his general conclusions about the nature and working of venom.

One of the conceptual tools for the analysis of experimental reports that I introduced in chapter 1 is the analytic distinction between narrative and argument. Larry Holmes initially developed this distinction through a comparison of writings by a contemporary of Fontana's, the mid-eighteenth-century naturalist René Antoine Réaumur, with a work authored by the early nineteenth-century physiologist François Magendie. Both scholars combined elements of narrative and elements of argument in their writings about experiment, Réaumur emphasizing narrative and Magendie argument.[14] Réaumur described series of linked experiments to investigate the digestive tracts of birds. He provided many details about the experimental conditions and circumstances of the various trials, such as at what hour the experiment was performed, the forms of the tubes the birds had to swallow and how these tubes changed shape inside the stomach, and so forth. Each modification of the experimental procedure appears motivated by the previous experiments, and at the end, the reader—now a fellow traveler compelled to agree with Réaumur—"feels that he or she has been with Réaumur through a scientific adventure." In this sense, the descriptions of the linked experiments are an effective argument for Réaumur's conclusions as well as a vivid story in their own right.[15] Holmes reminds us, however, that even those experimenter–authors who emphasize argument incorporate narrative elements in their reports. Their account might offer a succinct, cogent reconstruction of the logical (rather than the chronological) sequence of experimental trials so as to make a point, but to be convincing, it must still refer to the course of the investigation and must give the reader a sense of what was actually done.

Fontana's treatise is certainly not a narration of his research in the sense that it exactly follows the order in which the experimental trials were actually carried out. Nevertheless, his treatise has the flavor of a (very gruesome) scientific adventure story. The narration of the quest is interspersed with references to proper procedure, statements that are more general than the protocols for the investigation of the chemical nature of yellow fluid and of its effects on body functions. Such methodological concepts and views formed an integral part of the account of Fontana's experiments. Like Charas before him, Fontana made it clear that his treatise did not describe every single experimental trial he had actually performed. Like Redi, he put together narrative and methods discourse to advance and argue for a position.

The self-experiments that Fontana undertook to establish the chemical nature of venom illustrate how he integrated remarks about proper procedure into the account of his experiments. Tasting substances was one of the main modes of chemical assaying in the early modern period,[16] and Fontana took it on himself to taste the viper venom—"more than an hundred times." He knew from the writings of Redi and Mead that Redi's viper catcher and Mead and his friends had tasted it and that ingested venom was not fatal. The viper catcher had reported that the venom did not taste like much; according to Mead, however, it was acrid and pungent, as one would expect from a spiky salt. Fontana's methodological principles—he referred to "philosophical necessity"—required that he taste the venom for himself. He tasted both the liquid and the dry powder. He found the venom insipid, as did his brave servant, who had to swallow it as well—repeatedly, of course.[17]

Fontana described vividly and in great detail the taste and the feel that the venom left on his tongue: cold and insipid but not as insipid as spring water. Rather, it resembled "the almost insensible flavor of the fresh fat of animals."[18] Only if the undiluted venom remained in the mouth for a long time did it leave an unpleasant sensation, as if from something astringent. Nevertheless, Fontana declared himself unsatisfied with the evidence he had assembled. He went on to compare the taste of viper venom to that of the bee, wasp, and hornet. Again he performed self-experiments, and again he described all the sensations he experienced: the stinging and burning, the lasting pain. These experiments were done repeatedly and with variations, whereby Fontana took the venom from the stings of the insects or from the small vesicles that contained it. Finally, he fed viper venom to a dog. The animal kept licking its lips and begged for more. Evidently, it found the taste agreeable and was not in pain—another reason to conclude that venom was not acrid and fiery.

The argument against Mead is driven forward throughout the three chapters that deal with Mead's views. At the same time, the narrative preserves the sense of an investigative journey starting with observations and experiments. Along the way, the reader learns what measures Fontana had taken to ensure proper procedure. Fontana opened the discussion by stating Mead's initial position: that venom was an acid. He did not state his own contrary view, nor did he mention that Mead had in fact changed his position later. Fontana then presented his own experiments, which showed that venom was not an acid. Only then did he note that Mead had withdrawn his earlier claim but that Mead's book had reached him "too late" (too late, we might assume, to prevent Fontana from engaging in a set of experiments on the acidic nature of venom). He then proceeded to note that because Mead had changed his views, he himself could refrain from trying to clear up the apparent contradiction between the outcomes of his and Mead's experiments.

Fontana could have saved his readers—as well as the typesetters— much time had he simply stated at the outset that Mead's and his experiments both confirmed that viper venom was not an acid. But then the reader would not have known by which investigative pathway Fontana had arrived at his conclusions, and Fontana would have had fewer occasions to demonstrate how carefully he had traced all the circumstances and to expand on the proper procedure of experimentation.

When Fontana repeated Mead's initial simple color test, he found, contrary to Mead and to his great surprise (so he said), that the test paper did not turn red. Instead it yellowed—that is, it took on the color of the venomous fluid. As it "appeared extraordinary" to Fontana that "so learned a man as Mead should have been deceived in so easy an experiment" he repeated it again and "that nothing might be neglected, varied the experiment a hundred different ways."[19]

The notion of "many many" experiments was a leitmotif of methods discourse in Redi's letters. In Fontana's work, the leitmotif is the phrase "I varied this experiment in a hundred different ways."[20] It appears over and over again. In the extended critique of Mead, he could make this same point twice, first with respect to the color tests that both investigators had performed and again with respect to another issue on which he still disagreed with Mead—namely, the question of whether venom contained saline crystals with sharp points. In both cases, the reference to variations helped make the point.

What is more, Fontana told his readers exactly how he had varied his experiment. The contrast to Mead's account could hardly be any greater. Mead simply told his readers what he (and others) had found. Fontana described what he had done. To determine whether venom was an acid, he experimented with different forms of venom. He sometimes took it directly from the viper's tooth; sometimes he let the snake bite a piece of cotton. He used pure venom, diluted it, or mixed it with other body fluids such as saliva. But however he tried it, the paper did not turn red. He also could not observe the bubbles that, according to Mead, should appear on mixing venom with alkaline substances. Even the glance through the microscope did not give any evidence of effervescence.

Mead's second questionable finding—the microscopic observation of sharp crystals in viper venom—also could not withstand Fontana's scrutiny. In his search for saline crystals, Fontana even varied the instrument, using an English lens as well as a solar microscope, but as much as he tried, he could not see any crystals in the venom.

It is characteristic of Fontana that he offered an explanation (a rather unflattering one) for why Mead might have fallen into error. Mead might have looked at some bodies floating in saliva and mistaken them for salts—a mistake easily made by those who were "not very conversant in the use of the microscope" and "not well acquainted from habit with the shape of the different salts that are found in liquors."[21] Fontana also described—again, characteristically—two additional experiments he had devised to make himself "still more certain" and "to remove all doubt and suspicion on a matter so important and so generally adopted."[22] He peered down his microscope and watched how the drop of venom cracked and fragmented from the outside inward while it was drying. He repeated this several times, both with pure venom and with drops of venom diluted with pure spring water. Almost in passing, he also mentioned that he had had witnesses, two "celebrated Professors of the University of Pisa" who had assisted with the observations and agreed with the findings.[23] Only at the end of this comparatively long chapter (three times longer than the preceding ones) did Fontana announce that he was now sure that no salts were to be found in viper venom. He had "seen the theories founded on this principle, to explain the action of the viper's venom, fall and vanish before experiment, which proves that no salt, either acid, alkaline, or neuter, exists in this humour."[24] How could anyone doubt this conclusion?

A Hundred Different Ways of Making an Experiment

Fontana's other key insight—the explanation of the working of viper venom—is also supported by a detailed report of an investigative journey. In hindsight, it is obvious that Fontana relied heavily on Haller's distinction between two kinds of body parts—those that were irritable and those that were sensible. This distinction was one of Haller's major contributions to the study of life.[25] In his famous work on the irritable and sensible parts of animals from 1753, Haller derived the two concepts from physiological experimentation on animals: He called "irritable" all those tissues that contracted on being touched and "sensible" those parts that when touched transmitted an impression to the soul.[26]

Fontana's experiments showed that viper venom, just like other poisons such as opium, napel (*Aconitum napellus*—used in cooking as a meat tenderizer), and mephitic air, made muscles flaccid, livid, and putrid.[27] The body parts of people who had been bitten by vipers became paralyzed, again suggesting that irritability was compromised. Fontana procured "fifty of the strongest and largest frogs" and had each bitten by a viper, "some in the thigh, others in the legs, back, head, &c."[28] Some of the animals received a drop of venom through a wound Fontana had made with a lancet. Some frogs died very quickly, others after several hours, and others not at all. But there was one significant common feature: All animals suffered the loss of their muscular force after having been bitten or wounded and treated with venom. Stimulating and pricking the muscles and limbs of the dead animal produced no effect, even though in some cases, the heart still continued to beat. Like opium, napel, and mephitic air, viper venom made the body parts lose irritability, that "grand principle, both of voluntary and involuntary motions in the animal economy."[29]

This interpretation of the findings is clearly informed by Haller's ideas. What is more, Haller's work inspired Fontana's methodological views. Haller expounded several methodological rules for experimentation along with his experiments on sensible and irritable body parts. He did so not because he had taken an innovative approach to experimentation. He noted that he was "obliged to repeat and multiply" his experiments to prevent himself from "falling accidentally into any mistake." He declared that he had "examined[,] several different ways, a hundred and ninety animals." The strategy itself is not revolutionary. The novelty was in the results Haller advanced—because these results were so revolutionary, he had to be "very exact" in his proofs.[30]

The parallels between Haller's and Fontana's methods discourse are striking. Still, it would be rash to assume that Fontana simply borrowed the formulas offered by Haller or by Redi before him. Unlike the young Mead, who had couched his views on venom in the language of mechanical philosophy, Fontana ended up making Redi's and Haller's methods discourse his own. We will see in the next chapter that the French treatise on viper venom is not merely more methodologically reflexive and explicit than the Italian treatise but also much more specific about the various practical challenges Fontana encountered. The level of engagement with practicalities and the concern with identifying the sources of other people's errors suggests that Fontana's methodological views were shaped and sharpened by encounters with the complexities of experimental settings.

At the end of his first book on venom, Fontana announced: "Unless I am deceived, I have now, I think, happily terminated the controversies that have so long kept people at variance, on the mode of action of the venom of the viper."[31] This declaration was premature. As Fontana continued his studies on viper venom in the hopes of finding an effective antidote, a host of new questions opened up, and he continued to devise experiments. Because his research remained inconclusive, Fontana refrained from publishing any of it for some time, and he also delayed the publication of the French translation of the Italian book.

CHAPTER SIX

Thousands of Experiments

It was the publication of Balthazar-Georges Sage's study on volatile al-kali that was the final push for Fontana to publicize the results of his renewed endeavors to study venom. Sage presented volatile alkali (solu-tion of ammonia) as a remedy for viper bites, "canine madness," and all sorts of other maladies (1777). Fontana found that Sage's results did not agree with his own earlier findings. He reported to his brother that he had been in the process of bringing out the French translation of his book when "out came this book, full of miracles and resurrections, the product of a chemist, believed here to be a great chemist [...] I had to retrace my steps to examine anew this matter; without this no one would have be-lieved the experiments reported in my book."[1]

In his book, he framed his dismay a little differently, toning down his criticism. He voiced the concern that he might have been completely mis-taken and declared that he no longer knew what to believe. In addition, he expressed his general irritation with the fact that errors of previous re-searchers, as well as speculations that were not suggested or supported by experimental evidence, were in wide circulation.

Fontana's renewed engagement with the subject resulted in a massive study of more than 500 pages, published in French together with the French translation of the earlier, Italian work. It is a report of another investiga-tive journey that at first glance seems to be full of twists and turns, dead ends, and digressions; only patient and careful reading reveals the underly-ing logic of the argument. Among other things, the second book contains a new interpretation of the working of viper venom and an exploration of the "hidden causes" of the observable effects of venom poisoning. After relating countless experiments, Fontana presented the conclusion that it did affect the blood first and that it did so through an "obnoxious principle."

He also stood his ground against Sage and demonstrated that volatile alkali was not effective as an antidote to viper venom. Just like the Italian treatise, the new book is permeated by methods discourse. Fontana's methodological statements became much more pointed and elaborate. Not only that, but he even drew them together in a methodological essay, which introduces the French treatise.

The "Principal Methods" of Successful Experimentation

Fontana's methodological essay was at once a comment on the diversity of opinions about snake venom (some of which, Sage's among them, contradicted Fontana's own), an occasion to reiterate the general commitment to experimentalism, and an opportunity to express the principles of proper experimental procedure.

In chapter 1, we saw that some early modern experimenters produced essays and entire treatises exclusively devoted to proper experimental procedure and the new experimental method. Boyle's essays on unsuccessful experiments illustrated through numerous examples how many different kinds of contingencies and accidents experimenters might encounter. Precisely because of their accidental nature, these encounters defied any systematic treatment. Fontana, by contrast, took a systematic approach to methodology. He expounded three principles or "principal methods" of successful experimentation:

> The first is, to multiply the experiments exceedingly. It is almost impossible, in repeating them so many times, that one does not encounter the fortuitous cases that vary them, and that the final result of so many of them is not certain and constant.
>
> The second is, to vary them in a thousand ways, changing the circumstances as the nature and species of them may require, and giving them all the precision and simplicity they are capable of. This method supposes much greater talents and genius in the observer than the first, and there are few of these, even amongst the most skilful, who can boast of having invariably put it in practice.
>
> The third method is, not only to succeed in making experiments, decisive by their number, variety, and simplicity, but likewise to attain to a discovery of the source of the errours that others have fallen into.[2]

The first principle echoes earlier requirements that experiments be repeated several times, and the justification is similar to Redi's. The second principle resonates with the methodological views that Haller expressed. It describes a strategy of improving and simplifying an experimental setting that Redi and other early modern experimenters employed as well. In the early modern writings, this strategy is not explicitly brought up, whereas Fontana made multiple variations the key requirement for proper experimentation. Likewise, the third principle formalizes something that Redi practiced—Redi, after all, discussed Charas's erroneous ways—but that Redi did not explicitly require as part of proper procedure.

Undoubtedly, the disagreement with Sage was one motivation for Fontana to begin the second treatise with a methodological essay. At the time when the French treatise on viper venom was published, Fontana had just gone through a period of controversy in the context of another of his projects, the analysis of the quality of airs extracted from different waters.[3] Fontana designed a number of eudiometers and used instruments of his own design to measure air quality in Italy and London. As Jan Golinski points out, Fontana's instruments were very elaborate and complex and required considerable skill in their application.

The paper on air quality that Fontana published in the *Philosophical Transactions* does not explicitly state that experimental procedures had to be diversified, but he described series of experiments in which various circumstances, such as the materials of the test vessels, the shape of the vessels, the method of air extraction, and so forth, were recorded and systematically changed. When Fontana presented his instrument and method, he emphasized that instruments to measure air quality could easily produce incorrect results, being liable to various errors (the causes of which "sensibly alter this kind of experiments")[4] and extremely sensitive to variations of circumstances. Fontana added that these variations were so minute that even those observers of his experiments who were aware of their effect had been unable to notice it.[5]

At that time, several investigators debated the pros and cons of instruments for measuring properties of airs, and Fontana and others who took up and developed his approach were criticized because the instruments they used appeared too complicated, their difficulty in handling making them liable to error.[6] Against this background, it is understandable that Fontana made a conscious effort to defend his experimental approach and, to demonstrate that he had proceeded with the ultimate care and caution, had taken into account all possible circumstances and sources of error and had done his best to secure his results.

Modern readers of Fontana's methodological reflections will be reminded of recent work on exploratory experimentation. Friedrich Steinle has coined this expression as a corrective to what he calls the "standard view" in philosophy of science—the view that the role of experiments is to test scientific hypotheses.[7] This view is commonly associated with Karl Popper's *Logic of Scientific Discovery*. Steinle reminds us that experiments are not always performed to test theories. Drawing on examples from the history of electricity, especially on the work of Michael Faraday, Steinle characterizes a kind of experimentation that is performed to obtain empirical regularities in situations in which no well-developed theories are available.

Steinle further characterizes exploratory experimentation as a mode of knowledge generation; it generates a specific form of knowledge: new concepts. More important for the present context, exploratory experimentation comprises a set of strategies of experimentation among which are parameter variations, stabilization and simplification of experimental arrangements, exhibition of a phenomenon in the clearest possible way, and identification of concepts suitable for expressing the empirical rules governing the case. Steinle derives these strategies via empirical generalizations from detailed case studies, particularly from eighteenth- and early nineteenth-century research on electricity. In his view, however, the strategies are transhistorical. They are not bound to a specific historical period, such as the decades around 1800. Rather, they are typical for a specific epistemic circumstance—the absence of well-developed theories and hypotheses from a research field.

It was not Steinle's main goal to trace the methodological concepts and ideas as they were expressed and defended by past experimenters themselves. Indeed, Fontana's essay is striking because Fontana himself articulated several methodological principles related to parameter variation, simplification of experimental arrangements, and so forth that closely resemble the set of exploratory strategies exposed and abstracted by Steinle. Because theories and hypotheses about snake venom were available, Fontana did not do exploratory work to enter a new field, as Steinle's protagonists did. Fontana's position was under attack, so he must stake his claim—and he did so by, among other things, outlining the characteristics of proper experimentation. Throughout his second book on venom, Fontana placed much emphasis on the exploration of a setting, describing both the systematic variation of parameters and the simplification and purification of experimental designs. In the second principle of experimentation as the methodological essay presents it, however, these different practices are not explicitly distinguished; he simply and generally required "diversification" of experiments. Further conceptual distinctions of different patterns of variations gradually gained contours in the methods discourse of nineteenth-century investigators, and I will return to these distinctions.

Circumstances

Making sense of Fontana's second experimental project is a daunting task.[8] Fontana's endeavor carried him well beyond his original quest, something mirrored in his arrangement of the book. There is indeed an argument, and that argument is indeed brought to a conclusion, but its course is by no means easy to follow. In some parts of the book, volatile alkali is not even mentioned. Fontana did repeatedly state that his experiments had refuted Sage's hypothesis, but his explicit rejection of Sage's claim is buried within the first chapter of volume II of the treatise. Most of the argument against Sage is developed in the first part of the second treatise, but the discussion of volatile alkali extends into the first chapter of the next.

What begins as a systematic survey of the effects of volatile alkali on snakebite victims soon becomes a sprawling discussion of all kinds of issues related to the nature and working of venom: the effects of bites on different animals and in different body parts;[9] the chemical nature of the venom of vipers, bees, drones, and wasps; a determination of the quantity of venom necessary to kill animals and of the time it took to kill them; and, finally, a novel interpretation of the working of venom—or rather, a revival of the old idea that Mead had discarded: that the venom affected the blood (although not mechanically or chemically). Various narrative strands transgress the book's parts, chapters, and sections; sometimes, but not always, one section is devoted to one sequence of experimental trials. Here it is not necessary to follow Fontana through all the twists and turns of the story. Rather, I shall attend to methodological statements and reflections, examining how and at what junctions in the project Fontana brought up methodological issues. We will see that Fontana was long-winded for a reason: his account demonstrates how, exactly, his interpretation of the nature and working of viper venom is grounded in experimental findings.[10]

Fontana reiterated his general commitment to experimentalism in the second treatise. Usually at the beginning of larger sections, we find declarations such as the following: "Experiment alone may conduct us through the unknown paths of nature, and may lead us to new and unexpected truths. But at the very time that man, profiting by this torch, is making bold strides toward the truth, and soars as if he meant to govern nature herself, she stops him every moment, and by only discovering herself to him in part, seems afraid of being recollected; she thus continually reminds him of his weakness."[11] As before, the actual characterization of experimental procedures is much more specific than this formula. Throughout the text, the descriptions of experiments are intertwined with statements that invoke the principles that the methodological essay articulated. Although Fontana frequently noted that he had multiplied his trials, the emphasis is clearly on the second principle, the "diversification" of experiments. Take, for example, the very first chapter, right after the methodological essay. It is entitled "Whether the Volatile Alkali is a certain Remedy against the Bite of the Viper." The first sentence reads: "I deemed it necessary to examine this first question in the most circumstantial way, and therefore multiplied the experiments extremely, and diversified them very much. This is the only method that could lead to demonstration, and I flatter myself that my readers will be freed from all doubt."[12] The leitmotif from the first book permeates the second treatise as well. Fontana even turned against his own earlier work, claiming that he had not performed enough variations initially.[13]

For several reasons, it is illuminating to trace the many reappearances of the commitment to variations in the text. First of all, it is enlightening to read how Fontana explicitly justified variations of conditions and procedures. Moreover, we can see how the increased attention to variations of conditions and procedures gradually created new challenges to experimenter–authors. As more and more findings from series of experimental trials had to be considered, the readers had to be presented with increasingly complex experimental projects, and the organization of experimental reports became a matter of concern.

Early modern experimenters drew attention to contingencies and accidents that might affect the outcomes of experiments. Repetitions of experiments were the strategy of choice with which to deal with these contingent factors. Fontana, by contrast, demanded that circumstances surrounding experimental trials be examined. This move from "contingencies" to "circumstances" is not just a change of words.[14] It is an indication of a new conception of nature. In early modern writings, nature appeared fickle and "wanton," as Boyle called it, and Boyle emphasized that accidents and contingencies could not be treated in a systematic, methodical way. For Redi and his contemporaries, one principal task for an experimenter was to make sure that any "unseen hindrance" and freak accidents be eliminated from an experiment. Like Boyle, Redi did not call for an actual investigation of these accidents, just for enough repetitions of experimental trials to make contingencies inconspicuous. Fontana, by contrast, explicitly required that the circumstances of experimentation be investigated and assessed. All the measures that Fontana introduced to establish proper procedure suggest that nature was not "fickle" for Fontana. He was optimistic that through multiple variations of experimental settings, the circumstances surrounding and impinging on experiments could be identified.

Fontana described all the variations to his readers. The first chapter begins with experiments on effects of volatile alkali on sparrows and pigeons. Fontana reported:

> I had a pigeon bit in the leg by a viper, and instantly treated the part. At the end of a minute it fell forward, and could no longer support itself. In twenty seconds more it died. I had another pigeon like the first bit in the same way, but did not treat it. At the end of two minutes it fell forward, and in two minutes more it died.
>
> I had two other pigeons bit in the leg; one was treated, and the other not. The first fell at the end of three minutes, and died at the end of the twentieth. The other fell at the end of a single minute, and died likewise after the twentieth.
>
> Of two other pigeons bit in the leg, I treated only one. The one treated died at the end of forty hours, the other at the end of an hour.
>
> I had six other pigeons bit in the usual way. Three were treated, and three not. Those that were treated died at the end of 6, 22, 40, hours. The other three died at the end of 1, 2, 10, hours.
>
> I had two others bit in the leg in the usual way; one I treated, the other I did not. The treated one died at the end of eight minutes; the other at the end of two hours.[15]

This is the second set of experiments in the chapter. The first set, the one on sparrows, had similarly irregular outcomes. Fontana moved to pigeons because the effects were magnified, as it were, in the larger birds. The passage is remarkable for many reasons. First of all, it is reminiscent of Galenic trials to establish the efficacy of theriac. But there are other striking features: the verbose style of writing, the comparison of groups of birds, the repetitions (both the repetitive phrases and the descriptions of sequences of similar trials), and the presence of numbers.

It did seem clear that volatile alkali did not prevent death (all birds had died), but, as the numbers show, the times of death varied considerably—so much, in fact, that "reasonable conjecture" seemed impossible. The treated birds sometimes died after forty hours, at other times after eight minutes. Fontana reiterated his main methodological principle: "experiments must be multiplied, and the circumstances attending them examined more attentively."[16] Through what he described as "an infinite number of experiments, diversified in every possible way, and in which all the circumstances that accompanied them were rigorously examined," he established that a range of circumstances might have an influence on the experimental outcomes.[17] The time of death depended on the size of the viper, the force of the bite, and the amount of venom that was discharged with the blood that came out of the wound. Also, some vipers did not have any venom in their teeth.

The investigation of the circumstances attending the experiment was part of an extended effort to give the experimental design the precision and simplicity that the principal methods for successful experimentation required. Eventually, a change in the design of the experiment drastically reduced the number of circumstances impinging on the experimental outcome. Fontana severed the viper's head from its body. He then manually induced the venom into the wound. In this way, he could select the place of the bite; the compression of the jaw; and the quantity of the venom discharged. He could also prevent the venom from flowing out with the blood, as the tooth could be left in the wound. This design produced much more uniform—much more precise—results with both sparrows and pigeons. All sparrows died within five to eight minutes and all pigeons within eight to twelve: there were no exceptions.[18]

Numbers

The rule that close attention be paid to circumstances of experimentation is one distinctive feature of Fontana's methods discourse. Following that rule ultimately made the experimental design simpler and the findings more precise (so one would hope), but it created additional challenges for the investigator along the way. We already saw that many of Fontana's experiments yielded numbers. Fontana sought to obtain numerical data specifying quantities of venom as well as times of death and the weight of animals, using these as measures of the effectiveness of venom. When quantitative precision is a goal, discordant results become an issue. The variations among the times of death that resulted from similar quantities of venom evidently caught Fontana's attention.

The measurements of how much venom was necessary to kill an animal involved several practical challenges, which Fontana discussed in detail, dead ends and all. We have seen that some experiments were best performed with larger animals, in which the effects of the bite were magnified. In this context, however, it was "expedient" to use very small animals because they would die quickly and "to a certainty," making the data "less equivocal."[19] One practical problem was to find a way of measuring small quantities of venom. Fontana meticulously described the initial procedure he developed for this purpose, even though it was ultimately unsuccessful. He mixed four grains of the venom by weight with eight grains of water and spread it equally over a square inch of thin paper. He then cut the paper in halves and divided each half five more times so that he had two identically sized pieces of paper of each size. He pointed out that the procedure allowed even spreading of the venom, and he noted that he had checked this by comparing the weight of the pieces of the paper after they had dried. The muscles of ten sparrows were laid bare and the bits of paper attached to them. The results were disappointing: even though one should expect uniform results in the same animals, there was no relation between the time of death and the size of the paper (i.e., the quantity of the venom applied); pieces of paper of the same size gave quite different results. Repetitions of the experiment yielded even more irregular results. Fontana surmised that the problem was in the procedure: the reason for the irregularity might be because the paper did not discharge the venom evenly. He thus abandoned the method and adopted a new one.

In the second attempt to make the experiment work, he poured a given quantity of venom on a glass slide, which he then put on a scale. He dipped a tiny scoop ten times into the venom and established that the hundredth part of a grain of venom was thereby withdrawn. Repeated trials gave the same result, so Fontana concluded that with each scoop, the thousandth part of a grain of venom (or a sufficiently similar amount) was withdrawn. Again, Fontana mentioned a possible cause for discrepant results—the venom might evaporate—and accordingly noted that the experiment should be done very quickly, before the venom was diminished in weight by evaporation.

Although the quantity of the venom could thus be measured with satisfactory exactness, the experiments on sparrows using measured quantities of venom again gave irregular results (anything between a few minutes and three days). The irregularities might, Fontana surmised, result because more or less blood had flown out of the incision through which the venom was introduced. An experiment with two incisions and double the quantity of venom did not give any less irregular results.

Fontana then performed several similar trials with pigeons instead of sparrows, and these, too, failed to give uniform results. Early modern experimenters would have picked the "one best" or "one true" result from these outcomes, based on considerations about the specific features of particular trials, the quality of the instruments that had been used, and so on.[20] Fontana did not do that. If it was not possible to achieve complete uniformity of experimental outcomes, he offered approximate values or a range of measurements. His conclusions remained tentative "with some probability."[21] He noted that the thousandth part of a grain of venom was enough to kill a sparrow. Secondly, it took about four times as much venom to kill a pigeon. The sparrows weighed "somewhat less than an ounce each" and the pigeons "somewhat more than six ounces each." For the purpose of specifying relations between fatal quantities of venom and the weight of the animals bitten, Fontana rounded the measurements. Supposing that "sparrows weigh exactly an ounce, and pigeons exactly six," one could then calculate the quantity of venom required to kill larger animals: "an ox for instance, supposing it to weigh 750 lb, will be about 12 grains; and it will require nearly two grains and a half to kill a man, supposing him to weigh the fifth part of what an ox weighs, that is to say 150 lb."[22] The discussion ended on a Newtonian note. Fontana commented that his experiments suggested new hypotheses. Specifically, his calculation "takes for granted some new hypotheses more or less probable," such as the hypotheses that the action of the venom was proportional to its quantity and that the venom acted in the same way on animals of different species. Yet these hypotheses would require further probing by experiment.

Comparisons

Sometimes the attempt to identify the various circumstances surrounding experimental trials led to genuine and unexpected new insights. The investigation of whether tendons are susceptible to venom is case in point. To us, this seems rather an odd project, but Fontana's contemporaries would have been aware that Fontana engaged with a problem that had been prominent in Haller's treatise on irritability and sensibility. Fontana meticulously described the entire voyage of discovery. The details of this episode are particularly noteworthy because Fontana actually ended up refuting Haller's position, but, as in the case of Mead, only after describing the series of experiments that brought him to the point. According to Haller, tendons were insensible. To demonstrate this, he had exposed and pricked the tendons of the tibia and Achilles tendon in small animals. None of the animals had shown signs of pain. Haller told his readers that he had repeated these experiments "upwards of a hundred times since the year 1746, upon dogs, goats, rats, cats, rabbits, and various other animals, always with the same success." Not even a young man whose flexor of the finger Haller had touched with a pair of forceps had experienced any pain. Haller had also observed that the tendon never contracted when the muscle did; he had "seen it a hundred times." Tendons were neither sensible nor irritable. This finding marked a radical break with tradition. For centuries, exposure of the tendons had been considered extremely dangerous, as well as painful to the injured individual. But none of Haller's experiments had involved the fatal effects on the animals that traditional conceptions of tendons would imply. For Haller, the traditional beliefs were simply erroneous.[23]

Obviously inspired by Haller, Fontana set out to demonstrate that venom, too, had no effect on the tendons. He described at length how he had proceeded, and he kept the reader in suspense until the end of the section. Fontana stripped the tendons of three rabbits of skin, wounded the tendons with a venomous tooth, and covered the wounds with linen. The rabbits died. This was a big surprise because they should not have. Tendons are not endowed with vital principles (as Haller had shown), and they are disconnected from the blood and nerves, so a bite in the tendon should not cause any harm. Fontana thus assumed that the venom must have affected the animal in some other way—say, by entering the blood stream. To explore and ultimately exclude this possibility, he varied the experimental conditions in a number of ways. To avoid contact between the venom and the blood and nerves of the victim, he changed how the wound was inflicted, how the venom was inserted, and how the wound was dressed. To prevent the venom from seeping through the linen, he put a thin piece of lead between the folds. He instilled as little venom as possible to avoid spillover. He folded the linen sixteen times and dried and washed the wounds of three rabbits until "it was not possible for an atom of venom to remain within it."[24] He removed the linen immediately after having envenomed the wound and replaced it with fresh linen. He applied ligatures to the tendons of three rabbits and introduced the venom between the ligatures. But whichever way Fontana performed the experiment, it invariably had the same outcome: the rabbits died.

Not until he had described in grueling detail how he had examined each of the circumstances that might have produced the unexpected effect—and how each time he had failed to establish that this was the one—did he state his conclusion. The exposure of the tendon, not the venom applied to it, was the real cause of death. Fontana's final demonstration was brief and succinct. To demonstrate that it was fatal for an animal to have its tendon exposed, Fontana described a decisive experiment, similar in its design to those we encountered in the dispute between Redi and Charas. He used six rabbits, exposing the tendons of each. Two of the rabbits were wounded with a venomous tooth, two were pricked with a needle, and two had simply laid their tendons bare, without wounding or pricking. Again, all rabbits died. Fontana concluded that tendons were not susceptible to venom; exposure of the tendon was fatal for animals. The ancients had been right after all.[25]

Fontana did not develop this point, and he did not elaborate on his disagreement with Haller. He let the report of the investigations speak for itself, merely remarking that his results had implications for medical therapy, including the treatment of humans. He concluded the section on tendons with a reiteration of his methodological principles. He did not comment on the decisive experiment and its demonstrative force but only on the productive power of repetitions and variations: "The multiplied and varied experiments I made on the tendons, have been of very great use to me in the pursuit of my researches. If any doubt had remained on the subject; if I had not assured myself to a certainty that the bite of the viper on this part is not attended with any consequence; if I had apprehended that the venom could communicate itself to the animal through the medium of this substance; I should have a thousand doubts as to the parts on which the venom acts, in an animal that has been bit. No subject in nature is absolutely indifferent; and when such rare and extraordinary effects are to be examined in the animal body, nothing is to be neglected—nothing is to be deemed unnecessary."[26]

To evaluate the meaning and status of experimental strategies such as comparative experimentation, it is crucial to consider methods discourse in long-term historical perspective. Early modern experimenters applied comparative strategies similar to Fontana's. Redi did so when he wounded pigeons with broom straws to examine whether this intervention had a harmful effect on the animals. Charas performed the same kind of experiment. Mead did, too, when he compared unadulterated blood with a mixture of blood and venom. Even earlier examples of comparative experiments can be found in Galen's work and in medieval writings on drugs and poisons. Many similar examples of comparative experimentation can be found throughout Fontana's treatise. Like Redi and Charas, Fontana compared an animal that was injured, with venom then applied to the wound, and another that was injured but that did not receive venom. Like Mead, he compared the behavior of blood and the behavior of a mixture of blood and venom. Like Galen, he also compared animals that received venom and an antidote with animals that received only venom.

Fontana did not include the comparative strategy in his principles of successful experimentation. But he did single out comparative trials in his account by explicitly calling them "comparisons" or "comparative experiments." "Comparative experiments" even appear as an item in the Index (under "Nerves" we read: "—comparative experiments, by making simple mechanical wounds on the sciatick nerve"). He even stated that "I have made it a maxim, in almost the whole course of this work, to form comparative experiments, and only to compare those with each other, that were made at the same time and with the same circumstances."[27]

I noted at the beginning of chapter 5 that in the 1960s Fontana was hailed as the first venom researcher who used "adequate controls." We can now see more clearly why some later commentators on Fontana's endeavors could have come to the conclusion. Considering the weight that Fontana placed on comparative experimentation, and in light of the maxim that the experimenter was to form comparative experiments, it is understandable that Fontana has been considered a key figure in the history of control experiments. But we also begin to see why it would be misleading to describe Fontana's approach in these terms. One reason is that doing so would obscure the long tradition of comparative experimentation on which Fontana could build.

I will discuss control experiments in more detail in later chapters; here I will simply offer a few preliminary remarks. First, the very term *control* appeared only in the late nineteenth century. The systematic discussion of the structure of comparative experimental practices is also a nineteenth-century occurrence.[28] Second, the analogy between Fontana's trials and a modern controlled clinical trial with a treatment group and a control group strikes me as far-fetched even though Fontana sometimes compared small groups of animals that received different treatments. The comparison of groups rather than individual animals did not result from methodological considerations as they underlie a modern randomized controlled trial. Rather, by experimenting on small groups, Fontana followed his own methodological principle for successful experimentation. By giving several animals the same treatment, "fortuitous cases that vary experiments" are ultimately made inconspicuous and the comparative experiments become more decisive.[29] Fontana certainly was an innovator, but only a careful reconstruction of his methods discourse in historical perspective reveals what was novel in late eighteenth-century thought about experimental methods—namely, the acute awareness of the significance of systematic variation of experimental conditions as part of an attempt to identify causes and circumstances in a complex experimental situation.

Hidden Causes

We have already moved far away from the original issue that motivated Fontana's quest—namely, the efficacy of volatile alkali. In the last part of the treatise, Fontana turned to an even more fundamental problem in venom research: How exactly does viper venom work? This question was still wide open. Charas's Helmontian interpretation and Mead's mechanical explanation had both turned out to be unsatisfying. Fontana's earlier experiments had suggested that venom affected the irritability of tissues, but this explanation had also been abandoned. Twenty years later, he organized his experiments around a more elaborate theory of disease according to which venom possessed an active ("obnoxious") principle that moved through the body at great (yet not infinite) speed and produced identifiable lesions—both visible and hidden inside the body.

Fontana's experiments as described in the last part of the treatise thus dealt with intangible things. He admitted that we might never truly know the "active principle" of viper venom—just as the nature of gravity might forever remain obscure. But at least it was possible to find out more about what the venom principle did to the animal body, both in terms of visible lesions and in terms of internal alterations, and how quickly it acted.

This part of the treatise is organized much more tightly than the part that describes the experiments on volatile alkali and on the lethal quantity of venom. Fontana went into great detail telling the reader what trials had been performed, what attempts had been made to make the experiments simpler and more precise, and what efforts had been made to magnify the observed effects. Nevertheless, this part is anything but sprawling and meandering. It is organized in such a way that any reader who was tough enough to read through the gruesome descriptions with detachment could not but agree with Fontana's final conclusion: that the internal disease caused by viper venom was an affection of the blood and that venom acted by way of an obnoxious principle.

The account of these trials is both intriguing and repulsive. It is intriguing because Fontana devised ingenious experiments to investigate whether there was an "active principle" that was responsible for the lesions that venom produced and how fast this principle acted. It is also intriguing because he figured out ways to investigate the hidden lesions and because he discussed various methodological issues along the way. It is repulsive because he described in unrelenting detail the wounds he inflicted on the experimental animals and the numerous vivisections he performed to examine internal lesions and hidden effects. In fact, it was because of the great importance and the intricacy of these experiments that all the circumstances, tools, and procedures had to be meticulously reported. Fontana even described the syringe that he used to make injections, the way in which he moved the piston to fill it with venom, and the way in which the syringe was inserted into the blood vessel and emptied of its deadly content.

The first step in this part of the investigation was to refute the previous conceptions of venom-induced disease. From Fontana's findings, it can be inferred that the disease was not caused by a chemical or mechanical action. Fontana cut off the leg of a pigeon and had the leg bitten immediately, while it was still warm, palpitating, and bleeding. He then examined it for any signs of disease. Because it was important to have the amputated limb bitten as soon as possible after the bite—before the signs of life had disappeared—several experimenters had to work together. They had to follow a complicated choreography whereby one person restrained the animal and performed the operation while another handled the snake. Fontana reported that he had made a dozen experiments in this way. Never did it take more than three seconds between the amputation and the bite. In several trials, the bite and amputation took place within a single second. But none of the amputated legs showed anything "worthy of observation."[30]

Nevertheless, the experiments were repeated on the muscles of twelve frogs and afterward on more pigeons and frogs. But the results remained negative. For Fontana, it was thus clear that the venom did not produce any sensible changes in the amputated limbs. Venom did not act mechanically or by a simple mixture with body fluids. Nor did it react chemically with the humors of the leg. Such an exclusion of hypotheses might be "a step towards the truth" but was not enough to establish how and by what principles it did act.[31]

The external disease could be directly observed; it manifested itself in the visible lesions produced by the bite or venom injections, such as lividness and swelling. To determine the time it took for the venom to produce these visible symptoms, Fontana performed amputation experiments whereby the limb was cut off after the bite and the time of the amputation was increased in ten-second intervals. The legs were cut off at ten and twenty seconds, and so forth, after the bite. No change was perceived in the first trial. At twenty seconds, some symptoms of disease appeared, and in the remaining cases, the symptoms were fully developed. A ratio of five seconds showed the disease with certainty after twenty-five seconds. The action of the venom was not spontaneous; it took a certain time for the venom to take effect—"from fifteen to twenty seconds, or thereabouts."[32]

The internal disease was harder to detect. At first, Fontana simply took the time of death of the animal as a visible and measurable sign of the internal disease. To find out how long it took for the venom to bring about the internal disease, he again performed amputations. They were carried out just like the previous series, but instead of the time required for the external symptoms to show, Fontana measured the time of death. All things considered, Fontana stated that it took "somewhere betwixt fifteen and twenty seconds" for viper venom to produce the internal disease. Given that the symptoms of the internal and the external diseases appeared at the same time, Fontana concluded that the two diseases accompanied each other and were both produced by venom.[33]

Fontana then proceeded to investigate the hidden lesions and eventually even the hidden causes of the disease. No doubt, one reason for Fontana's care and attention to detail was that the series of experiments he described not only contradicted Mead's position but also showed that Mead had made the wrong experiments. Mead's experiments had shown that mixing venom with blood in vitro had no effect on the blood. Fontana's observations of the phenomena accompanying snakebite, however, indicated that the blood *was* involved in the disease.[34] In light of these findings, Mead's claims were simply not plausible. Fontana's new experiments were designed to show how venom and blood reacted *inside* the living animal. Vivisections revealed an immediate, noticeable effect. Of course, Fontana took the occasion to stress, yet again, that his test of Mead's experiment "shows how cautious we ought to be in the inferences we draw from experiments; and at the same time proves to us, that we know little or nothing, at least with any certainty, and without incurring the risk of being deceived, beyond that which is demonstrated to us by experiment alone."[35] As before, the reader learns how Fontana gradually adjusted the design of his experiments by varying several circumstances and constantly checking for other factors that might affect the outcomes of the trials.

Eventually, Fontana designed experiments to characterize the very cause of disease. Chemical tests in vitro had shown that venom was a gum, for it reacted just like gum Arabic: when mixed with spirit of wine, it gave a white precipitate. But, Fontana asked, does venom act on the blood "as a gum," or "as venom"? To answer this question, he compared mixtures of venom and blood and of gum Arabic and blood. The first mixture turned black and stayed fluid, but the second mixture remained red and coagulated. He concluded that "the changes that are wrought in the blood by the venom of the viper, are not the effect of a gummy principle, but of some other principle yet unknown to us, probably the very one that constitutes it a venom, and indeed we have hitherto been able to distinguish nothing more in this humour, than a gummy principle, and a venomous principle destructive of animal life."[36]

In retrospect, we might wish to say that Fontana did not, in fact, investigate the venom principle and the gummy principle as such. After all, he characterized both principles by their observable effects. From Fontana's perspective, however, it was at least established what the principle was *not*: it was not a mechanical or chemical agent, and it was also established that venom was a composite comprised of different agents. That was more than one step toward the truth.

Writing Analytically

"I feel that I have been too prolix," Fontana noted toward the end of the account of his venom research.[37] But because he had so many novel things to say, he felt justified in his chosen way of writing—he had chosen the "analytical" way, the detailed description of his work with ample information about actual procedures, circumstances, and failed trials. I, too, have been prolix in my presentation of Fontana's treatise because only a close reading reveals how methods discourse aids the argument and exactly how Fontana's way of writing compared to early modern experimental reports.

It would be a mistake to read Fontana's comment about prolixity as a late echo of the literary technologies that early modern experimenters were using to communicate their experiments. For Boyle and other seventeenth-century experimenters, "prolixity" had to do with the details, procedures, and contingencies surrounding a specific experimental trial. In the late eighteenth century, "prolixity" took on a different meaning. Fontana and his contemporaries conducted series of experiments—often, as in the case of Fontana's venom experiments, extremely complex projects. "Prolixity" now had more to do with the succession of steps of the investigative pathway. At this point, concerns about the prolixity of scientific writing had become threefold: How much detail should be provided about individual experiments? In how much detail should the series of investigations be reported? Should the organization of the report reflect the order of the investigation? More specifically, in exactly what order should data, thoughts, reasoning, and hypotheses be presented? At a time when experimenters were performing more and more extended series of experiments, this last problem had become especially pressing. Fontana's second treatise exemplifies a possible solution.

In philosophical discussions about scientific discovery and justification, it is often pointed out that prior to the nineteenth century, correct methods were thought to lead to correct outcomes. In other words, seventeenth- and eighteenth-century scholars adopted generative theories of scientific method, and methods of discovery were taken to have probative force.[38] Explicit accounts of the methods used in a particular investigation thus served as validations of the knowledge thus obtained.[39] Bacon's conception of his "new method" as it is presented in the *Novum Organum* exemplifies such a method of discovery with probative force. Notably, Bacon's "new method" is a method of reasoning successfully from findings that were obtained in successful experiments. The *Novum Organum* illustrates how best to arrive at knowledge about "form natures" (the most general properties of matter) via a systematic investigation of phenomenal natures. The reader learns how first to collect and organize natural phenomena and experimental facts in tables, how to evaluate these collections, and how to refine the initial insights with the help of further experiments. Through these steps, investigators will arrive at conclusions about the "form nature" that produces particular phenomenal natures. The point is that for Bacon, the procedures of constructing tables and evaluating the facts thus arranged leads to secure knowledge. In this sense, the procedures of construction and evaluation have probative force.

Newton, too, can be regarded as a proponent of a generative theory of scientific method. Newton's aim in the *Philosophiae Naturalis Principia Mathematica* was to present a method for the deduction of propositions from phenomena in such a way that those propositions become more secure than propositions that are secured by deducing testable consequences from them.[40] Newton did not assume that this procedure would lead to absolute certainty. Only moral certainty could be obtained for the propositions thus secured. But in his approach, propositions could be established and secured by showing that they followed from observed and experimentally produced phenomena.

For late eighteenth-century scholars, the generative theory of scientific method served as a guide to the writing of an experimental report. In his 1775 treatise *L'art d'observer*, for instance, the Swiss scholar Jean Senebier included an entire chapter on the question of how observers and experimenters should write about their empirical work.[41] Senebier demanded that investigators report their findings as well as the means by which those findings had been obtained. All procedures, all circumstances must be exactly described. The investigators also must indicate what they had seen well and what they had seen less well; and Senebier praised Haller for providing estimates of the degree of trust the reader could have in the facts Haller himself presented.

The degree of detail that was necessary in the description depended on the novelty of the phenomena reported. An investigator who describes, say, every aspect of the life of a horse, can keep it short; by contrast, an investigator who discovers a new being or who can illuminate the yet unknown sides of an important phenomenon ought to provide all the specific circumstances of the investigation.

Senebier also insisted that if these reports were to have any use, the author must bring order and method into the presentation of the findings. Otherwise the reader would not be able to appreciate how they were related, nor how they mutually informed each other. Notably, he stated that it depended on the purpose of the communication whether the systematic presentation of the findings or the path the experimenter had taken was more important. Both Newton's *Opticks* and his early letters on optics to the Royal Society were admirable works, but although the treatise on optics helped the reader understand the laws of optics, the letters showed how Newton had paved the way and how he had arrived "à la nature."[42]

Early modern experimenters sometimes drew a line between narratives of experiments and the interpretation of their significance (the "discourses" made on experiments, as Boyle put it). Boyle insisted that interpretations of experiments be kept separate from the narratives. Bacon offered concrete guidelines for the interpretation of the facts as they were presented in the tables. In the course of the eighteenth century, the question of how the experiments should be "discoursed upon" also was a matter of systematic discussion. Senebier devoted an entire section of his work on observation to the "analytic method."[43] He had in mind a method of inference from findings to ideas. The analytic method was based on a Lockean conception of knowledge, according to which abstract ideas originated in sensations. Senebier emphasized that the analytic method of inference—the progress from simple sensations to complex ideas—was the path to truth and at the same time the proper method of instruction.

I have not come across an explicit reference to Senebier's work in Fontana's writings, but as both researchers were in frequent correspondence about other issues,[44] it is highly likely that Fontana was familiar with his ideas. There are several striking parallels between Senebier's treatise and Fontana's own discussion of how to write scientifically about observations and experiments. There are some telling differences as well. For Senebier, prolixity and the analytic method were two different issues. Prolixity pertained to the description of the investigative pathway. It was a feature of accounts of pioneering investigations. It was discussed in the section that dealt with the communication of observations. The analytic method was introduced much later in the book, in the part about proper interpretations of empirical findings. It was introduced as a method of reasoning from the facts. For Fontana, by contrast, both prolixity and the analytic (or analytical) method were features of the account of the investigative pathway.

Fontana distinguished between "analytical" and "synthetic" methods of presentation, explaining that experimental reports written in the analytical fashion presented experiments in the order in which they were made. Reports presented according to the analytical method were very detailed and described not only the successful experiments but also the experimenter's errors. Fontana did not say much about the alternative—the synthetic method—except that in synthetic writing "mystery and reserve" abounded and that all the traces of the process of discovery were wanting.[45] Most likely he had in mind the mathematical notion of synthetic demonstration, which was preceded by a heuristic procedure by which solutions to mathematical and geometrical problems were discovered (whereby the heuristic procedure remained implicit). Fontana conceded that the synthetic method was more concise and perhaps even clearer than the analytical method, but (and here his comment on writing morphed into a comment on procedure) the analytical method was "the most certain, the most luminous, and the only one which leads immediately to a discovery." In a sense, Fontana's analytical method of presentation demonstrated the discoverability of his findings.

Some editing and rearranging was acceptable, and not all results need be presented. The idea behind describing the path of the experimenter was not that this path was necessarily the one right path. Rather, the idea was that correct appraisal of the findings required knowing how these findings had been obtained. When Fontana called for an analytical approach, he recommended it because it "inspires the reader with confidence, shows in what way the naturalist has searched into nature, and in what way she has answered to his researches."[46]

Because Fontana described not only the last, most improved version of an experiment and the final results of a series of trials but also the preliminary attempts at investigating a phenomenon, the exploration of an experimental setting, dead ends, new beginnings, and the problems that he encountered—as well as the steps he had taken to overcome those problems—the reader could get a sense of how he had arrived at designing and performing the final, definite experiments. He remarked that his readers could, and would, notice whether he had made a mistake when he drew conclusions from the facts that he had established. Because false observations gave rise to false theories, it was important to give his readers the means to spot those, too. The analytical method of presentation allowed them to do so.

When Fontana contemplated the "prolixity" of his writing, he was not chiefly concerned about the amount of detail in his descriptions of individual experimental trials; rather, he referred to the circuitous way in which he described the paths of his investigations. The question of prolixity was about whether and in what degree of detail the sequence of steps of the investigation should be described. Fontana took pride in having "endeavoured to be as exact as possible in the facts, and [in having] entered into a long detail on several of them."[47]

Fontana's contemporary, the Swiss geologist and meteorologist Jean-André de Luc, by contrast, despaired over his treatise on the atmosphere. He was at least as obsessed with details, circumstances, and investigative practices as Fontana was—so much so that he spent a large portion of his treatise on the atmosphere on discussions of the measuring instruments and procedures that were necessary for the project. The resulting tome, published after a delay of ten years, was a "ruin of a book." Unlike Fontana, de Luc ultimately failed to keep his meandering story together.[48]

Fontana often described successions of experiments when he could have presented aggregate accounts, and he reported what steps and measures he had taken to improve an experimental setting. Recall, for instance, the lengthy quote from the beginning of this chapter. Instead of writing "I had a pigeon bit in the leg by a viper. . . . I had two other pigeons bit in the leg. . . . Of two other pigeons bit in the leg, I treated only one. . . . I had six other pigeons bit in the usual way," he could have reported the total number of birds bitten and treated. Recall also the experiments on quantities of venom and on mixing venom with blood—in both cases, Fontana described chains of experiments that ultimately turned out to be useless for his intended purposes. His narrative preserves a certain chronological order in the sequence of experimental endeavors that are described, and at the end it leaves the persevering reader with a sense of scientific adventure (not to mention exhaustion, repugnance, and relief).

For Fontana, the telling of the journey of discovery served as a textual tool to assure the reader of the soundness of the findings. Already in the first book, Fontana used this technique to deal with Mead. In the second book, he used it in the report of the experiments that led to the refutation of Haller's position. Fontana could have stated at the outset of the section on tendons that his experiments proved Haller wrong, and he could have presented his findings in support of his claim. Instead, he led the reader down an investigative pathway that at the end permitted only one conclusion— that Haller must have been mistaken. Indeed, the second book as a whole is organized in this fashion; it is a painstaking report of the endeavor to refute Sage and to espouse an alternative position.

Thousands of Experiments

In the final paragraph of the supplement to the French treatise on poisons, Fontana declared: "A simple glance thrown on my work itself, will show how easy it is to be deceived in matters of experiment, even when many of them have already been uniform, and when one would the least expect: the possibility of being led astray. My experiments, on the side of truth, exceed the number of six thousand, and the observations I have interspersed through the work are at least as numerous. I know very well that the questions I have proposed and examined are likewise very numerous, and that there may be some few in the number, as I have observed on a former occasion, that have not been discussed with as many experiments as were necessary. But in spite of all this, I firmly maintain, that a few experiments will not be sufficient to destroy the great number I have made, and varied in so many ways, and that like contradictions will not be capable of making me change my manner of thinking."[49]

Much in this quote is reminiscent of the early modern experimenters and their methods discourse: the reminder that experiments might deceive, the modesty that the experimenter–author displayed, and of course the reference to numerous experiments. In his treatise, Fontana paraphrased several of Redi's statements (albeit without referencing the relevant passages of Redi's work), and he drew on the Tuscan literary tradition as well as on methodological ideas from the texts that were published in the context of the Cimento Academy. Fontana took the commitment to repetitions to the extreme, implying that thousands of repetitions might be necessary to gain confidence in an experimental result. The text is packed with claims that particular experimental trials were repeated numerous times or even a thousand times, that hundreds and even thousands of animals were used, and that the experiments were varied in a thousand ways. In the early modern debates, references to numerous experiments were particularly frequent at critical junctions—especially when opponents or rivals had to be challenged. Fontana added the supplement to his book because he had just encountered another researcher whose findings disagreed with his own. Clearly, "a few experiments" were not enough to prove Fontana wrong— the reference to many repetitions once again served as a weapon to disarm opponents in debates and disputes. And it was an effective weapon, as the reviews of his work show.

The numbers and statistics Fontana presented suggest that he did repeat his experiments to a degree that to twenty-first-century readers appears excessive. Fontana's first readers certainly took his statements literally. The author of the preface to the French translation—included in the English edition—pointed out that "one of the greatest merits of this work consists, not so much in the rare and numberless discoveries it contains, as in the luminous method with which the very important enquiries that are introduced in it are treated."[50] He added: "But what confidence ought not an author to inspire us with, who, after having said, *I have made more than 6000 experiments; I have had more than 4000 animals bit; I have employed upwards of 3000 vipers*; finds no difficulty in adding, *I may have been mistaken and it is almost impossible that I have not been mistaken!*—What a difference betwixt this authour and many others! betwixt opinion and certainty! betwixt ignorance and knowledge!"[51] Johann Friedrich Gmelin's review of Fontana's treatise, published in 1782, also praised Fontana for his patience and care, which Gmelin thought were evident from Fontana's having repeated his experiments thousands of times.[52]

Fontana's methodological notions are not simply paraphrases of early modern methods discourse, however, as close reading of his text reveals. Unlike Redi, he assumed that through multiple variations of the experimental

setting, the circumstances surrounding experiments could be identified—
"It is almost impossible, in repeating them so many times, that one does
not encounter fortuitous circumstances that vary [the experiments]"—as
the first principle of experimentation states. The commitment to hundreds
and thousands of repetitions, to variations, and to the identification of circumstances became a driving force for the entire project.

Fontana's methodological essay simply called for "diversified" experiments. In the descriptions of his experiments, we can—in hindsight—
identify different forms of "diversifications" to achieve different goals:
variations of experimental designs and procedures to establish whether
particular conditions are necessary or irrelevant for an effect—for instance,
the way in which a wound is dressed after an animal was bitten. Variations
of conditions could also serve to establish whether alternative conditions
could bring about the same effect—for example, whether the amputation
of a limb led to death when no venom was involved. Variation might also
mean amplification or reduction of a condition; the question here was how
such a change, such as in the size of the viper, would modify the effect. Finally, variations of experimental conditions could also serve the purpose of
bringing out the experimental effect in the best possible way—by using a
larger animal, the effects of the bite would be "magnified" in the larger organism, for example.[53] The obsessions with repetitions, diversifications, and
circumstances had palpable consequences. The large number of repetitions
that Fontana undertook to ascertain what really was going on in an experiment yielded discrepant experimental results. Because the uniformity of an
experimental outcome continued to be an ideal, the discrepancies called
for changes of experimental settings until uniformity of experimental outcomes was achieved. Sometimes the quest for uniformity did lead to a more
stable experimental design. Sometimes the quest for uniformity stimulated
changes in the direction of the experimental endeavor. At other times, efforts at unifying were not completely successful or entirely unsuccessful and
a result had to be extracted from a range of outcomes. Unlike the early
modern experimenters, Fontana did not pick the "one best" or "true" result from a range of outcomes. If it was not possible to achieve complete
uniformity of experimental outcomes, he indicated the range of measured
values. Even in the case of quantitative data, such as times of death, the final
result was an interval (from fifteen to twenty seconds, for example). The
gradual erosion of the ideal of uniform outcomes is an unintended consequence of the methodological principle of repetition.

The commitment to "diversification" of trials turned the attention from single experiments to series of experiments and from the uniformity of outcomes to differences. This opened up new questions, both questions of experimental design and related questions of conceptualization, interpretation, and, ultimately, reporting: What are the factors need to be varied? What do the differences between different experiments tell us? How much of this work should be reported? And how should the report be organized? Fontana's sprawling account is less vivid and definitely less amusing than Redi's or Platt's letters. It is not meant to entertain; it serves to demonstrate the discoverability of the findings. For Fontana, prolixity and analytic methods were no longer features of the presentation of one trial and the reasoning from facts to general propositions, respectively. They had both become features of the presentation of the investigative pathway. Only through careful interpretation of methodological statements do we realize that the resemblance between his and early modern writings on experimental trials is, in fact, largely superficial, and only then can we assess whether he really was a trailblazer for nineteenth-century experimenters.

CHAPTER SEVEN

Practical Criticisms

Chapter 6 presents Fontana's methods discourse both as a creative appropriation of a long tradition and as a product of the challenges he encountered in his own endeavors. In this chapter, Fontana's work becomes a vantage point for the study of nineteenth-century methods discourse. I begin with Silas Weir Mitchell, whose long-standing interest in venom research spanned the second half of the nineteenth century. Today, Mitchell is perhaps best known as a novelist and poet. Historians of medicine know him as an expert on nerve injuries and as the developer of the "rest cure" for the management of nervous diseases. But clinical neurology was only one of many of Mitchell's specialties. In biographies and obituaries, Mitchell is usually celebrated as a man of many talents—physician, educator, clinical neurologist, physiologist, poet, novelist, and medical reformer.[1] For the historian of methods discourse, Mitchell's work is of interest because it showcases the dynamics of medical and methodological thought during a key period in the history of the biomedical sciences, just like Mead's *Mechanical Account of Poisons* did for the eighteenth century.

The nineteenth century has become something of a historiographical embarrassment for historians of the life sciences. Much has been written on the development of the biomedical sciences from a number of historiographical perspectives. Histories have drawn on nineteenth-century medicine, physiology, and biology to exemplify the professionalization of the sciences, the rise of the research university, the role of teaching laboratories for the advancement of medicine and biology, the formation of large, state-funded research institutions, and so on. The historiography from the 1980s and 1990s was concerned with broad developments; older histories of the biomedical sciences from this period framed the nineteenth century in terms of the "rise of scientific medicine" or the "laboratory revolution."

More recent work in the history of the biomedical sciences has cast into doubt many of the long-standing historiographical concepts, generalizations, and grand narratives;[2] even the very notion of "scientific medicine" itself has been critiqued.[3] Instead, there have been calls for more localized, fine-grained, comparative studies of knowledge and practice, both in the laboratory and in the clinic.[4]

The study of methods discourse in mid-nineteenth-century venom research can do double duty. On the one hand, the comparison with late eighteenth-century experimental methods and methodological thought can give us a sense of the developments since the time of Fontana. On the other hand, such a study can offer a fresh perspective on mid- to late nineteenth-century biology and medicine. Venom research was situated at the intersection of several areas of biomedical investigation—including toxicology, physiology, medical chemistry, and therapeutics. Exposing continuities, resemblance, and innovation in the protocols, methodological conceptions, and commitments to experimentalism across these fields does justice to the complexities of local research environments while at the same time transcending these local contexts.

For many reasons, Mitchell's works on snake venom are particularly suitable for an examination of the transformations of biomedical knowledge, practices, and methods discourse. His medical career path was rather typical for a member of the medical profession in the United States. His work on snakes brought together different perspectives on venom poisoning, including chemical and toxicological studies of venom as well as clinical studies of venom poisoning and of the efficacy of antidotes. Mitchell produced his main works on venom in Philadelphia around 1860 and after 1885. He commenced his project at a time when many members of the American medical community began to debate the merits and demerits of "scientific" medicine and the value of laboratories and inscription devices. Mitchell made his last contributions to venom research when new findings in bacteriology began to impinge on nineteenth-century medical thought and practice.

The Life and Career of a Medical Doctor

Laboratory research was among Mitchell's main pursuits, but his opportunities were limited. In the mid-nineteenth century, when he was a medical student, laboratory exercises were not part of medical training. At that time, low-quality proprietary medical schools dominated the system of medical education. Teaching was largely based on lecturing and perhaps a few dissections.[5] Before the 1870s, medical men who wanted to pursue experimental research in physiology, chemistry, or toxicology were, at best, confined to small research facilities offering modest laboratory space if they did not want to work at home.

Mitchell did gain some laboratory experience while he was a student, mostly in chemistry. Mitchell's father was a chemist, and Mitchell's biographers note that Mitchell had been interested in chemistry from an early age.[6] Because of his family relationships, the student had access to a chemical laboratory, and he could also do some research in physiology.[7] After he had completed his medical training, Mitchell traveled to Europe, like other young physicians of his generation did. First he went to England, then to Paris, where he encountered Claude Bernard, the renowned French experimental physiologist, whose experimental work had drawn so many visitors to his "ghastly kitchen."[8] According to Mitchell's most recent biographer, Mitchell's French was so poor that he did not get much out of his visit. (In a letter to his family, Mitchell reported that he had been "greatly edified by hearing a lecture an hour long, scarce a word of which could I comprehend."[9]) Nevertheless, Mitchell's acquaintance with the French physiologist and especially with his work was formative for his entire scientific career, as we will see.

After his return to the United States, Mitchell practiced medicine in his father's surgery, where he also pursued some research. Throughout his career, Mitchell's access to laboratories remained limited, and even in later life, he could never devote all his time to laboratory research. His main source of income was always his surgery, and he conducted physiological experiments merely in his spare time. In his autobiography Mitchell wrote that in his early years he would spend every afternoon and evening doing research, often from 4:00 p.m. until 1:00 a.m.[10] Textual traces of Mitchell's time-consuming duties as a physician are scattered throughout his publications—in the accounts of his experiments, he repeatedly mentioned that his duties had kept him from continuing or completing projects or that he had been forced to leave an animal experiment unobserved for extended periods. One of Mitchell's biographers notes that in the 1850s, "the front office [of his surgery] was full of patients and the back office of rattlesnakes and guinea pigs."[11] In addition, Mitchell could also make use of a modest laboratory in the Philadelphia School of Anatomy's building.[12]

According to the older historiography, the development of nineteenth-century American medicine can be characterized as a steady move toward "scientific" medicine, a gradual acceptance—if slow and late compared to Europe—of the merits of laboratory-based and experimental medical research and animal experimentation combined with the introduction of practical training as an integral part of medical education. The rise of experimental, laboratory-based research in the United States was inspired, promoted, and facilitated by the activities of many young physicians who had been to Europe—particularly to Germany—and who sought to establish similar facilities and programs in the United States.[13] A number of them brought instruments home or later returned to Europe to buy laboratory equipment.[14]

The standard story is not wrong; it is just much too sweeping. Mitchell's career path shows that an explicit commitment to experimentation could very well be damaging for someone wanting to be successful in mid-nineteenth-century American medicine. Mitchell spent most of his professional life in Philadelphia. In the 1860s, he encountered strong resistance from many physicians who did not support laboratory work. During the 1860s, he attempted twice to get a chair at Philadelphia, once at the University of Pennsylvania, once at Jefferson Medical College. On both occasions he lost the competition (despite ample support from renowned experimentalists) to a medical practitioner who was not an experimentalist and had better social connections to the medical and academic establishment in Philadelphia.[15] "Philadelphia's 'lost' physiologist" was above all a busy physician with a medical practice.[16] Nevertheless, even though his attempts to become a professional physiologist remained unsuccessful, Mitchell eventually became a main player in the reforms of medical education and research at the University of Pennsylvania. In 1875, he was appointed to the board of trustees of the University of Pennsylvania, and in this capacity, he was involved in the reorganization of the medical curriculum, encouraged the shift to practical laboratory education, helped expand medical personnel, and promoted the construction of new laboratory facilities for the university.[17] And, of course, he also had his literary career.

Given his many professional duties and occupations, it is astounding how much research on venom Mitchell managed to accomplish. He began his experiments in the late 1850s and continued to do venom research, on and off, until the very end of the nineteenth century. He published a number of books and articles on the anatomy, physiology, and toxicology of poisonous snakes, in particular on the chemical nature of snake venom and its effects on organs, body fluids, and tissues. The results of his first studies were published in a book-length report, entitled *Researches on the Venom of the Rattlesnake: With an Investigation of the Anatomy and Physiology of the Organs Concerned*. The book appeared in the series of proceedings of the

Smithsonian Institution in 1860. The complementary text, "On the Treatment of Rattlesnake Bites, with Experimental Criticisms upon the Various Remedies Now in Use," appeared in 1861 in the *Medical and Chirurgical Review* and also as a separate publication in a slim volume produced by Lippincott, a Philadelphia publishing house.

During the Civil War, Mitchell did mostly clinical work, specializing in nervous diseases.[18] After the war, he briefly returned to venom research and did a few experiments to clarify some issues that had arisen from his earlier work. This led to another short publication.[19] But it was only around 1880 that Mitchell engaged in a new extensive project on snake venom. Edward Tyson Reichert, then a demonstrator of physiology at the University of Pennsylvania, assisted.[20] The scope of the later project is considerably wider than the earlier, covering the venoms of different kinds of poisonous snakes. The results were published in 1886 as another book-length report, *Researches upon the Venoms of Poisonous Serpents*, coauthored by Mitchell and Reichert. The publisher was again the Smithsonian Institution. Along the way, Mitchell published a few popular essays on snake bites in magazines for general audiences such as the *Atlantic Monthly* and the *Century Magazine*. This chapter deals with Mitchell's work from before the war.

New Audiences for Experimentalism

Michell's two initial works on snake venom were addressed to the two major audiences in mid-nineteenth-century American medicine: those interested in "elementary research" (as experimental investigations were sometimes called) and clinicians who had an interest in effective remedies and treatment options. In his Smithsonian essay of 1860, Mitchell advanced two overall points: First, venom was a composite of components, and not all of these components were toxic. Second, venom could produce "acute" and "chronic" diseases. In acute cases, death occurred rapidly, within minutes, and in these cases, respiration and the heart became enfeebled. In secondary or chronic poisoning, death occurred only after several hours, and postmortem dissection showed changes in the blood. The slender book on the treatment of snakebite from 1861 is a treatise on therapeutics. It discusses the efficacy of antidotes, in particular "Bibron's antidote," a remedy for rattlesnake bites that the French zoologist Bibron was promoting.[21] Both writings contain descriptions of a number of experiments.

The reader also finds an explicit commitment to experimentalism. In the preface to the Smithsonian essay, Mitchell declared: "The conclusions arrived at in the pages of this Essay, rest alone upon experimental evidence. That in so varied and difficult a research, it may be found that I have sometimes been misled, and at others erred in the interpretation of facts, is no doubt to be anticipated. I began this work, however, without preconceived views, and throughout its prosecution I have endeavored to maintain that condition of mind which is wanted in experimentation, and that love of truth which is the parent of rational inferences."[22] In the introduction, he assured his readers that when his experimental evidence had been inconclusive, he had always admitted it—"thinking it better to state the known uncertainty thus created than to run the risk of strewing my path with errors in the garb of seeming truths."[23] In Mitchell's treatise on therapeutics, the statement regarding the merits of experimentalist approaches in medicine is not quite as explicit. But we do find a brief commitment to rational therapeutics. The declared goal was to arrive at a *rational* understanding of the treatment of snakebite, which meant, to Mitchell, a quantitative, experiment-based assessment of the efficacy of treatments.

Mitchell's version of experimentalism does not appear particularly original or novel. One would not be surprised to find such passages in an eighteenth-century text on physiological experimentation; in fact, Fontana wrote very similar things. What is relevant is not the content of Mitchell's commitment; what is relevant is that he made it at all. Surely in the nineteenth century the value of experimentation was no longer in dispute?

Once again, it is important to remember that two of the key questions about the commitment to experimentalism are the question of the context in which the commitment was made and the audience to which it was directed. Redi conversed with fellow naturalists and with aristocratic patrons. Mead—the young Mead—sought to promote Newtonianism in a community of physicians. In Mitchell's case, two debates are relevant: the debates within the medical community about the merits and demerits of scientific medicine and, increasingly, the debates both within the medical community and between members of the medical profession and members of the public about vivisection. The concept "scientific medicine" had different meanings for different people, it was deployed in different contexts and for different purposes, and researchers who said of themselves that they pursued "scientific" medicine did quite different things.

We already saw that Mitchell took a certain professional risk when he explicitly identified himself as an experimentalist. Although Mitchell's experimentalism does not sound very radical, at least in hindsight and in comparison with Fontana's, it appears much more so if we take into account that the kind of medicine that many members of the American medical community considered properly "scientific" was observation-based medical practice. From this perspective, the promotion of experimentation must have appeared as a rather drastic move to some. In the first half of the nineteenth century, experimental research was often vehemently criticized for being speculative, mystifying, and thus unscientific. It was certainly not uncritically accepted as a sound basis for therapeutics.[24]

Those practitioners who advocated medical reforms and laboratory-based research did so to distinguish and distance themselves from diverse groups both within and outside the medical community. They had to defend their views within the community because several of its members did not see any practical value in laboratory-based physiology. These people argued that clinical judgment was not aided and might even be distorted if the clinician relied on the insights gained in physiological experiments.[25]

The commitment to experimentalism was also a provocation to those who were concerned about the use of animals in medical research. Such concerns were growing during the 1860s. Mitchell's early essays on venom became a reference point in these debates. John Call Dalton, a physiologist at the renowned College of Physicians, presented Mitchell's work on snake venom as an exemplary case of a new style of medical research that had displaced the old practices of seeing, touching, listening, and dissecting. In an address delivered at the New York Academy of Medicine in 1866, he commended his friend's work for "the clearness and elegance of its style, the abundance of its material, and the precision of its results. It illustrates very distinctly the value of strictly elementary researches as a necessary preliminary to those of a more practical nature."[26] That address was, in fact, an address in defense of vivisection in response to a particularly forceful antivivisectionist attack on the medical community.[27]

Dalton was a great admirer of Claude Bernard; he had attended Bernard's lectures on experimental physiology in Paris as a young man. Like Bernard, he supplemented his own physiology lectures with vivisections for the purpose of demonstration. Dalton relied on Mitchell as an ally in his battle against antivivisectionists who called for restrictions and regulations to prevent what they saw as useless and unethical cruelty in medical teaching and laboratory research.

Mitchell himself did not make strenuous efforts to defend vivisection. He gained his insights on snake venom at the cost of the lives of numerous rabbits, dogs, and pigeons, a fact that he did not hide. But unlike his friend Dalton, who publicly, repeatedly, and expressively defended the use of live animals in physiological experimentation, he rarely explicitly addressed the issue. When he did so, in his article in the *New York Medical Journal*, his comments suggest that he had little patience for the sentiments of antivivisectionists and made clear that his research was not their business. He praised Dalton's "eloquent defence of vivisection," and he frankly stated that he took responsibility for "a large expenditure of the lives of birds, dogs and rabbits." He was responsible to his own conscience and "to the Maker, who has endowed us with the will and the power to search into the secrets of His universe," definitely not to those "many ignorant and well meaning persons, who have recently sought to take away from us the chief aid of the modern physiologist."[28]

Mitchell's 1861 treatise must have appeared particularly suitable for Dalton's purpose because it illustrated how animal experimentation could be used to assess the efficacy of treatments. Nevertheless, it is one thing to advocate a program, a system, or a reform of medical practice in public lectures and addresses and to promote new programmatic goals, such as "scientific" medicine or "rational" therapeutics in order to build a community, to distinguish oneself from one's predecessors, or to defend one's ideals. It is quite another thing to turn these larger goals into novel kinds of approaches to biomedical research or to use these approaches in one's research.

Dalton, at least, maintained that Mitchell's approach to biomedical experimentation was innovative. He congratulated Mitchell on his methodological acumen. Commenting on Mitchell's study of antidotes to rattlesnake venom, Dalton declared: "It would be difficult to find a medical treatise which should illustrate more fully than this the judicious caution and reserve which guide the physiological experimenter, and which enable him to avoid the sources of error that lie in his way. The constant employment of comparative and counterexperiments, and the frequent variation of the methods employed, indicate the care and faithfulness of the investigations, and inspire a well grounded confidence in their results."[29] We will see that Dalton's characterization of Mitchell's methodological ideals misses the point in a rather interesting way.

Eclectic Protocols

Mitchell stated that initially he had simply wished to evaluate the potency of Bibron's antidote, a concoction of "five drachms of bromine, four grains of iodide of potassium, and two grains of corrosive sublimate." He had hoped to assess the value of the drug by procuring a few snakes, having them bite animals, and giving the antidote to the victims. But matters turned out to be less simple than he had expected. "After destroying many animals and attaining only negative results," he embarked on a more systematic project to acquire the information that he needed: information about the nature of the venom and how it was conveyed to the victims, as well as about the precise effects it produced in them.[30] The two treatises from 1860 and 1861 present the results of this endeavor.

The Smithsonian essay was written for readers who had an interest in experimentation. Mitchell's short treatise on the treatment of rattlesnake bites addresses the readers of the *Medical Chirurgical Review*—an audience of practicing physicians. The Smithsonian essay referred those readers who wanted to know about rattlesnake venom "from a purely medical point of view" to the essay in the *Medico-Chirurgical Review*, which, in turn, referred to the Smithsonian essay those readers who were curious about the details of Mitchell's experiments.[31] In the introduction to the Smithsonian essay, Mitchell presented his project as a continuation of a research tradition stretching back to the ancients, even though he worked on rattlesnakes, not on vipers. Mitchell kept referring to Fontana's views and experiments, and Redi, Mead, and even Celsus appear in the footnotes. But the questions he asked and the conceptual framework and protocols he used owed much more to contemporaneous questions and concerns in medical chemistry, pharmacology, and pathology than to seventeenth- and eighteenth-century investigations.

The protocols in the Smithsonian essay show that for Mitchell, the uniformity of experimental conditions became a prime concern. Mitchell had a more regular supply of resources than most of his predecessors. In the early modern period, the meat of vipers was an essential ingredient of theriac, a remedy for snakebite and various other health troubles. For their experiments, investigators often made use of the numerous vipers that had been collected for the production of theriac. Mitchell had a much smaller supply, which he obtained from various locations, mostly in the Alleghenies,[32] but he kept the snakes not only to have venom readily available for his experiments at all times and also to be able to provide a regular food supply and a well-known environment for the animals. He was thus able to offer general information about the habits of rattlesnakes in captivity. And he did even more to ensure the uniformity of experimental conditions. He used known quantities of venom in his experiments, which he injected; and in the experiments on treatment, he did all experiments on dogs of large size and used quantities that had previously been determined to be fatal.[33] In the essay on therapeutics, he even turned his concern with the uniformity and stability of experimental conditions into an "experimental criticism." Not only did he emphasize that stable and uniform experimental conditions had to be created, but also he stressed that fallacies might result if experimenters did not pay attention to this issue.

Paying attention to protocols—to the design and procedural methods of experimentation—proves useful for situating Mitchell's work within early nineteenth-century plant and animal chemistry, pathology, and physiology of the nerves while at the same time making us aware of connections among research fields that discipline-oriented histories might miss. Mitchell investigated whether venom was a composite substance, what its proximate principles were, and whether it was possible to isolate the physiologically active principle from it. The concepts of the "proximate" or "immediate principle" and the "active principle" were key concepts in early nineteenth-century chemical analysis, particularly within pharmacology. Proximate principles are those chemical components that can be extracted from animals and plants. Mitchell's friend Dalton defined proximate principles in his textbook of human physiology. A proximate principle is *"any substance, whether simple or compound, chemically speaking, which exists, under its own form, in the animal solid or fluid,* and which can be extracted by means which do not alter or destroy its chemical properties."[34] Active principles formed a subgroup of immediate principles— namely, the physiologically effective substances of plants.[35]

In the course of these researches, a recognizable pattern of experimentation emerged, indeed a *protocol* of identifying active principles through a series of steps. An exemplary case of successful isolation of an the active principle from plants—well known at the time—was François Magendie and Pierre Joseph Pelletier's research on ipecacuanha, a plant that was known to induce vomiting. To isolate the active principle of the substance— the "emetine" or vomitive matter—the researchers studied its physical and chemical properties and its physiological effects and compared the effects with those of the substance in its original form.[36] Soon after Magendie and Pelletier had announced the active principle of ipecacuanha and after they and other researchers had identified several more of these principles, these principles became known as salifiable plant bases, or "plant alkalis." Similar investigations were carried out on *strychnos* plants and other plant groups. The discovery of alkaloids in plant chemistry transformed medical science in the second third of the nineteenth century.[37]

In all these cases, the isolation and chemical study of the active principle is combined with animal experiments to establish the effects of the isolated chemical agent. Claude Bernard, Magendie's student and a model
for Mitchell, followed the very same steps in his famous study on curare,
the arrow poison.

Mitchell's early investigations on alkaloids fall squarely within the scope
of the endeavors to isolate active principles of plants. Together with his
friend, army surgeon William Hammond, he extracted the alkaloids from
corroval and vao, two varieties of woorara (curare), and tested their effects on the animal organism in animal experiments. In the resulting paper, the authors explicitly refer to Pelletier's earlier analyses of curare.
They described the physical characteristics of the two substances and the
extraction of the "poisonous principle" through water and alcohol. They
identified the principle obtained from corroval as an alkaloid and proposed to name it "corrovalia." Finally, they referred to experiments that
showed that the substance was highly toxic.[38]

It is exactly this protocol that we can find in the chemical portion of
Mitchell's Smithsonian essay. The common experimental techniques for
the extraction of principles were filtering or precipitating. Mitchell described how by boiling venom, decanting the liquid, filtering the coagulate,
and washing it with cold water, he obtained an albuminoid substance and a
fluid. Injections of each into the breasts of pigeons showed that the fluid
was deadly but the coagulum innocent. Mitchell also found that treating
the fluid with strong alcohol yielded a second precipitate that was highly
toxic, as animal experiments established. This nitrogenous body was the
active toxicological element or essential principle. In line with common
practices of naming active principles—such as strychn*ine* from *strychnos*,
morph*ine* from sleep-inducing opium—he named it "crotaline" (from *crotalus*—rattlesnake).

The second part of the Smithsonian essay deals with the diseases caused
by snakebite. Here, we find two other protocols—the familiar animal experiments in which different species of animals received venom either
through bites or through injections in various body parts as well as experiments that were inspired by nervous physiology.

Chapter 6 showed how the design of Fontana's amputation experiments
and vivisections embodied eighteenth-century conceptions of life and
disease—not in the sense that his experiments were hypothesis-driven
but in the sense that the experimental designs appear plausible in light
of certain ideas about life, disease, and vital and obnoxious principles. In

the pathology portion of his book, Mitchell followed toxicological tradition when he classified snake venom as "septic or putrefacient" poison. In Matteu Orfila's seminal work on toxicology, snake venom is one species of septic or putrefying poisons, another is "exhalations from burying-grounds, hospitals, prisons, ships, privies, marshes, putrid vegetables, and stagnant water," and yet another is "Contagious Miasmata, emanating from pestiferous bodies, or bales of merchandize coming from a place infected with the plague."[39] The consequences of snakebite are conceptualized as disease in terms of early nineteenth-century pathological theories, and Mitchell stressed that the "natural history" of this disease, its symptoms and lesions, had to be described in detail. First and foremost, the effects of venom on blood had to be investigated. The study of blood was one of the key concerns in mid-nineteenth-century pathology. Blood was analyzed chemically and examined microscopically in both normal and diseased conditions, and experimenters hoped to modify the condition of blood in animals so as to simulate its condition in certain diseases.[40] Symptoms were minutely described, and in most cases, postmortem dissections were performed to learn about lesions. Mitchell studied the symptoms and lesions caused by snake venom in several animal experiments, most often with dogs. Like Fontana, Mitchell preferred to experiment on bigger animals because symptoms and lesions were magnified, as it were. Reed-birds, for instance, were so small that they almost always died instantly when bitten by a rattlesnake; they were thus useful as indicators for the presence of venom but unsuitable for the study of its physiological effects.

The occurrence of changes in blood became the demarcation criterion between acute or primary and chronic or secondary poisoning. In acute or primary poisoning, no alteration of blood takes place. In chronic or secondary poisoning, the changes are very pronounced: coagulation, potential damage to blood disks, and fibrin disappearance in vitro with mixtures of venom and blood. All these processes were well known in contemporaneous hematology as concomitants of disease.

Mitchell's expressed aim was to illustrate both "constant" and "exceptional" lesions accompanying the disease of venom poisoning. He presented his results in the form of tables, which list essential information about each trial—which body part was wounded, time of death, symptoms such as convulsions, internal lesions, and—last, and most important for the characterization of the disease produced by the venom—the state of the blood in each trial ("loosely coagulated," "chiefly uncoagulated," and so on). To the reader, the table highlights that there is no constancy in the

lesions except that the blood is affected. The last of the columns includes
everything from "blood perfectly fluid" to "coagulated well," for example.
The table related to the poisoning of rabbits contains so much information
that it is too big to be squeezed onto a single printed page. It is split in two
parts, a table of symptoms and a table of lesions. Table I lists the twitching, jerking, gritting of teeth, the convulsions and the labored breathing.
Table II describes the appearance of the inner organs and the character
of the blood. Again the reader is left with the impression that the only
general conclusion that can be drawn is that based on the information
provided, no generalization about the venom disease can be made.

   Although the pathological perspective dominates the second part of
Mitchell's treatise, there are a number of experiments on physiological
effects—namely, the study of venom on the sensory and motor nerves of
frogs. The source of inspiration is obvious: Following Galvani's discovery
of animal electricity, numerous experimenters repeated Galvani's experiments to explore muscle and nerve functions.[41] Mitchell investigated the
twitching of frog muscles, using venom instead of galvanic stimuli. Mitchell did not comment on the design; indeed, he probably did not have to.
He merely informed the reader that the frog was "prepared as if for use
for a galvanoscope," which must have been clear enough information for
the mid-nineteenth-century physiologist.[42]

   The purpose of these experiments was not to produce and study a disease. Rather, the experiments were designed to study the effects on specific body systems (such as nerves). Experimental changes had to be such
that the body system of interest was isolated, stimulated, and the effects
(such as twitches) were recorded.

   The section does not quite fit with the rest of Mitchell's project—no
diseases, symptoms, or lesions are described, no substantial conclusions
drawn from the experiments. Precisely for this reason, the section is telling.
It shows that mid-nineteenth-century biomedical experimenters could
(and did) draw on a common stock of protocols. Tracing protocols across
contexts and time periods helps in appreciating what is novel, what is part
of a long tradition, and how different scientific fields are connected.

Comparisons and Counterproofs

The recurrent theme of late eighteenth-century methods discourse was
variation, or "diversification." Mitchell, by contrast, only occasionally men-

tioned variations of experimental conditions. This certainly does not mean that he did not care to perform variations. Dalton for one commended him on "frequent variation of methods employed." But Mitchell himself did not discuss this practice, at least not nearly as much as the late eighteenth-century experimenters did. Only once or twice did he explicitly note that the repetition of an experiment with some variations in the setting was a crucial strategy for proper experimentation, and when he did so, he sounded like an echo of Fontana. Having described a particularly intricate experimental design, he noted: "it is still desirable that these experiments should be repeated, with every possible modification; since, as I have endeavoured to show, this, like all other portions of our subject, is girt about with such difficulties as may well baffle the most careful."[43] The methodological term that stands out both in the Smithsonian essay and in Mitchell's essay on the treatment of snakebite is the term *test* or *check*. The word *test* is used in several ways, depending on the context. "Test" can simply mean "examine further," as in the phrase "the irritability of the motor nerves in the sciatic trunk was next tested." In chemical experimentation, "test" means an indicator for a chemical substance. Even small birds can be used as "tests"—as indicators—for the presence of venom. The term *test* can also be used in the sense of "probe." "To probe" or "to test" means to determine the existence or quality of something— for example, the acidity of a solution. Some common procedures for doing so were described in chemistry handbooks.

From the perspective of the history of methods discourse, the most intriguing instance of "test" or "check" is the test that involves comparative experimentation. For example, when Mitchell determined the toxic effects of a boiled mix of venom and water or an otherwise chemically altered venom, he secured a small quantity of pure venom as a "toxicological test" of the activity of the pure venom against which the toxicity of the mixture was determined.[44] Sometimes he also referred to such a procedure as a "check experiment." Conversely, the results of experiments with combinations of venom and chemicals were "checked or tested" by giving the chemical substance alone to an animal. In this way, one could establish whether a strong reagent alone was fatal. When Mitchell studied the effect of a mixture of venom and alcohol on a pigeon, he noted that a "check experiment" had been performed to see what happened if the pigeon was injected only with alcohol.[45] When Mitchell investigated the effect of venom on muscles, he punctured exposed muscles with dry clean fangs whose ducts had been stopped with wax and compared the time and intensity of the twitching with the effects of an injection of venom through the fangs.[46]

Mitchell made the performance of a comparative experiment a condition of proper experimental procedure. He did so in a critical comment to study some recent experiments on the effect of venom on plants, a description of which he had found in the *St. Louis Medical and Surgical Journal*. The author who was the target of Mitchell's criticism had used a lancet to inoculate healthy vegetable plants with snake venom. The plants were found withered the next day. Mitchell pointed out that the methodology was lacking; indeed, he found the accounts in the *Medical and Surgical Journal* "so very limited, and so wanting in statement of details, that it is difficult to accord them any great value as scientific evidence."[47] To be able to assess the validity of the results of experiments other investigators had performed, readers had "a right to demand every possible knowledge as to the temperature and season, the size of the plants, the amount of venom employed, *and the effect of wounding similar plants to the same extent, but without the use of the venom*."[48]

Mitchell's emphasis on comparisons did not remain unnoticed. Dalton specifically mentioned Mitchell's "constant employment of comparative and counterexperiments." We have seen in earlier chapters that the practice of comparative experimentation had long been common in the experimental sciences and that there are even a number of premodern precedents for this practice. Fontana made it a maxim to perform comparative experiments to assess the validity of experimental results. However, only in the first half of the nineteenth century do we find more explicit and critical analyses of the structure of these comparative experiments. At that time, comparative experimentation was explicitly conceptualized in books on the logic of science (or natural philosophy). Here, the model was physical experimentation. In John Herschel's *Preliminary Discourse of the Study of Natural Philosophy*, for example, comparative experimentation is introduced in Baconian terms as an example of "crucial experiments." Like Bacon, Herschel ascribed special status to these kinds of experiments. In addition, he had quite concrete things to say about their structure, and he also offered a general characterization: "We make an experiment of the crucial kind when we form combinations, and put into action causes from which some particular one shall be deliberately excluded, and some other purposedly admitted; and by agreement or disagreement of the resulting phenomena with those of the class under examination, we decide our judgment."[49] Herschel then proceeded to discuss in detail the rules that should guide the search for causes through comparative experimentation.

A few years later, John Stuart Mill introduced the "method of difference" as the very mark of experimentation. Mill found inspiration in Herschel's methodological discussion as well as in another renowned treatise on method, Auguste Comte's *Cours de philosophie positive* (1830–42).[50] Both Comte and Mill characterized physical experimentation—the ideal type of experimentation—as an act of placing an object in definite artificial conditions. (Observation, by contrast, is a noninterventionist, passive process, as Herschel also noted.) The method of difference is a comparative method, whereby the experimenter compares two states of affairs: the preexisting state in which the phenomenon of interest is absent and the new situation in which that phenomenon has been introduced. If an effect occurs only in the second situation, then one can conclude that the phenomenon that the experimenter introduced is the cause of the effect in question.

In the most basic sense, the method of difference thus involves the comparison of an artificial situation with the natural course of things. In actual experimental inquiries, the comparison will usually be between two artificial situations, in one of which the phenomenon of interest is present and in the other of which it is absent. In any case, it is critical for the successful application of the method of difference that all conditions of the initial state be fully known to the experimenter. The method of difference is definitive; it is, according to Mill, the only method by which one can ever "arrive with certainty" at knowledge about causes.[51]

Mitchell's and Dalton's comments on comparative experimentation, already quoted, resonates with these more abstract discussions of the methodology of experimentation. Yet we should not jump to the conclusion that experimenters in the life science "applied" Mill's methods. Dalton, for one, was likely drawing on another source: Claude Bernard.

Bernard, too, had something to say about the structure of comparative experimentation, but he gave the issue a new twist. He introduced the terms *comparative experiment* and *counterproof* as technical terms, and he explicitly contrasted the two. He insisted that comparative and counterproof were different things—the first relevant for physiology, the second for physics.

Bernard's 1865 monograph *Introduction to the Study of Experimental Medicine* is the sum total of Bernard's reflections on the proper methods of physiological experimentation. In it, Bernard outlined a methodology of physiological experimentation that carved out a special niche for comparative experiments in physiology. Bernard acknowledged that experimentation in biology was less definitive than experimentation in physics. In this he followed his compatriot Comte as well as the Englishman Mill. Bernard agreed with Mill about the importance of the "method of difference" for physics, but he insisted that the method of difference was rarely applicable in physiology, because the experimenter was never in a situation in which all the experimental conditions and effects were fully known. In this respect Bernard was in tune with Comte, who maintained that biological phenomena were so complicated and complex that they offered "almost insurmountable impediments to any extensive and prolific application of such a procedure [experimentation] in biology."[52]

Comte suggested two ways in which experiments in biology should proceed. The artificial phenomenon could be introduced either into the environment of the organism or into the organism itself. The second option was highly problematic and not very informative, because the "natural sympathy" of the organs, their harmony, would obstruct experimental practice. Comte favored the first kind of experimentation, which made experimenters "better able to circumscribe, with scientific precision, the artificial perturbation we produce; we can control the action upon the organism, so that the general disturbance of the system may affect the organism very slightly; and we can suspend the process at pleasure, so as to allow the restoration of the normal state before the organism has undergone any irreparable change."[53] Bernard, however, was more confident about the power of comparative experimentation. Even though he agreed that strictly speaking the method of difference could not be applied to physiology, experimentation in the life sciences could lead to more definite results than Comte had assumed. The hallmark for proper procedure is the "comparative experiment."

Bernard's concept of comparative experimentation in physiology was his answer to the complexity of life and the intricacy of the living organism he encountered as experimenter. In Bernard's methodological discussion, a conceptual tool specifically for the practical purposes of experimental physiology emerged.[54] Bernard explicitly contrasted the comparative experiment with another, more definite kind of check or test: the "counterproof." The counterproof is connected to experimental reasoning in physics. It is the successful application of the "method of difference." The method of comparative experimentation is a pragmatic counterpart of the method of difference as it were, a method that takes into account that life, living organisms, and their environments are too complex to be fully understood by the experimenter.[55]

Like many other medical men, and in stark contrast to advocates of Newtonian medicine, Bernard insisted that methodological ideas from physics could not simply be transferred to the life sciences and that bio-medical experimentation was profoundly different from physical experi-mentation. Experimenters could never make rigorous experiments on living animals "if we necessarily had to define all the other changes we might cause in the organism on which we were operating. But fortunately it is enough for us completely to isolate the one phenomenon on which our studies are brought to bear, separating it by means of comparative ex-perimentation from all surrounding complications." Comparative exper-imentation reaches this goal by adding "to a similar organism, used for comparison, all our experimental changes save one, the very one which we intend to disengage." Suppose an experimenter wished to study the result of section of a deep-seated organ. Getting to that organ would require in-juring other surrounding organs. What is the effect of the section itself, and what is collateral damage? To avoid confusion about causes and effects, the experimenter must perform the same operation on another animal without actually performing the section. "Comparative experimentation in experimental medicine," Bernard pronounced, "is an absolute and general rule applicable to all kinds of investigation, whether we wish to learn the effects of various agents influencing the bodily economy or to verify the physiological rôle of various parts of the body by experiments in vivisection."[56]

Bernard even insisted that because of the variability of experimental animals, comparative experiments must be performed on one and the same animal, whenever possible. At this point, it is obvious that the *Introduc-tion to the Study of Experimental Medicine* was the product of Bernard's reflections on his own experiences with physiological inquiries, because Bernard's toxicological work contains some illustrative examples of the practice of comparative experimentation on one and the same animal. To investigate the influence of curare on the irritability of muscles, Bernard experimented on frogs. He found the irritability of frogs to be so variable that it was not illuminating to conduct comparative experiments on two different frogs. The comparison had to be performed on one and the same animal. Bernard described in text and image how the frog had to be pre-pared in such a way that only one of its hind legs would be affected by cu-rare. In this arrangement, the irritability of the "normal" and the poisoned leg could be directly compared.[57]

The divergence from Mill's methods is crucial. Bernard did not as-sume that in actual physiological experimentation, all causal factors are knowable and can be isolated and individually manipulated, nor that the experimenter has full power over the experimental setting. In other words, the notion of comparative experimentation is a retreat from Mill's ideal. As is well known, Bernard was a determinist at heart, and he firmly believed that organisms were governed by general laws.[58] Still, the introduction of the concept of comparative experimentation is an explicit acknowledgement that in practice, ideal experimental conditions could only be approximated. Comparative experimentation was seen as a realistic workable counterpart to Mill's ideal, not an elaboration of it. Mill's methodology could not address the most pressing problems of scientific experimentation in the life sciences—the complexity of living things. Bernard tried to cope with what Ronald Fisher would later call "the [experimenter's] anxiety of considering and estimating the magnitude of the innumerable causes by which his data may be disturbed."[59]

It is hard to say whether Dalton was aware of, and whether Mitchell could have been aware of, the subtleties of Bernard's position. Mitchell, of course, knew Bernard's physiological writings, especially his work on toxicology and the studies on curare. Bernard's *Introduction to the Study of Experimental Medicine*, however, appeared a few years after Mitchell had completed his main project, just before Dalton gave his address in defense of vivisection. Mitchell did not make a distinction between "counterproofs" and "comparative experiments." His tests or checks involve a comparison along the lines of the method of difference, and we have seen how important it was for Mitchell to obtain uniform and stable experimental conditions.

Dalton's "Bernardian" reading of Mitchell's methodology as "constant employment of comparative and counterexperiments" is not entirely appropriate (if we read "counterexperiment" as "counterproof"). Although Mitchell's methodology favored comparative experimentation, it did not reflect the methodological distinction that Bernard made between comparative experiments and counterproof. However, the important point is not whether Dalton interpreted Mitchell correctly or whether Mitchell had read Bernard correctly but rather that there was now an ongoing discussion about the very structure of proper experimental procedures. This discussion was not completely detached from scientific pursuits, but the more abstract reflections about experimental practice were presented in venues other than scientific articles and proceedings. Herschel's *Preliminary Discourse of the Study of Natural Philosophy*, Comte's *Cours de philosophie positive*, Mill's *System of Logic*, and even Bernard's *Introduction*

to the Study of Experimental Medicine are not—at least not primarily—
outlets for the presentation of research outcomes. They are primarily
(methodo)logical treatises. In these discussions, the difference between
the practice of identifying causes in complex, complicated real-life experimentation and the structure of causal reasoning in ideal experimental
situations came to the fore. The experimenters in the life sciences tried to
develop strategies for the former, whereas the authors of methods treatises concentrated on the latter.

Experimental Criticisms

Apart from the general commitment to experimentalism, the Smithsonian essay contained very little explicit methodological discussion. In the
essay on therapeutics, by contrast, methodological issues are explicitly
addressed, as the full title of the work indicates: *On the Treatment of Rattlesnake Bites, with Experimental Criticisms upon the Various Remedies
Now in Use.* Mitchell's notion of "experimental criticism" is ambiguous,
however. It means both the experiment-based critique of common beliefs
about treatment options and the critique of experiments themselves. The
reader finds detailed discussions of the efficacy of two chemical agents
that were used for the treatment of snakebite: iodine, which was promoted by the Chicago physician David Brainard, and Bibron's antidote.
The efficacy of both agents had been tried in experiments before, but according to Mitchell, these experiments had been poorly designed.

 The practical criticism of experiments on antidotes involved a discussion of two "fallacies of experimentation": "1. Fallacies in regard to the
use of antidotes of all kinds, arising from want of exact knowledge as to the
secretion of venom, and the mode in which the serpent uses its fangs and
ejects the poison. 2. Fallacies as to antidotes, arising from want of information on the natural history of the disease caused by the venom."[60] The
experimental criticism as Mitchell presented it is an intriguing conglom-
erate of factual and methodological critique of snake venom research.
In part, Mitchell chided his fellow researchers because they did not correctly understand basic facts about snakes. Nor did they understand the
mechanisms of different kinds of antidotes. This lack of understanding
also led to badly designed experiments—experiments with indeterminate
quantities of venom, for example. The efficacy of antidotes could only be
assessed if similar cases were compared, and this condition was not easy

to ensure. The venoms from snakes of different age, size, or vigor might not be alike. And even if two snakes were alike in these respects, and even if the amount of venom contained in their fangs were the same, it did not mean that their bites were alike, too. In fact, the mechanisms of the bite differed so much that "the danger of the bite is utterly unequal."[61]

The most important experimental criticism had to do with proper accounting. In many known cases of snakebite, the disease caused by venom was, in fact, not fatal for human victims. Like the proverbial cold medicine, which is guaranteed to cure a cold in seven days, an antidote to snake venom might be much less potent than advertised. To assess the efficacy of a particular treatment, it was not enough simply to assume that snake bites are always fatal and then to estimate how successful an antidote was by counting how many of a group of treated individuals died. Instead, one must investigate the natural history of the venom-induced disease, including the question of how many individuals actually died from it.

Mitchell's criticisms are particularly interesting in view of the discussions in the first half of the nineteenth century about the value of numbers, statistics, and quantification in medicine.[62] In the medical community more generally, this discussion intensified in the early nineteenth century as a debate about the merits and demerits of the so-called numerical method that the French clinician Pierre Charles Alexandre Louis had developed. The numerical method was a quantitative tool for the systematic comparison of therapeutic outcomes in groups of patients. It involved collecting instances for the success and failure of a particular treatment.[63]

Louis compared groups of patients who all had the same disease—angina tonsillitis or pneumonitis, for example. These groups of patients were bled on different days, the effect of the treatment was recorded, and the average duration of the disease for the different groups of patients was calculated. There were three problems that Louis hoped to address in this way. First, the conditions for treatment and therapeutic assessment were complex and not always completely uniform. Second, strictly speaking, no two patients were alike. They differed on features such as the severity of the disease, age, strength, stature, and so forth. Third, there was the possibility of error of judgment on the side of the therapist—it was extremely challenging for the observer to make precise judgments about the time of the onset of the disease, its termination, or its severity. Drawing on Laplace's work on probability, Louis insisted that all these issues could be addressed by comparing the responses to treatment of large groups. He reasoned that in large groups, the similarity of conditions "will necessarily be met with, and all things being equal, except the treatment, the conclusion will be rigorous." In the assessment of groups, "*the errors, (which are inevitable,) being the same in two groups of patients subjected to different treatment, mutually compensate each other, and they may be disregarded without sensibly affecting the exactness of the results.*"[64]

The reasoning underlying here is different from the reasoning that motivated many variations of experimental trials. Notably, it is directly opposed to Fontana's approach to experimentation. As we have seen, Fontana assumed that through systematic variations of experimental conditions it was possible to identify all the circumstances that potentially affected experimentation. Louis, by contrast, called for the treatment of large groups, because he assumed that these factors could not be reliably identified.

The fate of the numerical method is well known. Many members of the nineteenth-century medical community were rather critical of Louis's approach and of statistical approaches more generally. Bernard in particular was very much opposed to statistics, and his critique of the numerical method was biting. In the *Introduction to the Study of Experimental Medicine*, Bernard reiterated some of the criticisms against the numerical method that Louis himself had already anticipated in his treatise on bloodletting. The first requirement for the proper use of statistics was that "the facts treated shall be reduced to comparable units."[65] But in clinical contexts, this requirement could not be met. The records were not reliable, diagnoses were obscure, the causes of death were often carelessly recorded, and so on. Because there was too much variety among the individual patients, averages were meaningless. Bernard also abolished statistics in experimental physiology, drawing attention to a controversy that his own research had helped resolve. Experimenting on the spinal nerve, some experimenters had found that the anterior spinal roots were insensitive but others that they were sensitive, and Bernard asked: "Should we therefore have counted the positive and negative cases and said: the law is that anterior roots are sensitive, for instance, 25 times out of a 100? Or should we have admitted, according to the theory called the law of large numbers, that in an immense number of experiments we should find the roots equally often sensitive and insensitive?" Of course not. Such statistics would be "ridiculous." The proper scientific way to deal with the case would be to establish the exact conditions in which spinal roots are sensitive and insensitive.[66] Statistics was wanting, because statistical approaches did not provide any insight into causes. The correct approach to physiological inquiry was the experiment-based search for causes, according to Bernard—and we already saw how he thought he could address the problem of individuality through his conception of comparative experimentation.

Even those members of the medical community who welcomed quantitative approaches to the study of disease were quick to criticize Louis, because he justified his method along the lines of the law of large numbers, and the populations he examined in his trials were rather small. Others criticized the advocates of the numerical method because it led the physician away from the individual sick patient. Yet others had qualms about the method itself, like Bernard did: it was based on the wrong assumption that all patients were alike. As William Coleman has shown, there were a few attempts in the mid-nineteenth century to improve on Louis's approach and to introduce statistical tools that were more appropriate to the very small numbers of subjects that were used in therapeutic trials (as well as in physiology). But the medical community at large did not welcome these tools. Therapeutic trials on groups of patients continued to be pursued, but there was little methodological discussion and reflection on these trials. The discussion intensified again only around 1900, and it was then that statistical tools came to be accepted in the medical community.[67]

Mitchell's essay on treatments of snakebite is remarkable because it outlines the conditions for adequate tests of the efficacy of antidotes along the lines of the numerical method. The "experimental criticisms" Mitchell put forward in the 1861 essay on the treatment of snakebite resonate with some of the issues Louis discussed. Mitchell's conceptualization of experimental animals as "diseased" made it possible for him to frame the method for the assessment of animal experiments in those terms. We have seen how concerned Mitchell was with uniformity and stability of experimental conditions. At the same time, he emphasized the individuality of the "patients"—the snakebite victims—he considered in his experiments. This becomes most obvious in the 1868 study on rattlesnake bites, which supplemented the earlier essays. Mitchell remarked: "The puzzling factor of individuality perpetually comes into the equation with elements of doubt, so that it is only by multiplying results that we can reach the requisite amount of assurance that in a large given number of animals the individual cases of unusual resistance are not likely to interfere with our conclusions."[68] For Mitchell, the reason why many animals needed to be treated was that animals were individuals; they all reacted slightly differently to the treatment with venom. In this sense, they were like sick patients receiving treatments.

Unlike Bernard, Mitchell assumed that for the correct assessment of the power of an antidote and for the evaluation of the success of treatments, some application of statistics was required. In particular, the mortality rate of snakebite victims must be known. One reason why some of the experiments in the Smithsonian essay were so remarkable was that they showed that some species of animals were more resilient than others. Dogs, for instance, were much less likely to die than were rabbits and pigeons, and in half the experiments on the effect of rattlesnake venom on dogs that Mitchell reported, the bite victims recovered. (The survivors did not necessarily escape their fate, though: The large spaniel who was recovering from a bite in its shoulder "was so well on the ninth day, that it was used for another purpose."[69]) Contrary to popular opinion, it was also not true that rattlesnake bites were always fatal for humans. Referring to another table in the Smithsonian essay, which showed that several snakebite victims had recovered after treatments with oil, alcohol, iodine, and other substances, Mitchell commented drily: "either all treatment [. . .] is successful, or else [. . .] the greater part of the cases must have survived under any form of medication."[70] What seemed to be a successful treatment might as well be a case of recovery by natural causes. In any event, "[t]he mere fact of their surviving can assuredly be no test of the value of a plan or treatment."[71] Because there were no comprehensive statistics about the fatality of rattlesnake bites, it was not possible to make any concrete assessments of the efficacy of treatments.

The results of Mitchell's own experiments with Bibron's antidote were discouraging: Of nine canine snakebite victims that were treated with the drug, only two survived. Given that roughly every second victim survived without any treatment, this number is not too impressive. Experiments with other agents were equally disappointing. If done properly, these experiments showed that those treatments were in fact ineffective.[72]

Some New Arrangements in the Narrative

Fontana composed his celebrated treatise on viper venom in the late eighteenth century. Mitchell put together his reports about eighty years later. We have already seen about the changes in the content of methods discourse, especially the quite fundamental changes in the protocols and the shift of emphasis from variations to "checks" in the methodology. Even the commitment to experimentalism, although quite similar in its wording, had new significance in the context in which Mitchell made it. One might also expect quite a bit of change in the organization of experimental reports between the late eighteenth and the nineteenth centuries. Indeed, historians of scientific writing have argued that the changes in this period were particularly profound. They find in the second half of the nineteenth century the roots of the modern, modular form of scientific writing—that is, the familiar division into the sections "Introduction," "Methods," "Results," and "Discussion." These historians maintain that in the course of the nineteenth century, methodological reflections disappeared from experimental reports and that the reports (especially the articles in scientific journals) became increasingly less reflexive and more standardized. The alleged decrease in reflexivity and the increase in standardization are taken as indications that methodological issues in the experimental sciences were largely settled.[73]

This view strikes me as problematic and misleading in several respects. Mitchell's essays exemplify a transitional stage—they do exhibit some new features that we recognize in hindsight as anticipating organizational elements of later scientific writings but they were in many ways quite traditional. It is obviously not the case that methodological issues were largely settled. There were quite intense discussions in the mid-nineteenth century about all sorts of methodological issues—about the numerical method and quantification, about statistics, about the structure and epistemic force of comparative experimentation, and so on. Even though there might have been fewer discussions of methodological issues within experimental reports, the methodology of experimentation clearly was still a topic of concern.

At least for the most part, accounts of experiments were still organized sequentially. Protocols were described along the way, not in separate methods sections. Like Fontana, Mitchell combined narrative and methodological statements to form an argument. For instance, like Fontana, Mitchell described how particular experimental settings were gradually improved to make the findings more secure. One such sequence of trials explored the influence of heat on the destructive power of venom. In these experiments, the venom was heated to successively higher temperatures and then injected into the breasts of small birds. The report includes a table showing the results of a series of ten experiments. Death occurred after longer and longer times, and the bird that received venom heated to 212 °F survived the injection. The table thus suggests that the venom gradually lost its power when heated and had become completely inactive at 212 °F, and Mitchell stated this.[74] In the next paragraph, he declared that his conclusion had been mistaken. Only then did he state that he had re-examined the issue after some time. The new experiments had demonstrated that the results as they are shown in the table were misleading because they falsely suggested that heated venom was less dangerous than venom at room temperature. He went on to describe the causes of the error into which he had fallen: He reminded his readers that boiling caused coagulation of a part of the venom and that the active portion of the venom was in the fluid, not in the coagulant. He had been forced to work with very small quantities of venom, which meant that the amount of fluid venom was minute. During the boiling process, most of that fluid would cling to the test tube and would thus not be injected. This was the reason why at higher temperatures death occurred more slowly or not at all. The remedy was straightforward: The experiments had to be done with larger quantities of venom. Indeed, in the second series of experiments, all birds died "with the usual symptoms."[75] In sequences like these, the order of presentation matters for the argument. The narrative describes the gradual stabilization of an experimental arrangement, culminating in the specification of a causal relation (or, as in this case, the absence of one). The point of the description of the experiments resulting in the misleading table is not simply that Mitchell was an honest man, honest enough even to admit that some of his experiments had failed. Rather, by showing how one could explain what happened when small quantities of venom were boiled, Mitchell strengthened his own position. The outcome of the failed experiment made perfect sense if it was considered in light of his idea that venom was a composite of a fluid and a precipitating part.

This manner of writing is quite similar to late eighteenth-century experimental reports. What is different is that the flow of the narrative is occasionally broken up and that Mitchell did not leave the reader in suspense, at least not to the same extent as Fontana did. Both of these new arrangements are indicative of a move away from the main goal that Fontana pursued in his writing—to demonstrate the discoverability of his findings to the reader. The new arrangements indicate a move toward making it easier for the reader to grasp the significance of the findings. The data Mitchell obtained in his experiments are sometimes (not always) presented in tables, which are typographically separated from the main text.

In a table, descriptions of actions and findings are necessarily terse and succinct and are often represented by a number, space in the columns being limited. Mitchell's table has seven columns. The first gives the number of the trial. The second states the number of fang marks: one or two. The third lists body parts bitten: Thigh. Breast. Back. Leg. The fourth column presents the "duration of life from the time of the bite," times ranging from two hours to nine hours, ten minutes. The symptoms and lesions are also just briefly indicated. The fifth through seventh columns contain pithy characterizations of visible symptoms and alterations of organs and body fluids.

Mitchell's tables could not remove the uncertainty about what overall conclusions might be drawn from the empirical findings—and, of course, tables are not a nineteenth-century invention. But the introduction of tables had a quite profound effect on the organization of Mitchell's argument, as the emphasis shifted from describing the way in which data were produced to the synoptic presentation of evidence. This benefits the reader. Producing and filling in the table forced the author to be systematic and comprehensive and to abstract from the actual chain of events. Information presented in tabular form can be taken in at a glance; comparisons between trial outcomes are quick and easy to make. The format of the table, the spatial limitations set by the rows and columns necessitated the introduction of salient categories ("locality of the bite," "state of blood") and descriptive concepts ("uncoagulated," "loosely coagulated"). In this sense, the presentation of relevant information in tabular form is a significant divergence from the narrative format of the investigative journey that we find in other parts of the Smithsonian essay.

In contrast, Mitchell's 1868 essay on the toxicology of rattlesnake venom, the supplement to the earlier essays, presents a narrative consisting of a string of linked experiments performed to answer a particular research question. Mitchell asked: Why is it that venom is not poisonous when ingested? Why does it not act on the mucous surfaces of the stomach? The discussion begins with a hypothesis—either the venom is altered by the gastric juices, or it is incapable of osmosis. The narrative covers a series of experiment to determine which of the alternatives is correct. The story is driven by inconclusive results that needed to be clarified and by new questions that these experiments opened up. Mitchell held his readers' interest through remarks such as "To relieve myself of doubt, I made a second set of experiments . . ." or "I resolved my difficulties by the following experiment . . . ."[76] The readers can follow along and are drawn into the story by Mitchell's account of inconclusive outcomes, open questions, false leads, new leads, and failed trials.

This report closely resembles the eighteenth-century accounts of investigative journeys. Mitchell's readers also learned that even though the question was clear, there was no distinct, decisive *experimentum crucis* that could unmistakably identify one of the two alternatives as correct and the other as false.[77] Much intricate and challenging experimentation was needed for the experimenter to reach the point at which such a decision could be made with some confidence.

CHAPTER EIGHT

Controlling Experiment

Roughly twenty years passed until Mitchell returned to venom research. Together with the then demonstrator of physiology at Philadelphia, Edward Tyson Reichert, he conducted new experiments on the composition and action of rattlesnake and other venoms. By 1880, when Mitchell embarked once more on venom research, several universities and medical schools had acquired laboratory facilities, including the University of Pennsylvania. In the preface of the 1886 study *Researches on the Venom of Poisonous Serpents*—again published by the Smithsonian Institution—the two authors sign as "S. Weir Mitchell, Edward T. Reichert. Physiological Laboratory of the University of Pennsylvania."[1]

Some of those physicians who had gained experience in European laboratories were now training their own students. Several larger universities had acquired laboratories. Instruments such as thermometers, sphygmographs, kymographs, and photographic cameras (often imported from Europe) produced new kinds of records of phenomena of life, health, and disease, and the study of microorganisms was beginning to transform theories of disease and therapeutics. Mitchell and Reichert's work bears traces of all these developments.

Mitchell was heavily involved in the administration of the College of Physicians in Philadelphia, including (from 1886 onward) several years as president. He encouraged experimental research and sought to obtain funding for researchers. Together with Henry Pickering Bowditch of Harvard and H. Newell Martin of Johns Hopkins University, he took the lead in the founding of the American Physiological Society in 1887.[2] The society was devoted to the encouragement of research in physiology (broadly understood), and it admitted only those who already had a record of research in physiology, histology, pathology, experimental therapeutics, or hygiene.

At first glance, it seems that nothing much had happened between the publication of the first and the second Smithsonian essay, at least with respect to the general approach and the methods used to study venomous snakes. The second essay is broader in scope—different species of snakes are considered—but *Researches upon the Venoms of Poisonous Serpents* also begins with investigations of the chemical composition of snake venom, and the main part of the book describes animal experiments to study the action of this substance. On closer look, however, one finds a number of quite dramatic changes in the preferred kinds of protocols, in the way in which experimentation with animals was conceptualized, and in the methodological tools Mitchell and Reichert brought to bear on these experiments. Venom served as a means to explore specific body functions, and the changes thus produced were recorded with novel kinds of recording devices. Venom was tested systematically with a battery of common chemical testing procedures. Venom was also being scrutinized for any microorganisms that it might contain, and these microorganisms were studied with the help of novel culturing techniques.

Like the first Smithsonian essay, the second one exemplifies larger developments in the biomedical sciences, including a distinct shift from the investigation of experimentally induced diseases, symptoms, and lesions to experimental manipulations of physiological functions, increased use of measuring and recording devices, systematic studies of bacteria cultures, and a preoccupation with experimental controls.

I noted in the previous chapter that the few experiments on body functions that were mentioned in the first Smithsonian essay seemed to be strangely disconnected from the main part of the project, the study of symptoms and lesions in snakebite victims. In the second essay, physiological studies take center stage. Descriptions of experiments with bacteria and venom now play the part that the physiological experiments played in the first—bacteriological experiments are reported, but they do not seem to have a significant role in the main thread of the argument, the discussion of the effects of venom and venom components on cardiovascular functions.

Precisely because Mitchell and Reichert were no great innovators methodologically, their work is informative. It is a good measure of how established physiological experimentation and instruments had become in the 1880s and how quickly new bacteriological methods and practices were becoming common features in pathological laboratories and in biomedical experimentation. The second Smithsonian essay showcases how the latest trends in biomedicine were taken up by those whose investigations were not, as it were, on the cutting edge.

Experimentalism, Scientific Medicine, and the Laboratory

"[A]s regards scientific medicine, we are at present going to school to Germany," the U.S. Army Surgeon and librarian of the Surgeon General's Library John Billings stated in an address at the International Medical Congress in London 1881. Even in the United States, many people now believed that it was "no impediment to an original investigator to have to devote a moderate portion of his time to giving instruction either in the laboratory or in the lecture-room."[3] In the 1850s and 1860s, not all members of the medical profession in the United States embraced the ideal of experimentation, and many were in fact quite critical. In the 1870s and 1880s, the professional attitude of the majority of American medical men had become less hostile toward experimentalists, at least within academia. Billings's statement also shows that for him, at least, scientific medicine was linked to the laboratory. In the 1860s, "scientific medicine" had meant all things to all people. To an extent, this was still the case in the last decades of the nineteenth century. Calls for scientific medicine continued to be made for a host of reasons: to denigrate the advocates of the numerical method, to ridicule the practice of homeopathy,[4] to protect the profession against interference by antivivisectionists, and to stimulate reforms in medical education.[5] Sometimes, the proponents of scientific medicine simply listed a number of accomplishments that, according to them, qualified as "scientific," and often the advocates of new systems of healing also frequently claimed that their systems followed the principles of science.[6] Depending on the goals, the call for scientific medicine could mean more and better laboratory facilities, experimentation with precision and recording instruments, or "rational therapeutics," and often the practitioners combined their call with concerns about the competitiveness of American medicine. Mitchell's friend Dalton advocated experimentation on animals, including vivisection. For Billings, it was the connection with physical science and the use of precision recording instruments such as the sphygmograph and the kymograph or measuring instruments such as the thermometer that made medicine scientific.[7] Mitchell's colleague Wood advocated "rational" or "scientific therapeutics," by which he meant that experiments with drugs upon lower animals and upon healthy human beings were the "rational scientific groundwork for the treatment of disease."[8] Other advocates of medical reforms argued that chemistry was crucial for the understanding of the nature of disease, for therapeutics, and for pharmacology. The call for scientific medicine often, but not always, included a commitment to experimentalism. Almost all advocates of experimental research in biomedicine expressed such views in lectures and addresses that were printed in medical and popular journals, as well as in the introductions to their medical treatises and textbooks.

Some decades ago, many historians described the transformations of nineteenth-century biomedicine as a "laboratory revolution." Billings and other advocates for scientific medicine did draw connections between scientific medicine and laboratory research and instruction, but *laboratory revolution* was not an actors' term.[9] Recent (and not so recent) work on the history of scientific experimentation and instruments before 1800 shows that none of the practices that have been associated with the nineteenth-century "laboratory revolution" were particularly novel in the second half of the nineteenth century. Small-scale laboratories—for instance, alchemical ones—existed long before the nineteenth century.[10] Michael Faraday published a manual for students in chemistry laboratories in 1827.[11] One of the most detailed analyses of the concept "laboratory style" traces this research style even to the mid-seventeenth century. Ian Hacking has characterized the "laboratory style" as a form of investigation that constructs and uses apparatuses intended to isolate phenomena and create new ones. According to Hacking, the laboratory style of research is exemplified by Robert Boyle's air pump experiments.[12]

Historians have shown that only two features of late nineteenth-century research laboratories were unprecedented. The first is the size of some of the new laboratories and the scale of activities that these facilities enabled. Large-scale laboratories made it possible to acquire and put to use costly and high-quality instruments; provided the investigators with space, time, and resources (animals!) and facilitated collaboration and the division of tasks among researchers.

Another related feature of nineteenth-century laboratory research was due to the rapid industrialization and urbanization of nineteenth-century society and culture: New machines and the reorganization of nineteenth-century society offered new technical resources for the conceptualization and realization of experiments. All kinds of recording devices using photography and graphic recording techniques were employed in physiological experiments and for the observation of body functions. These devices included steam engines and small motors, novel kinds of galvanic cells and other machines that were adapted from contemporaneous industrial technology. Emil du Bois-Reymond's laboratory in Berlin is perhaps the most impressive institutional realizations of the project of "organic physics."[13]

In the United States, however, many research facilities were of modest size, and not every medical school had a laboratory, let alone all the devices available to the researchers in the Prussian capital. Several doctors who returned to the United States from their studies abroad were successful in their efforts to establish at their home institutions laboratory facilities, practical training, and curricula along the lines of the German model. The big German-style research laboratories in America opened in the 1870s and 1880s.[14] The institutional expansion in America really only began around 1900 with the opening of big medical schools at Johns Hopkins University, Harvard, Pennsylvania, Chicago, and Michigan and culminated in the founding of the Rockefeller Institute in New York.[15]

The programmatic calls for scientific medicine often included calls for animal experimentation, including vivisection, the use of precision instruments, quantification, and explanations of physiological and pathological processes through chemistry and physics. On the occasion of the opening of the biological laboratory at Johns Hopkins University in 1876, H. Newell Martin specified the requirements for successful scientific work in physiology, placing the emphasis of his speech on the European achievements in experimental physiology. Martin insisted that biomedical research was properly scientific only if it was based on the physical and chemical sciences. The researcher "must have a fair knowledge of mechanics, experimental physics, and chemistry; he ought to (I would almost again say he *must*) be able, besides English, to read at least French and German with facility—assuredly, if he cannot, he will labor with much toil and sorrow—and the more mathematics he knows, with the present rapid importation of quantitative ideas into biological science, the better for him."[16] Martin proceeded to illustrate these familiar proclamations with concrete and specific examples from experimental physiology.[17] Finally, Martin also stressed that research in his laboratory would be undertaken in close collaboration with the hospital, implying that experimental physiology could have direct effect on therapeutics and that pathological observation could validate experimental research.

We know, however, that we need to be careful not to overestimate the number of people who were actively pursuing experimental research. The American Physiological Society, for instance, comprised several reform-minded, research-oriented physicians, but it was very small compared to the American Medical Association, with more than 4,000 members.[18] Although experimentalism, "rational therapeutics," and experimental physiology were promoted and defended by several medical men, historians have identified only few active vivisectors in the United States. In 1883, Mitchell's friend and colleague Horatio C. Wood estimated that vivisections were performed in just five American cities: Baltimore, Boston, New York, Philadelphia, and Easton, Pennsylvania.[19]

Mitchell's work did move to the laboratory—in the 1880s, he and Reichert could make use of the physiological laboratory at Philadelphia. Yet he was never part of a large-scale "urban lab," and of course he had to see to his surgery. But the availability of laboratory space did leave traces in the Smithsonian essay (literally, as we will see). Like Mead and Fontana before him, Mitchell appropriated new trends in medicine for his venom research. In the 1860s, experimental physiology invaded his pathological studies of rattlesnake venom. In the 1880s, the new bacteriology encroached on his research on poisonous serpents.

Lab Notes

The twenty-odd years that passed between the first and the second of Mitchell's research projects on snake venom made a difference in the design of the project. Although the overall questions remained the same—the chemical composition of the venom and the effects on the body—the conceptual framework for the design and interpretation of experiments with venom had shifted as a result of broader developments in the field. Chemical experiments still helped identify active principles in venom. But ideas and aims related to remedies and treatments are largely absent from the experimental part of the second essay, as is the language of lesions and symptoms. The protocols for the design and interpretation of a number of the animals experiments with venom were now drawn from contemporaneous experimental physiology: Experiments were performed to investigate specific body functions in isolation (cardiovascular functions in Mitchell's case) and to establish how these body functions were altered by the influence of venom. The chapter entitled "Pathology" in the second Smithsonian essay uses elaborate protocols from the new bacteriology along with traditional comparative experiments, which were common in older research on spontaneous generation.

Like the earlier Smithsonian essay, the second essay discusses the physical and chemical analysis of venom. The chemical part of the book contains

Mitchell and Reichert's lasting contribution to venom research, the new conception of the toxicologically active principle of venom as a composite. In his autobiography, Mitchell conjured up a classic "eureka moment," here not a dreamy vision of a coiled-up snake but the sight of a tattered rug. The rug was "made of a rope, one corner of which was partly torn off and slightly resembled a serpent. Suddenly there came into my mind the idea that the poison of serpents must be double and not a single poison."[20] The Smithsonian essay contains nothing of this. Instead, Mitchell and Reichert presented the new chemical analysis in terms of contemporaneous protein chemistry. Using then-common methods of analysis (and securing the advice of a local chemist), Mitchell and Reichert found that the toxic component of the venom contained two albuminoid substances, which they identified as "peptones" and "globulins." The globulins could be further resolved in three different principles. They named them "water-venom-globulin," "copper-venom-globulin," and "dialysis-venom-globulin," whereby the names reflected the three chemical processes by which the globulins could be isolated.[21] Venoms from different kinds of snakes differed both in their proportion of globulins and in the proportions of globulins and peptones.

The results of Mitchell and Reichert's new analysis regarding the chemical composition of venom were novel, but the overall approach was not, although it had become more refined. To investigate the fluid, Mitchell and Reichert could draw on a battery of analytic tools that medical chemists now had at their disposal. For instance, the tenth edition of John Attfield's manual of medical and pharmaceutical chemistry, written for students in Great Britain and the United States, contains more than 600 pages describing tests for the most common chemical substances that a student might encounter, including globulins, peptones, and alkaloids.[22] On the basis of such tests, Mitchell and Reichert established the composite structure of the venom.

The perspective of the toxicological investigation has shifted from the study of venom-induced diseases to the study of venom-induced disturbances of normal physiological functions. As a consequence of the new chemical analysis of venom composition, the physiological project became extremely complex. Not only did the investigators have to examine the effects of pure venom, but the effects of each of its elements—the peptones and each of the three kinds of globulins—had to be studied as well. And to do this exhaustively, each component had to be tried on different body systems, tissues, and fluids. The middle chapters of Mitchell and

Reichert's essay, chapters VII through IX, are devoted to this Herculean task.

The conceptual framework for the toxicological investigation had completely changed. Mitchell and Reichert no longer studied venom-induced diseases and the symptoms and lesions associated with them. They no longer described the effects of the venom on the entire living organism or on blood and tissues. Instead, the reader finds experimental setups designed to isolate specific body functions in the animal. External agents were employed to disturb the normal physiological functions, and instruments recorded the alterations that were thus caused. Like other agents used in physiological experimentation, such as electrical and mechanical stimulation, venom was an external agent that altered certain "normal" body functions—respiration, pulse, blood pressure—in distinct, measurable ways. The expectation was that the comparison of observations and measurements of different modifications induced by venom—such as the acceleration or deceleration of the pulse of a rabbit that had been injected with venom or different venom components—would elucidate the "normal" effects of the venom. Mitchell and Reichert studied functions in venomized but otherwise intact animals as well as in animals whose pneumogastric nerves (nerves that supply the lungs and stomach) were cut and on whom sections of the upper spinal cord had been performed. Such sections isolate the heart from the nerve centers and thus permit separate investigations of the effects of venom on these body systems. (To be kept alive, the animal requires artificial respiration.)

The traces of contemporaneous physiological research are clearly visible. The changes in body functions were measured with up-to-date recording devices, a manometer for recording the blood pressure and a revolving drum equipped with "Marey's tambour" for recording the rate of breathing.[23] These experiments did not result in vivid descriptions of changes in the texture and coloration of tissues and organs. They produced numbers: heartbeats, pressure measurements, and counts of breaths per minute, presented in numerous tables.

Had antivivisectionists ever got hold of Mitchell and Reichert's book, they would have been horrified by the relentless accountancy of death that is displayed in the right-hand column of each table.[24] The tabular information provided in the first Smithsonian essay gives at least the impression that all the pain and suffering is for a purpose. The reader learns something about the symptoms and lesions produced by snake venom, and there is a promise perhaps of future treatment. In the second Smithsonian essay, we merely find records of a succession of deaths. The experiments are really "elementary researches" into physiological processes—the deaths of the animals mark the limits of the perturbations that body systems and functions can withstand. Typically, the remarks read: "Injected intravenously 0.015 gram dried venom of the Crotalus horridus dissolved in 1 c.c. distilled water." At 9:40 minutes: "Convulsions." At 10:10 minutes: "Dead." Or: "Injected intravenously 0.004 gram dried venom of the Ancistrodon piscivorous dissolved in 1 c.c. distilled water. Convulsions." At 1:30 minutes: "Injected as above. Struggles." And at 4:00 minutes: "Killed by pithing."[25]

The physiological part of the essay is a bit of an oddity. On the one hand, it is meticulous almost to obsession in its presentation of the masses of data and in its systematic, comprehensive coverage of the effects of pure venom and of all four physiologically active venom components on healthy animals and on animals with sections of nerves and spinal cord. On the other hand, the project remains strangely incomplete. Mitchell and Reichert decided to treat venoms of different species of snakes as physiologically and toxicologically equivalent. They did mention that each of the experiments should have been carried out with several different venoms but noted that they did not have enough supplies and resources to do so. Moreover, the measurements were not made at regular intervals. For instance, in experiments 5, 6, and 7, the experimenters measured the effects of pure venom on arterial pressure. In experiment 5, the measurements are taken at minute markers 0:10, 0:20, 0:30, 0:40, 1:00, 1:20, 1:40, 1:55, 4:00, 7:00, and 8:00. In experiment 6, measurements are taken at 0:05, 0:10, 0:20, 0:30, 0:40, 0:50, 1:00, 1:10, 1:20, 1:30, 1:40, 3:40 5:40, 7:40, 9:40, and 10:10. In experiment 7, measurements are taken at 0:05, 0:10, 0:20, 0:30, 2:40, 3, 5, 6:30, 7:30, 8, 8:30, 9:00, and 10:00.[26] Although the report contains pages and pages of such numbers—almost all of them sequences of measured values dependent on time—there is no evidence of data processing. The (few) comments on the data remain qualitative. No significant conclusions were drawn from this time-consuming and taxing endeavor. In fact, the authors state at the outset that clear conclusions are not possible: "In all of our observations we find that the results produced in animals, under apparently the same conditions and by using the same doses, vary very greatly; sometimes the pulse is quickened from the first and remains beyond the normal until death ensures, sometimes there is a primary diminution followed by an increase, at others there is a diminution which continues until death. The pulse is generally found to vary much in frequency. These facts all suggest that the action of the pure venom is of a complex nature."[27] Indeed, what little commentary there is typically reads like the following: "Twenty experiments were made with pure venom upon normal animals; in three the pulse-rate was diminished and remained below normal, in two there was a primary increase followed by a diminution, and in one of these the pulse-rate afterwards went above the normal, while in another there was a primary diminution followed by an increase."[28] The outcome of the experiments with *Crotalus horridus* and *Ancistrodon piscirorus* were equally inconclusive. Analysis and interpretation is left to the reader, and one gets the impression that no definite general conclusions can be drawn from the experimental outcomes.

The insights presented in the chapter following the physiological discussions, chapter X, entitled "Pathology," are also preliminary at best. To the modern reader, the chapter is an exemplification of how late nineteenth-century investigators attempted to make experiments and observations comparable across different contexts. A number of established techniques and procedures are mentioned, such as "the methods of Sternberg and Koch," "Sternberg's glass-bulbs," and the like.[29] This is common in other fields, too. In bacteriology manuals, we find Sternberg's bulbs, Miquel's bulbs, Lister's flasks, Aitken's test tubes, Pasteur's apparatus, and so forth.[30] In experimental physiology, researchers used Marey's tambour or Ludwig's kymograph.[31] Microscopy manuals list countless staining and fixing agents, typically with names attached—there is Thiersch's Carmine Fluid, Beal's Carmine Fluid, and so on.[32] But above all chapter X exemplifies the transformation of disease theories in the late nineteenth century.

Mead conceptualized the plague as a disease that was caused by a poison, the "plague contagion." For Mead, contagion was a hypothetical nonorganic, subvisible material substance. His conception of contagion was well compatible with early modern miasmatic disease theories according to which diseases (such as mal'aria) were caused by bad air. It was only in the second half of the nineteenth century, that the (modern) conception of specific, identifiable material causes for diseases gradually took shape.[33] The mid-nineteenth-century investigations of disease-causing agents had their roots in early nineteenth-century research on microorganisms, their classification, and their biological agency.[34] At that time, microscopists drew parallels between disease processes and processes of putrefaction and fermentation, and we saw traces of this conception in the first Smithsonian essay, where Mitchell classified snake venom as "septic or putrefacient" poison. In 1844, the German anatomist and pathologist Jacob Henle published a widely influential article on miasmas and contagions. In it, he suggested that infectious diseases require a "substance" that would be discharged from the diseased body and transferred to a healthy one. This article marked a fundamental reconceptualization of disease, because the bodily disease became clearly separated from its cause.[35] In the last decades of the nineteenth century, the causes of disease that Henle had hypothesized could be made visible with the help of the microscope and staining and culturing techniques. Organic entities— germs—were identified as specific causes of disease.

In their research on pathology, Mitchell and Reichert used both the protocols that were associated with traditional research on micro-organisms and spontaneous generation and the protocols that were emerging from research on bacteria. Mitchell and his coworkers[36] examined whether venom caused putrefaction in organic substances. They compared test tubes filled with sterilized bouillon with or without rattlesnake venom, which were either sealed or left open, and studied the liquid for evidence of putrefaction. All samples except the sealed, sterilized bouillon showed putrefaction, but in the samples that contained venom, the putrefactive changes occurred more rapidly. Related series of experiments showed that more rapid putrefaction also occurred when muscle tissue was treated with rattlesnake venom and when the first experiment was repeated with moccasin venom.

Inspired by the new bacteriological research, Mitchell and Reichert also examined the bacteria contained in venom samples. The novel bacteriological conception of infectious diseases was embedded in a not-so-new set of practices of investigation. Recall that the identification of active principles required the isolation of the agent, its chemical analysis, and its transfer to a living body to establish its effects. Late nineteenth-century bacteriologists followed a similar protocol to identify disease-causing agents. According to Robert Koch, a microorganism could be considered a cause of disease if it was present in a diseased animal, could be isolated from the animal and grown in pure cultures, and produced the disease when a healthy animal was inoculated with it. These requirements for determining causes of disease later became known as "Koch's postulates."[37] One of the first expressions of these requirements can be found in Koch's famous lecture on the etiology of tuberculosis of 1882, which was immediately printed in the *Berliner Klinische Wochenschrift*.

Just a few years later, Mitchell and his coworkers used the same protocol—if in a slightly different manner than intended by Koch. The section "culture experiments" describes how Mitchell and his coworkers isolated the bacteria they found in venom, transplanted them to flat glass vessels covered with jelly to obtain "pure cultures," and inoculated the venom-micrococci into animals.[38] They examined whether the bacteria they had found in venom were pathogenic—they were not, as it turned out. When the microorganisms were inoculated into healthy rabbits, rats, and pigeons, no animal died, there were not even any lesions similar to those produced by venom. Despite their negative outcome, their experiments on the role of microorganisms in venom poisoning were timely enough to be included in the treatise.

Control and Complexity

In chapter 7, I examined the debates in the mid-nineteenth century about the merits of Mill's methods of inquiry for biomedical experimentation. The strategy of comparative experimentation was introduced as a pragmatic, realistic alternative to the ideal search for causes as it was encapsulated in Mill's methods of inquiry. Other striking features in nineteenth-century methods discourse suggest that concerns about complex and hard-to-manage experimental situations were growing. One particularly interesting sign of this struggle with complexity is the rise of the term *control*. Trying to gain control means trying to find ways of managing complexity and variability in an experimental situation.

The concept of "control" is puzzling. It was only during the last decades of the nineteenth century that the term (*experimental*) *control* or *control experiment* began to appear in experimental reports. On the other hand, many instances of control experiments *avant la lettre* have been identified. Fontana has been described as "the first" to use controls, and some historians have drawn attention to much earlier episodes. The earliest instances for "control experiments" have been found in antiquity.[39] Seventeenth-century experiments on spontaneous generation have been cited as instances of control experiments.[40] A challenge from van Helmont to the humorists has been presented as control experiment, and so on.[41]

In his 1954 essay *The Nature and History of Experimental Control*—still the most in-depth analysis of the history of experimental control[42]—Edwin Boring helpfully distinguished among three meanings of "control" in the context of experimentation: control in the sense of restraint (keeping conditions constant), control in the sense of guided manipulation (causing an independent variable to vary in a specific manner), and control in the most general sense of check or comparison. This distinction is instructive in that it helps us to see that the third, broad conception of control underlies many attempts to identify "the first" control experiment in history. It is the element of comparison that some historians appear to have in mind when they date the origin of control experiments back to the early modern period or even to antiquity. In this general sense, however, all experimentation is controlled experimentation. Boring notes, very plausibly, "Control in the sense of a check or comparison, the original meaning of the word, appears in all experimentation because a discoverable fact is a difference or a relation, and a discovered datum has significance only as it is related to a frame of reference, to a relatum."[43]

Boring and other historians have suggested that Mill's work on the methods of inquiry from the *System of Logic* should be regarded as the origin of the narrower methodological notion of experimental control. His method of difference should be interpreted as the first philosophical conceptualization of controlled experiments, even though he did not use the actual term *control*.[44] This interpretation needs to be amended, however, at least for the life sciences. In experimental contexts, the notion of control was very frequently used in late nineteenth-century bacteriology. But I have been unable to find any reference or connection to Mill's methodological thought in this literature. If Mill was mentioned in medical and biological journals, he was mentioned with reference to his political and economic writings. If we look for instances of "control" in bacteriology, we find certain recurrent practical procedures of comparison, but little of Mill's rigorous experimental reasoning. In bacteriological experiments, for example, test tubes containing bacteria are routinely compared with "controls," that is, with test tubes that do not contain bacteria.

These kinds of comparisons are frequently described in Mitchell and Reichert's essay. When they injected a mix of venom and tannic acid (which was known to react strongly with albuminoids) into the breast of a pigeon, they compared the outcomes to "a control experiment in which 1.5 c. c. saturated solution was injected into the breast of a pigeon."[45] In one of the experiments on putrefaction, moccasin venom was mixed with sterilized bouillon in a test tube. Six test bulbs filled only with pure, sterilized bouillon were kept as a "control experiment."[46] When the investigators examined how venom affected the shape of red blood disks, the effects were compared with the effects of other liquids or reagents "tried in control experiments."[47] When the effect of venom on the movements of spermatozoa (taken from a live rabbit) was examined, "[s]pecimens treated with the venom were examined side by side with control specimens."[48]

This type of experiment exemplifies the definition of "controlling experiment" from the *Century Dictionary and Cyclopedia* of 1897: "Controlling experiment, in *chem.*, a corroborating or confirmatory experiment." The definition is illustrated with an example from W. R. Bowditch's book *Coal Gas*: "For a *controlling experiment*, the gas may be passed for a short time through the alcoholic ammonia alone."[49] The experiments just mentioned and the definition exemplify a practical procedure that is the practical counterpart of the formal system of reasoning Mill presents.

What, if anything, does the appearance of the term *control* signify? It may be tempting to interpret the emergence of the term *control* as an echo of the rapid industrialization that affected all aspects of nineteenth-century society, a reflection of the advancements in technology, engineering, automation, mass production, and the transformation of social organization that industrialization entailed. In the late nineteenth century, the United States became the dominant technological nation, and new technologies and systems of production gave rise to new concerns with order and control.[50] Historians of physiology have shown more concretely how much the organization of the big nineteenth-century physiology laboratories resembles a factory, and how many experimental designs and instruments were borrowed directly from engineering contexts. Older work in history of physiology drew a direct link between the rise of industrialization and the "technological attitude" of physiological experimenters, which was seen to be manifest in the manipulation and creation of events in experiments. In this perspective, knowledge becomes identified with the control imposed by the "investigator–technologist"; knowledge "equals control," and to some, even Bernard appeared as a representative of the new "technological" attitude to experimentation.[51]

However, as we saw, Bernard emphasized how very difficult it was to manage the causal conditions of physiological experiments—much more difficult, in fact, than in physical experiments. Bernard's emphasis on comparative experimentation does not represent an optimistic technological attitude. On the contrary, it indicates a concern about *lack* of control over the experimental situation.

It is clear that the term *control* does not *originate* in the technology–engineering domain. In the first half of the nineteenth century, it was used predominantly in sociopolitical contexts—to refer to the management of (potentially unruly) people and institutions, administrative officers,[52] schools,[53] railways,[54] inventories, armies,[55] expenses and prices,[56] and—last but not least—oneself.[57] James Clerk Maxwell's 1868 communication to the *Royal Society* illustrates this point very nicely. The paper is considered a hallmark of *control theory* in engineering, but it does not contain the notion of control. The title of the paper is "On Governors," Maxwell's "governor" being "a part of a machine by means of which the velocity of the machine is kept nearly uniform."[58] Thus, we may speculate that in the nineteenth century, the sociopolitical notion of control migrates to both engineering and life sciences. We may further speculate that the rise of *control* as a term in the life sciences does signal a shift in methodological thought, but not the optimistic one that Boring and others have identified (*viz.*, the systematization of the experimental search for causes that makes the findings more secure). It signals a new attitude to the epistemic status of the findings resulting from such a comparison, which is much more pragmatic and less optimistic.

Bernard's notion of comparative experimentation was an acknowledgement that the ideal experimental situation required for the application of Mill's method of difference cannot be attained in practice. Likewise, the practice of directly checking against control animals and control substances can be seen as a retreat from an ideal, the ideal of a secure standard of comparison. In the early nineteenth century, it was common practice to evaluate observations, measurements, or experimental outcomes against invariable, fixed reference points for comparisons across local contexts. All sorts of things were used as standards, not only conventions regarding units of measurement but also particularly accurate observations of nebulae, the "most probable" (!) proportions between the jaw and dental elements of a skeleton, the quantity of tannin in oak bark, or the quantity of zinc lost in the decomposition of water at the Volta electrometer.[59] As different as these "standards" may appear, they all are (presumed to be) nonlocal, fixed, invariable reference points for comparisons. But as the items on this list suggest, some of these standards turned out to be less fixed and immutable than would be required for a transcontextual point of reference. Experimental controls appear as a pragmatic alternative to universal standards of comparison; just as comparative experiments were a pragmatic, realistic alternative to counterproofs.[60] The word *control* became common in the experimental sciences when it turned out that experimental situations were too complex to be completely controlled.

The term *control* might have suggested itself to some because it was "in the air," but the shift in physiology from a methodology of "tests" to a methodology of "control" indicates, above all, a *loss* of confidence and trust in the certainty of experimental results. The gradual replacement of the terms of *check* or *test*, which have the ring of definiteness, with the term *control*, which signifies an effort to maintain order, indicates that the complexity of the experimental situation no longer appeared fully manageable. Some of the methodological innovations from the nineteenth century can be understood as compensatory measures along these lines. Even the medical literature had grown so vast that it had to be pruned.

A Deluge of Scientific Papers

In the late eighteenth century, scholars like Fontana and Senebier made explicit comments about how to write on experimentation. Late eighteenth-century experimenters were encouraged to be prolix and to describe their entire journey of discovery. Following the investigative pathway would enable the reader not necessarily to see for himself but at least to judge for himself whether the findings from this expedition could be relied on. The underlying assumption was that experimenters were trustworthy people—the detailed report of their investigations enabled the reader to judge the soundness of the project, not the veracity of the investigator.

Those medical men in the 1880s who commented on medical writing did not discuss whether the order of research should be reflected in the logic of arguments. Rather, we find complaints and expressions of concern about the implications and consequences of the changing market for medical literature. The expanding book and magazine market had repercussions on scholarly and literary writings—not all of them welcomed.

For Mitchell, the expanding book, magazine, and journal market created opportunities. As a literary writer, he took advantage of the new periodicals aimed at wider, educated audiences. His popular essays on snake venom research were published in popular magazines, such as *Century Magazine* and the *Atlantic Monthly*. He published several of his short stories in the latter magazine as well. The popular essays on rattlesnakes appeared shortly after the two Smithsonian essays and contain some information about his research. Of course, the account of his investigation takes a quite different form. The rattlesnake is now a "gallant-looking reptile" armed with "death-giving juices."[61] Both of the popular pieces offer colorful descriptions of snakes and of the "wonderful machinery" of the bite, how the "coiled death" lashes out, the neck bending back, mouth wide open, fangs exposed, the points entering the skin: "Quick as thought the lower jaw shuts on the part, deeper go the fangs, and, the same muscle which closes the jaw compressing the glands, the venom is injected among the tissues which the fangs have pierced."[62] Only the second popular essay offers a description of the experiments Mitchell had performed and of the methods he had used. In the first popular essay, Mitchell simply asked his readers "to accept the proposition without being troubled with the proof," as a full explanation of the "modern means of studying the effects of poisons" would take too much space—which might have been a polite way of saying that the description of these experiments was too difficult for his readers to understand, or perhaps an attempt to avoid confrontation with antivivisectionists.[63] By the time he was preparing the essay for *Century Magazine*, Mitchell obviously presumed that his readers had at least some basic knowledge of and appreciation for science. He now described in some detail the chemical composition of venom and the experiments by which the chemical agents could be identified. On the other hand, this piece appears much more visually attractive than the first popular essay thanks to several illustrations of snakes, coiled up in their natural habitat, watching their prey, and striking.

Along with the new outlets for popular science writing, the medical literature had also grown vast, creating new challenges to the researcher both as writer and as consumer of medical texts.[64] In his address at the big International Medical Congress in London, John Billings not only commented on the state of "scientific medicine" but also offered hard data to illustrate the growth of medical literature around 1880, both in the United States and worldwide. The figures are impressive. For instance, in 1879 "the total number of original articles in medical journals and transactions which were thought worth noting for the Index Medicus was a little over 20,000. Of these there appeared in American periodicals 4,781; in French, 4,608; in German, 4,027; in English, 3,592; in Italian, 1,210; in Spanish 703; in all others 1,248."[65] Billings, founder of the *Index Medicus*, announced that based on his search in bibliographies and catalogues of the main medical libraries, he estimated that medical literature was growing at the rate of about 1,500 volumes and 2,500 pamphlets yearly. But he was not too pleased with the current state of medical writing—those "cheap and dirty volumes of modern days, with their scrofulous paper and abominable typography."[66]

The poor quality of the printed items was a minor trouble compared to the poor quality of the content and organization of the texts. Billings implied that late nineteenth-century editors of journals and proceedings should be, but were not, very powerful gatekeepers for their profession. It was mainly left to readers to assess the quality of the materials that were printed—an almost impossible task amid the flood of printed materials. For Billings, it was clear that trust in experimental outcomes could not be based on trust in people and in the veracity of their reports. Experimenters were not trustworthy just because they were scientists. Although those in the investigator's immediate surroundings usually knew whether that person did good and reliable work and could thus form an appropriate judgment about that person's writings, others had "no clue to the character of the author." That was a problem, for, as Billings pointed out, there were those "books and papers from men who are either constitutionally incapable of telling the simple literal truth as to their observations and experiments, although they may not write with fixed intention to deceive, or from men who seek to advertise themselves by deliberate falsehoods as to the results of their practice." Because their closer colleagues would be aware of this, these researchers would "find it necessary to send their communications to distant journals and societies in order to secure publication."[67] Obviously, if editors did nothing to prevent publication of these items, gullible readers might be led astray in their judgments.

Billings chided editors because they did not take the responsibility—at least not sufficient responsibility—to evaluate the content of the manuscripts they were offered, "a responsibility which cannot be altogether avoided by any formal declaration disclaiming it."[68] But reliability was only one thing. The other big problem was ease of access to information. Writers had a responsibility here, too: they ought to present their research in a properly organized fashion.

No doubt Billings would have been quite impatient with de Luc's, Réaumur's, and Fontana's sprawling treatises, and especially with Fontana's call for prolixity in experimental reports. In the late nineteenth century, the ideal of quick access to information had trumped the late eighteenth-century notion that the telling of the journey of discovery could inspire the reader with confidence. According to Billings, authors of articles or at least the editors of transactions or societies had to make sure that the articles they published carried a properly informative title instead of a heading like " 'Clinical Cases,' 'Difficult Labour,' 'A Remarkable Tumour,' 'Case of Wound, with Remarks.' " Otherwise it was almost impossible to identify all those items in the deluge of papers that were really relevant and had to be acknowledged in one's work. Researchers had to realize that their readers were "for the most part hard workers, busy men, who have a right to demand that their library table shall be provided with properly prepared materials, and not with shapeless lumps."[69]

As a scientific author, Mitchell did a better job than many, but the second Smithsonian essay is certainly not a model of stringent, elegant, and concise scientific writing. On the contrary, the essay looks as if it was put together in haste—not as a well-crafted product of a writer who had literary talent. Leafing through Mitchell and Reichert's Smithsonian essay, one can find some new features in the organization of the materials, but, like the methodological innovations, these features do not appear to be the result of a conscious stylistic choice on Mitchell's part. Rather, they appear as a pragmatic choice of someone who had many other things to do as well.

Chapters I–VI of the second Smithsonian essay read like Mitchell's earlier scientific writings, containing reports of a number of experiments, a few tables, and a good deal of interpretation. The themes of these chapters and of most of the sections are properly identified by informative section headings (such as "The Activity of Venom when applied to Serous Surfaces"). Within the sections are descriptions of procedures for each experiment. The rest of the book is much less developed. Chapters VII–IX, the reports of the physiological experiments, are markedly different from the preceding chapters. They contain the records of 174 experiments but little else. The data from these experiments are simply given as lists or in tabular form. There is only a minimum of additional information and almost no interpretation—even less than in the tables of 1860. Chapter X, entitled "Pathology," is a bit of a hodgepodge of several loosely related issues and is remarkable chiefly for the two sections on putrefaction and culture experiments.

Because the style of the middle chapters is so different from the others in the rest of the book, one may assume that they were penned by Reichert, who simply copied large parts of his laboratory notebook. Mitchell certainly had a research agenda, but he had little time for systematic experimentation thanks to his countless other obligations as a physician, university administrator, and literary writer. Reichert was a professional physiologist and demonstrator, and presumably he had regular and frequent laboratory access.

Information about Reichert is hard to find. Even though he became the chair of physiology in 1886 and remained in that position for twenty-four years, he left few traces. He did publish a book on hemoglobin crystals that had a modest effect on protein research.[70] Nevertheless, most commentators regarded him as a rather minor figure. Alexander Abbott, bacteriologist and hygienist at Philadelphia, had fond memories of the lunch club that convened daily in Reichert's laboratory and of Reichert's artistry with the skillet.[71] In George Corner's history of the School of Medicine of the University of Pennsylvania, Reichert is introduced almost in passing as "mediocre lecturer" and "industrious investigator."[72]

Judged by the essay he coauthored with Mitchell, Reichert was very industrious indeed. The middle chapters are an exemplary piece of accountancy. They contain pages after pages of experimental data, the experiments numbered consecutively, the data organized as plain tables with three columns for times, counts of the animal's pulse (or blood pressure, or respirations), and "remarks." The remarks are sparse—a couple perhaps for each experiment. The pages read as if they were taken straight from the laboratory notebook. There are a few details about the quantity and kind of venom and sometimes a hint about the behavior of the animal during the trial ("struggles," "breaks loose").

There is no narrative in the middle chapters of the essay. There is no information about the research activities or explanations or justifications of any aspect of the experiments that were performed. The experiments on cardiovascular functions are grouped by theme ("injections of copper-venom-globulins," "injections of dialysis-venom-globulins") but there is no internal sequence within the group that would take the reader through the material.

The same is true for the pathological chapter. Here we do not even find an overall theme or motivation; the chapter records explorations of all sorts of issues related to bacteriology. Some of the novel approaches that Mitchell and Reichert described seem to have been included just because they were new opportunities that suggested themselves for exploration.

Mitchell's work on snake venom leaves the toxicologist, physiologist, and pathologist with loose ends and open questions, but to the historian of methods discourse it is instructive even so. If nothing else, it shows yet again how important it is to put scientists' discussions of methods (or lack thereof) in context and historical perspective. At first glance, Mitchell and Reichert's report of their physiological experiments suggests that the criticisms of experimentalism and the concerns about reliable methods had been quelled, as do their references to standardized procedures and instruments. But this is not the case. What the essay does show is that toward the end of the nineteenth century, specialized communities had begun to form within the larger biomedical community. These smaller groups shared not only protocols, tools, and methods but also certain recognizable methodological strategies. Such protocols, tools, methods, and strategies provided a pool of resources and opportunities that other experimenters could draw on in their research. One can trace the migration of those protocols, approaches, and tools across the different subfields (and the practice of naming things conveniently helps). Sometimes such an appropriation of resources can lead to innovations; at other times, it goes nowhere.[73]

CHAPTER NINE

Unobservables

At the age of 70, Mitchell still took a keen interest in venom research. More than ten years after his study with Reichert, he published another paper on the interaction of blood and venom, which resulted from the collaboration with Alonzo Stewart.[1] In this short paper, the authors linked their venom research to contemporaneous immunology, in particular to Calmette's and Fraser's demonstration that the repeated injection of small doses of venom into animals could protect these animals from the effects of up to fifty times the lethal dose. In 1899, Mitchell suggested to the new chair of pathology at Philadelphia, Simon Flexner, that the application of the tools of immunology to venom research might be productive. Inspired by Mitchell, Flexner and his assistant Hideyo Noguchi embarked on a project to examine the action of venom on blood and tissue, which probably went well beyond anything Mitchell had envisaged.[2]

In hindsight, Flexner and Noguchi's venom research does not appear to be a momentous contribution to toxicology, pathology, or biochemistry (certainly less so than Fontana's work). My choice to focus on Flexner and Noguchi's venom research is motivated in part by their investigations' having been encouraged and guided by Mitchell. But many other scientists had at least an equally large share in the study of snake venom. When Mitchell began his project, a community of venom researchers did not exist. Before 1900, finding out about the composition and action of snake venom was for the most part the concern of particular individuals, and it is quite easy to identify protagonists of snake venom research for a given period, such as the late eighteenth century. As snake venom research became part and parcel of the larger research program of immunology, an increasing number of people became involved in the investigations. When

Flexner and Noguchi took up their venom research, they could—and did—engage with a sizeable group of specialists. They could draw on a rapidly increasing stock of new findings and a stream of research reports and discussions of techniques that were being published both in established and in new journals. Venom research had become an international endeavor. Flexner and Noguchi referred to, and engaged with, scientists from Europe, the Americas, and Asia. Snakes and venoms were sent and exchanged across continents and oceans, and articles on venom studies were published in scientific journals with international readerships. The bibliography of Noguchi's 1909 monograph on snake venom comprises eleven pages in two columns listing more than 200 items, mostly published after 1880.[3] A host of researchers had become involved in the topic.

It is worthwhile to probe Flexner and Noguchi's work, because Flexner was in the center of, and was even a leading figure in, the institutionalization of early twentieth-century American medical and biological sciences. He was connected to all the institutions and organizations that were central for the history of late nineteenth-century American medicine and biology more generally: The Marine Biology Laboratory at Woods Hole, the biology laboratory at Johns Hopkins University, the American Physiological Society, and the Rockefeller Institute for Medical Research. Through his teacher William Welch and the other researchers in the biological laboratory at Johns Hopkins University as well as at the MBL, Flexner became familiar with the latest developments in bacteriology, toxicology, immunology, and experimental embryology. His work is instructive for the historian of methods discourse because it elucidates developments in venom research, in the methodology of biomedical practice, and in the institutional organization of the fields of biology and medicine more generally. Flexner and Noguchi could draw on a much broader pool of resources than Mitchell had available—conceptual, methodological, and (as we will see) material.

A Not-so-Typical Career in Medicine

In the late nineteenth century, infectious diseases were reconceptualized as diseases produced by pathogenic microorganisms. The body's defense mechanisms were explained in entirely new ways. As a consequence, new kinds of drugs and new forms of treatment appeared possible. The field of bacteriology—and, along with it, immunology—emerged as an area of specialization for medical experimenters. In the 1870s and 1880s, new staining and culturing techniques for microscopy helped bacteriologists see and describe the disease-causing agents whose existence several mid-nineteenth-century pathologists had assumed. In the 1880s and 1890s, numerous findings about properties of bacteria, their actions, and their interactions with other bodies contributed to the formation and stabilization of the fields of bacteriology and immunology, following research by Robert Koch and Louis Pasteur on microbes as disease-causing agents and therapeutic vaccination.[4]

Meanwhile, another fundamental problem was turned into a problem of experimental research: the problem of biological organization. Comparative anatomy and morphology had provided detailed descriptions of the complexities of life. Experiments with developing embryos were expected to reveal how organized life forms could come into being. To some degree, the fields of biology and (medical) physiology and pathology were growing apart, and a number of subfields emerged within the biomedical sciences, complete with societies and specialized professional journals: experimental embryology, bacteriology, immunology, and so on. Historians of American science have shown that the American Physiological Society encouraged laboratory and experimental research (also animal experimentation and vivisection) in physiology, histology (including pathological histology), experimental therapeutics, and hygiene.[5] New biological institutions, notably the Marine Biological Stations at Woods Hole (or Woods Holl, as it was called then) and elsewhere were mostly concerned with evolution and heredity, embryology, cytology, and genetics.[6] The new Rockefeller Institute (founded in 1901) was an institute specifically for medical research with the expressed aim to combine laboratory research with clinical research.[7]

We know that during the process of gradual institutional separation of medical departments and biological research stations close personal and intellectual ties and commitments remained among researchers in the more "medically" and in the more "biologically" oriented institutions and organizations. The APS, for instance, remained small and exclusive (performing and publishing original research was a condition for membership), but it maintained close bonds not only with the medical community but also with the American Society of Naturalists (founded in the mid-1880s) and the smaller, affiliated groups springing from it.[8]

Perhaps the most pivotal center for American biologists was the Marine Biology Laboratory. Established in 1888, the MBL was at first a summer resort for academic biologists.[9] It offered summer courses, a library, and basic resources for research projects to American biologists, as well as living organisms for systematic study. It facilitated interaction and communication among American biologists and thus helped build and strengthen the community of professional biologists, but it also hosted medical researchers—Flexner and Noguchi among them. By the mid-1890s, more than sixty investigators and more than 160 students from dozens of institutions and organizations met at Woods Hole every year.[10] Some of the most important strands of research in twentieth-century American biology originated or were pursued at Woods Hole, including Edmund Beecher Wilson's cytological research, Thomas Hunt Morgan's work on *Drosophila* genetics, and Jacques Loeb's experiments with embryos.

At the universities, biological and medical research and teaching continued to be intertwined as well. The ambitious new research laboratory at Johns Hopkins University, for instance, was nominally a "biology" laboratory. The head of the institute, Martin regarded physiology as a key part of biology and placed his own emphasis on physiological research. The laboratory he established at Johns Hopkins was equipped for it, especially for research on the heartbeat, Martin's own subject of interest, and he pursued experimental research with his students.[11] At the same time, the students also took courses in morphology (embryology and comparative anatomy) as part of their graduate training. Subjects related to morphology were represented by the second instructor, William Keith Brooks.[12] Brooks also ran a successful summertime seaside laboratory at Chesapeake Bay. Students could (and some did) participate in Brooks's summer courses at the Chesapeake Zoological Laboratory.

In the early 1890s, when Flexner was a student at Johns Hopkins University, the work in the laboratory became more and more medical as a consequence of both the decline of Martin—who had become addicted to alcohol and thus was no longer able to perform his professional duties— and the increasing influence of William Henry Welch, the chair of pathology and first dean of the Medical School. Even Welch did not completely sever the connection with biology. His own work focused on the experimental and clinical study of abnormal body functions and activities.[13] But he kept abreast of biological research, especially on evolution. In his address before the Fourth Congress of American Physicians and Surgeons in 1897, for instance, he discussed the concept of adaptation. This address shows that Welch was familiar with the research in experimental embryology by Wilhelm Roux, Hans Driesch, Jacques Loeb, and others.[14] In his address, Welch discussed the extent to which pathological processes could be understood to be adaptations due to variation, natural selection, and heredity. He argued that these processes could ultimately be explained in terms of physiological processes or mechanisms. He also stressed that "those new fields of experimental research called by Roux the mechanics of development of organisms, and also in part designated physiological or experimental morphology" could help shed light on the general question of how living organisms responded to changing external or internal conditions.[15]

In the 1890s, Welch's student Flexner negotiated his own career path between biology and physiology. The younger brother of reformer of medical education Abraham Flexner (the prominent author of the *Flexner Report on the State of American Medical Education*, commissioned in 1908 by the Carnegie Foundation for the Advancement of Teaching) had not come equipped with the systematic scientific training that Martin, the first director of the biology laboratory had insisted be prerequisite for all of his students. Flexner left high school early to work in several jobs and eventually at a local pharmacy.[16] He was sent to attend lectures at the Louisville College of Pharmacy, where he took courses and laboratory classes for a couple semesters.[17] During the eight years he was working at his brother's drugstore, he read various medical books and journals and practiced microscopy, but he never obtained a formal university degree, except an M.D. from the Medical Institute at the University of Louisville—not a very reputable institution. Nevertheless, Flexner was admitted to Johns Hopkins University. He took up his medical studies at Baltimore in 1890, well after the opening of the biology laboratory but still years before the opening of the Medical School.

During his first years at Johns Hopkins, Flexner received his training "on the job," as it were. He spent a great deal of time in the laboratory. He did some microscopy, attended Welch's lectures and demonstrations in pathology, and helped him and the assistant professor in pathology, William Councilman, with routine laboratory tasks and autopsies. Several of his teachers and instructors had ties to European laboratories.[18]

In 1893, Flexner himself took a trip to Europe and became acquainted with the organization of German pathological laboratories. His trip was comparatively short. He visited several different places—Strasbourg, Freiburg, Heidelberg, Berlin, Vienna, Leipzig, among others—rather than spending an extended period in one laboratory as many other American visitors did. According to his biographer, Flexner was impressed with the resources the German laboratories offered but not with the organization of the workforce. He thought that in his own laboratory back home, he had more freedom to choose the topics and projects he wished to work on.[19]

Flexner's career path was steep. He published numerous papers while at Baltimore. Among other things, he conducted detailed investigations of the lesions produced by vegetable and animal toxic agents.[20] In the mid-1890s, he spent a couple of summer months at Woods Hole, where he could make use of Jacques Loeb's laboratory. While at the MBL, he investigated the nervous system of *Planaria*, and he lectured to the MBL community on bacteriological poisons—toxalbumins—and their actions on tissues and organs.[21] Despite his lack of formal academic credentials, he became a professor of pathology at the University of Pennsylvania in 1899 and, just a couple of years later, he played a leading role in the founding of the Rockefeller Institute for Medical Research in New York. In 1901, he became a member of the first board of directors, and a few years later he became its first director, a position he held for more than thirty years.[22]

Almost immediately after he had taken up the chair of pathology at the University of Pennsylvania, Flexner began working on snake venom. He had already left his mark on bacteriology—he had identified a bacillus causing tropic dysentery and named it *Shigella flexneri*. He accomplished this feat on a trip to Japan and the Philippines in 1899 as chair of a special commission of Johns Hopkins University. The overall aim of that commission was to apply the latest tools of bacteriological research to study the causes of tropical diseases.

Flexner's assistant, the Japanese microbiologist Hideyo Noguchi had come from the Institute for Infectious Diseases in Tokyo. He had worked with Shibasaburo Kitasato, who in turn had worked with Emil von Behring in Robert Koch's laboratory in the early 1890s. Flexner met both Kitasato and his assistant when the commission was passing through on its way to thc Philippines. Noguchi, who was eager to pursue a career in the West, followed Flexner to Philadelphia to work in his laboratory.[23]

Just before Flexner moved from Philadelphia to the Rockefeller Institute in New York, he and Noguchi published two substantial papers on venom research. It was the first extended research project for both. The collaboration on this topic lasted until after 1903. At that time, both researchers had moved to the Rockefeller Institute. Around 1904, Flexner had begun to concentrate on other issues. Noguchi continued to pursue venom research, and Flexner, too, remained interested and involved in his progress. The two researchers published a few more papers, some coauthored, some singly authored, on various aspects of venom action, as well as a book-length treatise on snake venoms.[24]

Experimentalism and the Bacteriological Revolution

The people associated with the new medical and biological institutions and societies continued to voice their support for scientific medicine and reforms of medical education in countless speeches and addresses. As in the previous decades, they called for a number of things, such as an unflagging commitment to physics and chemistry, reliance on measuring and recording instruments, and practical training for students. But now there was also a more overtly expressed concern about the competitiveness of the American medical profession than in the earlier calls for scientific medicine. Moreover, the reformers endorsed collaboration between laboratories and hospitals.

Ten years after Martin's hopeful outline of the promises of experimental physiology, in an address on the occasion of the tenth anniversary of Johns Hopkins University, Flexner's teacher William Welch underlined that laboratory work and training positively affected therapeutics. Like Martin, he stressed that the "higher purposes" of medical education would be attained "only by the establishment of well equipped laboratories and by the foundation of hospitals."[25] Also, the medical department had to be part of a university, because some branches of medical study belonged to the philosophical rather than the medical faculty. Here was a compelling reason why those independent medical schools that many university professors abhorred really were detrimental to the profession. In 1898, speaking at the Johns Hopkins University commencement, Welch belittled the old proprietary medical schools and their emphasis on book learning. He proudly stated: "We have broken completely with the old idea that reading books and listening to lectures is an adequate training for those who are to assume the responsible duties of practitioners of medicine. Anatomy, physiology, physiological chemistry, pathology, bacteriology, pharmacology and toxicology are taught during the first two years by practical work in the laboratory, and in the last two years disease is studied in the dispensary and at the bedside, not merely as it is described in books."[26]

At that time, however, experimental physiology was no longer the primary reference point for interactions with the broader public, trustees, and administrators. As the century drew to a close, there was a recurrent theme in the popular addresses and speeches of the members of the medical community: the key role of bacteriology for the prevention and treatment of diseases.

Many historians of medicine have described the rise of bacteriology as a "revolution" just like the "laboratory revolution" (and in fact intimately linked to that revolution).[27] Late nineteenth-century bacteriology has been considered "revolutionary" because it centrally involves laboratory experiments: the systematic manipulation of pathogens, of experimental animals, and of their environments. Notably, unlike "laboratory revolution," "bacteriological revolution" is an actor's term. References to laboratory research in bacteriology and immunology had become a common theme in public lectures and addresses, both to illustrate how scientific medicine had already become and to demonstrate that the new medicine had already led to important practical applications in disease prevention and treatment. Time and again members of the medical community reiterated the point that the new bacteriological findings promised immediate rewards for the clinic.[28]

References to bacteriology also bolstered the arguments to quiet those antivivisectionists who were still campaigning against the use of animals in biomedical research. In 1896, Henry Pickering Bowditch spoke about the value of research for medicine at the Annual Meeting of the Massachusetts Medical Association, defending the use of animals for medical research. He specifically referred to bacteriological research as one area where animal experimentation had led to therapeutic successes.[29] The promises for prevention and treatment of disease of early bacteriological research provided the members of the medical community with additional arguments for their cause. Nevertheless, the very fact that so many researchers felt the need to speak out on behalf of scientific medicine suggests that they were still concerned about their career options, their reputations, and the future of biomedical research more generally. And indeed, the *Flexner Report* of 1910 identified only a few medical schools that provided appropriate training and admitted students based on adequate entrance exams. Overall, it lamented the sorry state of American medical education.[30]

Flexner—the other Flexner, Simon—and Noguchi pursued snake venom research at the beginning of their professional careers, some years before they both moved to the Rockefeller Institute. At that time, neither of the two wrote much for wider audiences, nor did either of them give many talks to the broader public. Flexner's public lecture at the MBL addressed a community of biologists and was published in the collected volume *Biological Lectures Delivered at the Marine Biological Laboratory of Wood's Holl*— hardly an outlet for a general audience. Those of Flexner's public speeches and addresses that were printed in journals and proceedings for broader audiences appeared for the most part after 1920—and they addressed members of the wider medical community (at commencements, for example) or members of professional organizations such as the AAAS rather than nonmedical audiences. Unlike Mitchell, Flexner did not write for popular magazines.

Flexner made the most pronounced commitments to experiment-based medical research and education much later in his career, however, when he was an established teacher of the medical profession and received frequent invitations to official functions.[31] In an article in *Scientific Monthly* of 1933 entitled "Triumphs of Experimental Medicine," he explicitly compared the medical experimenter studying the conditions of disease with the chemist and physicist who "seek to discover by experiment the nature of the phenomena of matter."[32] Now, research on hormones was for medicine what bacteriology had been a few decades earlier.

There is nothing particularly remarkable about Flexner's own speeches except perhaps how deftly he framed his own research on snake venom in terms of the "revolutionary" changes of late nineteenth-century medical research. In 1903, just after the two papers on snake venom had come out, Flexner delivered an address to the Medical and Chirurgical Faculty of Maryland at their annual meeting. He described how much toxicology had contributed to the understanding of infectious diseases. In doing so, he drew attention to the latest progress in the investigations of the properties and alterations of blood. The bulk of the article dealt with his and Noguchi's achievements and with the potential that studies of agglutination and hemorrhage might have for the treatment of thrombosis.

Tissues, Cells, and Body Fluids

Mitchell and his coworker used a number of protocols that they took over from other fields of medical and chemical research and adopted for their purposes: the several steps of investigation to identify "active principles" of organic substances, the study of symptoms and lesions characteristic of diseases, and the instrument-aided recordings of body functions. In the 1880s, Mitchell and his coworkers had already begun to adapt the tools and practices of bacteriology for venom research. Like the bacteriologists, they studied the growth of microorganisms in vitro in various media. The protocol also served to identify disease-causing agents—by forced infection, as it were: the microbe must be identified in a diseased body, must be cultivated and grown in the laboratory, and must be transferred to another healthy organism, in which it must produce the disease. In Mitchell and Reichert's laboratory, this procedure was used to investigate whether the bacteria contained in venom were disease-causing agents.

The protocols of bacteriology and immunology were integral parts of Flexner and Noguchi's overall project. When Mitchell suggested to the two young investigators that the immunological perspective might be fruitful for the study of snake venom, he directed them to a flourishing area of research that featured established procedures and theoretical frameworks. Some key experiments in bacteriology and immunology had become models that guided further inquiry. One of these was Emile Roux and Alexander Yersin's demonstration that the diphtheria bacillus produces a poison (toxin) and that this toxin caused the disease. Studies of diphtheria and other infectious diseases such as cholera indicated that toxins from bacteria could trigger a defense mechanism in the living body if the organism was inoculated repeatedly with cultures of bacteria. Such an immunization process induced the production of "antitoxins" in the blood serum, which could protect the body from the disease that the bacteria would have caused in a body that was not immunized. When blood serum was transferred from an animal that had been immunized into the body of another unprotected animal, the protective power of the antitoxins remained intact, and the second animal was also protected from the disease. In 1890, the German Emil von Behring and the Japanese researcher Kitasato (Noguchi's teacher) demonstrated this first for the toxin of tetanus bacilli, and then Behring did the same for diphtheria toxins.

The new findings stimulated countless experiments to characterize the mechanism of the immune reaction. The fluids and corpuscles involved in the immune reaction were isolated and in part even created in the laboratory, both in vitro and in the living animal. The investigators followed characteristic steps: They triggered the production of specific defense agents through repeated injection of "foreign" tissue extracts and body fluids in laboratory animals.[33] This practice gave a new meaning to repetition in experimentation. The repetition of the injection did not bring about—and indeed was not intended to bring about—the same effect. Repetitions were performed to stimulate the production of antitoxins, which were the ingredients of curative serum.

During the 1890s, several researchers produced in rapid succession contributions to a more detailed account of the phenomenon of immunization.[34] Paul Ehrlich's work showed that the immune reaction could be triggered not only through bacterial toxins but also through plant toxins.[35] The German Richard Pfeiffer found that the bacteria-destroying power of cholera serum involved dissolution (lysis) of bacteria. Other experiments indicated that serum had "clumping" or "agglutinating" power—bacteria added to serum would clump together. The Belgian researcher Jules Bordet, Ehrlich's nemesis, showed the similarity between the dissolution of bacteria and the dissolution of "alien" blood corpuscles (blood corpuscles from a different species).[36] Blood serum could produce hemolysis even before it was immunized. Bordet's investigations also suggested that the defense mechanism of the body by which bacterial and plant toxins, bacteria, and alien blood corpuscles were destroyed required a mediating or trigger substance. This trigger substance was contained in the cell-free serum and could be easily destroyed by heating the serum.

In immunological experiments, blood, blood corpuscles, blood serum, immunized serum, bacteria, and so on were mixed, centrifuged, heated, cooled, and diluted. The resultant fluids were checked for evidence of clumping or dissolution and tested for toxicity. What sounds like a largely exploratory endeavor was in fact heavily informed by theoretical ideas. Drawing on their theories of the immune defense and the results of their experiments, investigators established how many "principles" or agents must have contributed to the effects that could be observed and how the agents worked together. Bacteriological and immunological experiments were indirect in the sense that the agents that were required to trigger the dissolution and clumping of bacteria and blood corpuscles were not directly observable. Their presence was inferred from the observed effects of the experiments with substances that were thought to contain them.

Ehrlich's own experimental practice illustrates this indirect approach. In one of the early contributions to the study of lytic action, he and his coworker Morgenroth attempted to isolate the various agents that were involved in hemolysis. They used blood serum from a goat that had been immunized with sheep blood. The immunized serum thus dissolved solutions of fresh sheep blood. They then used goat serum that was heated (and therefore presumably free of the "something" that acted as trigger substance) and mixed it with sheep blood, and—as was expected—no lysis occurred. Finally, the heated serum was mixed with fresh serum (known to contain the trigger substance). This restored the lytic function. Commenting on these experiments Ehrlich and Morgenroth wrote: "In haemolysis [...] we are therefore forced to assume [!] the existence of two substances. One of these, specific and quite resistant (stable), we shall call the *immune-body*, following Pfeiffer's nomenclature. The other, normally present and highly labile (unstable), we shall for the present term *addiment*."[37] (Ehrlich would soon label the second substance "complement.")

Ehrlich's infamous "side-chain" theory provided the theoretical framework for much of Ehrlich's experimentation. The theory Ehrlich developed explained how toxins acted on body tissues, how antitoxins were produced, and how antitoxins protected the body from the action of the toxins. According to Ehrlich, all aspects of the toxin–antitoxin relation could be explained through chemical bonds among molecules via so-called side-chains. From the outset, Ehrlich's theory was very complex, and it became increasingly more so as he modified it to fit every new finding into his approach. For my purposes, the theory as such is less important than its playing a constitutive role in experimentation. The elements of the immune response are identified as putative—yet invisible—causes for observed effects.

Perhaps I seem to have strayed quite far from the topic of venom research—but venom research was expected to contribute to bacteriology and immunology in several ways. For example, one major task for bacteriologists was to determine more precisely the bactericidal power of blood, serum, and various organic tissues. In this context, it was instructive to establish how the presence of venom affected the bactericidal power of organic substances. Did the addition of venom to a mix of bacteria and body fluids or tissues decrease or even destroy the bactericidal power of these organic substances? Moreover, and even more important, venom also produced lysis and agglutination of blood and bacteria, the very phenomena that were the center of interest in contemporaneous immunology. But venom was easier to investigate, because there were no cell divisions or changes of virulence as they were in bacteria cultures.[38] The mechanism of venom poisoning served as a model, as it were, for the immune reaction. It was a convenient tool for the study of these mechanisms.[39]

In the footnotes and introductions to their publications, Flexner and Noguchi acknowledged the interest Mitchell had taken in their work and the guidance he had given. In the text itself, however, the two investigators presented their studies as "an integral part of the work on haemolysis and bacteriolysis, which is now attracting so much attention among bacteriologists and pathologists"—a much more appropriate characterization of the intellectual context of their work.[40] After just a couple of paragraphs, it becomes obvious to the reader that Flexner and Noguchi's work was a great deal more indebted to new immunological research than to Mitchell's investigations. The two authors drew heavily on the current literature, and their texts are full of the technical terms of immunology; they even coined a few themselves.

Flexner and Noguchi never doubted that Ehrlich's theory was the correct one. They took it on, lock, stock, and barrel, as a framework for their venom research. Ehrlich's concept of the body's immune response formed the background against which they performed their experiments. In their first publication on snake venom, the authors—having dutifully acknowledged Mitchell's initial suggestions—not only situated their research in the broader context of bacteriology and immunology but also explicitly identified Ehrlich's theory of the body's immune response as the theoretical framework within which they operated and borrowed its protocols for experimentation. Venom action could be analyzed as a combination of several components and mechanisms as theorized by Ehrlich.[41] Ehrlich's theory did not have the status of a set of hypotheses that were yet to be confirmed. The authors repeatedly pointed out that their results supported Ehrlich's explanation of the immune response, and they developed certain aspects of this theory, but the theory as such was never explicitly put to test. Rather, it constituted the very entities that the experimenters manipulated.

One of the tasks the two scientists set for themselves was to design experimental settings that enabled them to mix isolated components of venom with isolated components of blood (corpuscles, serum, trigger substance, or complement), so that the interactions among these components could be studied "purely." According to Ehrlich's theory, blood consisted of corpuscles and serum with intermediary body and complement. If blood was mixed with venom, lysing and clumping of blood corpuscles occurred. It was known that agglutination happened before lysing took place. It was also known that hemolysis required two components working together—Ehrlich's intermediary body and complement, which were contained in blood serum. But what exactly were the roles of the individual components in these processes? To answer these questions, Flexner and Noguchi had to overcome daunting practical challenges: how could they isolate the components that were assumed by Ehrlich's theory, and how could they study their actions and mechanisms?

To study "pure" agglutination without subsequent lysis, the two scientists experimented with serum-free blood corpuscles. To obtain these, the blood corpuscles were washed several times, thereby washing removed serum from the corpuscles and hence the factor that was necessary for lysis.[42] That it did so was confirmed by the expected outcome of the experiment: when the washed corpuscles were combined with venom, no lysis occurred.

Studying the action of pure complement was less straightforward, because this substance was notably difficult to isolate. As was known from Bordet's and others' work, it was easy to remove the thermo–labile complement from serum—one had only to heat it. But it was not so easy to treat serum in such a way as for pure complement to remain. Flexner and Noguchi produced pure complement by an indirect procedure, treating at low temperatures washed corpuscles of a particular species of animals with serum of another species. According to Ehrlich's theory, the corpuscles would bind with the serum-intermediary bodies. But because lysis did not occur at freezing temperatures, the complement could be separated from the serum-intermediary body–corpuscle complex by centrifuging the mixture.[43] Blood corpuscles could then be mixed with the "pure" complement thus obtained. As expected, no lysis occurred.

But when venom was subsequently added to the mix of pure complement and (serum-free) blood corpuscles, the blood corpuscles dissolved. Because lysis required complement and intermediary bodies and the intermediary bodies in this mix could not have come from the blood (the corpuscles had been washed), Flexner and Noguchi concluded that venom must contain intermediary bodies and that the venom-intermediary bodies produced lysis. The finding that venom contained intermediary bodies was considered a major one at the time, as evidenced by the 1915 edition of the *Reference Handbook of the Medical Sciences*.[44]

The complexity of the project increased further as the researchers described experiments with venom components and then also with snake blood components. Flexner and Noguchi's second article, published in the *Medical Bulletin* of the University of Pennsylvania, and shortly after reprinted in the *Journal of Pathology and Bacteriology*, offered a more exhaustive analysis of the composition of venom and the interactions of venom and blood components and extended the analysis to the mechanisms of toxin-induced hemorrhage and proteolysis. This paper also addressed head-on the differences in the actions of venoms from different kinds of snakes, an issue that had been mentioned before but that had not been systematically pursued.[45] Later papers discussed in more detail various agents of lytic action, such as cytolysins (1904), hemolysins (1904, 1906), and "tetanolysins" (1906).

Flexner and Noguchi compared the interaction of venom with snake blood and snake serum and compared the results with the results from experiments with venom and blood from other animals, the potential victims of snakebite. They found—interestingly but not unexpectedly—that venom was much more effective when it combined with complement from an "alien" species than when it combined with complement from snakes. In keeping with Ehrlich's theory—as well as with Ehrlich's habit of adjusting and modifying his theory to accommodate new findings—the two researchers explained their findings by concluding that venom contained two kinds of intermediary bodies, those that could combine with snake complement and those that could combine with the complement in other animals' blood, and that snake venom contained more "heterocomplementophilic" intermediary bodies.[46] Snake serum intermediary bodies, by contrast, were "isocomplementophilic": they united with the snake complement rather than with foreign complement.

We do not have to follow Flexner and Noguchi in their entire quest. The point is that in all these experiments, they, just like their fellow immunologists, aimed to establish the existence and agency of substances that were not directly observable. They had to find ways of identifying these substances and of assessing their agency, all while stabilizing very complex experimental environments. These methodological challenges linked their work with contemporaneous medical and biological experimentation in bacteriology and even in seemingly remote fields, such as experimental embryology.

Standards and Thresholds

Boyle, Fontana, and many other scholars who lived prior to the nineteenth century are known as people of many talents, so we are inclined to consider their scholarly activities from multiple perspectives. In contrast, the nineteenth century, especially the second half, is known as the period of emerging disciplines and professional specialization. However, even though the experimenters in pathology, experimental embryology, bacteriology, immunology, and experimental physiology had quite different aims and research questions, they had many similar methodological concerns: How could one stabilize and standardize experimental conditions? How could unwieldy experiments be controlled? What was the most reliable procedure for the identification of invisible causes for observable effects? And what was the proper relation between theoretical assumptions and experimental outcomes? If we want to understand the nature and significance of methods discourse in late nineteenth-century biology and medicine, it would be a mistake to limit the analysis to one specific medical or biological field. Comparing how scientists in different disciplines were dealing with methods-related issues reveals novel and surprising connections among those fields.

One of these shared methodological concerns is the concern with standardization. Historians of science have shown how the standardization of experimental objects, practices, and instruments functions as a means both to reduce the complexity and variability of experimental situations and to decrease the amount of trust in others that was required in scientific research.[47] Immunological research played a key role as a driving force for standardization practices not least because of the increased involvement of the pharmaceutical industry in drug research and therapeutics.[48]

Toward the end of the nineteenth century, experiments in immunology became more quantitative as researchers aimed to determine thresholds and minimum or maximum effective doses of drugs, poisons, and experimental substances more generally. The notions of units and thresholds became common tools for the general and comparative characterization of the activity of chemical or biological agents. In their pathbreaking investigation of the effect of tetanus bacilli on brain tissue and spinal marrow, for instance, The German researcher August von Wassermann and the Japanese Kanehiro Takaki operated with a "test solution of poison" [*Test-giftlösung*]. This fluid contained a concentration of tetanus bacilli that was lethal for a specific kind of animal of a specific size (white mice of 15 g). The procedure to establish the strength of the test solution was a biological assay, because it determined the efficacy of the serum with respect to the reaction of organisms to organic matter. Doses of this standardized solution (the lethal dose and two, three, and ten times the lethal dose) were mixed with emulsions of brain and spinal marrow substance and injected into mice to examine whether the brain and spinal marrow had antitoxic capacities (they did).[49]

By 1900, the concept of standard was ubiquitous in immunological research, largely as a result of the gradual commercialization of drug

development and testing.[50] Standards—specifications of units and procedures—became crucial as researchers and pharmaceutical industries
sought to produce immune sera that worked reliably. Standardized biological assays were indispensable for the mass production of immune sera
because they allowed comparisons across laboratories. Ehrlich's research
became much more quantitative as he turned to the problem of determining the strength of immune sera for commercial purposes, especially
diphtheria-curative serum.[51] A British commission had found great variation in the strength of the sera that were commercially available in Britain
and on the Continent. An article in the *Lancet* of 1896 reported the results
of a range of tests showing an "astounding variation in the quality of the
serum supplied by different makers and the puzzling manner in which the
antitoxin from certain sources varied from time to time."[52] Ehrlich took
it upon himself to develop a definitive standard for the strength of diphtheria sera, whereby the *efficacy* of a certain quantity of the serum was
fixed—in other words, standardization ensured that the same quantities
of serum always had the same efficacy. His seminal paper "The Assay
of the Activity of Diphtheria-Curative Serum" was published in 1897.
The "standard" that Ehrlich proposed serves as a unit for comparison
across laboratories, but it is not a physical quantity like the kilogram. It
is a capacity or potency. In the context of his work on diphtheria, Ehrlich
sought to determine the "minimal lethal dose" of diphtheria toxin as a
unit of activity—namely, the minimal dose necessary to kill guinea pigs
of a certain size within a particular time. For Ehrlich's contemporaries (as
for historians of science today), it was one of Ehrlich's most important
and lasting achievements that he was able to determine standardized units
for the production of curative serum. Ehrlich himself noted later that his
initial ambition had been "to introduce measures and figures into investigations regarding the relations existing between toxine and anti-toxine."[53]

Not all projects in immune research became part of commercial enter-
prises, and not all researchers were immediately involved in the commercialization of immunological research, but the notions of standard, standardization, thresholds, and minimal doses became conceptual resources
for many around 1900. In the 1880s, Mitchell hardly ever used the terms
standard or *standardization* except in a few references to a "standard"
laboratory solution or agent. Flexner and Noguchi, by contrast, made
ample use of the concepts "standard," "threshold," and "minimal lethal
dose." At the outset, they determined the minimal lethal dose of the original venom that they were investigating. They did so by testing the effects

of venom in guinea pigs of the "standard" size that Ehrlich had used in his research on diphtheria antitoxins and that Wassermann and Takaki had used to define their test solution of poison. They treated the minimal lethal dose (M.L.D.) as the dose that killed a guinea pig weighing 250 grams[54]—in their case, in twenty-four hours. Twice this dose caused death after two or three hours; three times the M.L.D caused death after thirty to forty-five minutes.

They also made ample use of the concept of minimal effective doses in their systematic study of the capacities and activities of the different venoms and venom and blood components.[55] One of the main outcomes of Flexner and Noguchi's research was the demonstration that venoms from different kinds of snakes contained toxic components with different capacities: hemolytic, hemorrhagic, agglutinating, and so on. Cobra venom was mainly a neurotoxin, rattlesnake venom was chiefly a hemorrhagin, and snake sera were hemotoxins. They showed this by determining all kinds of minimal active doses and comparing the results. They determined the M.H.D. (minimal hemolytic dose), the M.A.D. (minimal agglutinative dose), and the M.Hr.D. (minimal hemorrhagic dose).

Unlike Ehrlich's immune serum, these standards did not become reference tools for a pharmaceutical market, nor even for the wider community of venom researchers. They were really only fixed points of comparison for Flexner and Noguchi's own investigations. As such, they were quite useful. Based on these standard of comparison, the relative hemolytic and hemorrhagic power, and so forth, of different kinds of venom could be expressed quantitatively—at least approximately.

The "comparative quantitative determination" had severe limitations, however, because both the venoms that were tested to determine the standard and the conditions of experimentation were variable. Flexner and Noguchi noted that the figures they had obtained were "relative only, as the particular specimen of venom according to the animal yielding it and other external causes, will be subject to variation."[56]

Securing Control

The endeavors to determine standard units and thresholds show, again, that variability—of experimental conditions, of organisms, of biological specimen—was one of the main challenges to the nineteenth-century experimenter. In the 1880s, increased efforts to control unwieldy experiments were the flipside of the encounters with variations, instability, and complexity. Around 1900, we continue to find many calls for measures of control with the aims of reducing the variation and complexity of experimental objects and conditions and making experimentation more secure.

Many of these calls for control are about "securing control"—that is, about keeping the experimental conditions stable. Medical men pointed out that laboratory experiments were advantageous because, unlike clinical investigations, they offered "controlled conditions." William Welch, then Dean of the Medical School at Johns Hopkins University, made this point in his Huxley Lecture on immunity in 1902 (delivered at the Charing Cross Hospital in London).[57] Flexner noted that animal experiments in an artificially designed environment made it possible "to secure that control of experimental conditions that alone can make biological experiment accurate and advance logical and not a thing of chance."[58] Charles Whitman, the director of the MBL, even proposed a "biological farm," which would make it possible to investigate life "under conditions that secure most favorable control for experimentation and study."[59] This notion of control refers to the endeavor to produce and maintain order in an experimental situation. Only experiments performed under controlled conditions allow reliable inferences.

Then there are examples for so-called control experiments that involve comparison against a fixed or determinate reference point. There are numerous instances of such experiments in the milestone papers in bacteriology and immunology. In a series of experiments to make pigeons insensitive to rattlesnake venom, the outcomes of repeated inoculations of pigeons were compared in "control experiments" with the effects of inoculations of "fresh pigeons."[60] Mice injected with curative serum to immunize them against tetanus and diphtheria were routinely compared with "control mice," which were injected with tetanus and diphtheria bacilli.[61] To test the bactericidal properties of the brain and spinal marrow, the effects of injections containing a mix of brain matter and tetanus toxin were compared with "controls," both with injections of emulsions of liver, kidney, spleen, bone marrow, and serum from the same animals and with an injection of tetanus toxin without any additions.[62] Flexner and Noguchi employed this concept of control just as everyone else did. In their study of the bactericidal properties of venom they compared experiments with venomized serum and with nonvenomized serum. They found that in the former media, bacteria grew much more rapidly than in the latter.

And outside bacteriology and immunology, the same kinds of control experiment were done as well. For instance, in the detailed descriptions of the countless experiments through which Loeb studied the behavior and development of marine organisms, the effects of an intervention in the organism or its environment were compared in control experiments [*Kontrollversuche*] with the undisturbed organism.[63]

From today's perspective, the frequent reference to control signals a concern about how to establish causes in a reliable way. It is tempting to read the calls for control of experimental conditions as attempts to create an experimental design and environment that approximates Mill's ideal experimental situation and the descriptions of control experiments as a practical *application* of the method of difference. At the time, however, concerns about causal reasoning and concerns about controls were two different issues.

Control experiments like the ones already mentioned exemplify the "controlling experiments" that function as "corroborating or confirmatory" experiments per the dictionary definition I quoted in chapter 8. That dictionary also offers a separate definition of "control-experiment," which is an experiment "made to establish the conditions under which another experiment is made."[64] In Flexner and Noguchi's study of the influence of venom on bactericidal properties of serum, for example, this is exactly the point. One project was to investigate how the addition of venom to blood serum affected the capacity of normal blood to destroy bacteria.[65] Flexner and Noguchi added bacteria—bacillus anthracis, coli, and typhoid bacilli—to samples of normal blood and to blood mixed with venom and counted the number of bacteria after certain time intervals. They found that in those blood sera samples that did not contain venom, bacterial growth was prevented—this was the regular course of things in normal blood sera. In the mixtures, by contrast, bacterial growth increased after an initial small drop in numbers. These results suggested that venom did indeed have a destructive effect on the bactericidal action of blood serum.[66] But there was also the possibility that the presence of nutrients affected the speed of bacteria growth. It was not possible to remove hemoglobin completely from blood serum with the available techniques of centrifugalization. If hemoglobin acted as a nutrient for bacteria, bacterial growth in the venomized mixture could be caused by the hemoglobin that still remained in the serum, and not by the destruction of its bactericidal power. The control was designed to exclude this possibility: Flexner and Noguchi purposely added a nutrient (peptone) to the mixture of bacteria and serum "as a control."[67] In this series of experiments, the bacteria grew at a slightly greater rate, but far more slowly than when venom was added to serum, which indicated that the increase in nutritional value of the mixture did not have a marked effect on the speed of bacteria growth.[68] The increased growth rate in the initial experiment was really due to the destruction of the serum's bactericidal power. In these kinds of situations, the control experiment is performed to ensure that the experiment of interest really works the way it is assumed to work and that no confounding factors interfere.

Causes

In late nineteenth-century medicine and biology, the concept of cause was explicitly brought into methodological debates, although discussions about controls and discussions about causes were not necessarily connected. In fact there were two different kinds of questions that were raised about causes: Researchers across different fields were discussing the nature of causes and also the ways in which reliably to identify causal agents and causal relations.

Concerns about causes were raised and critiqued in similar ways by researchers in fields as different as experimental embryology, bacteriology, immunology, and pathology.[69] Pathologists, for instance, debated whether all, some, or no diseases were caused by external, tangible pathogens.[70] They also discussed what criteria should be applied to specify the causes of particular diseases.[71] Immunologists tried to identify the agents involved in the mechanism of the immune response and to characterize their role.

Researchers working on disease causation and on the body's defense mechanism concentrated on the adult, fully developed organism and thus accepted biological organization as a given.[72] Embryologists carried out experiments on eggs and embryos to elucidate the process of development.[73] These researchers hoped that by manipulating embryos and their environments they would be able to establish what factors caused and directed the development of the embryo. They debated—intensely and passionately—whether there was a specific category of cause that produced formation and organization in living beings, and whether the process of biological organization and its formation could be turned into a problem for experimental research.[74] The larger question of whether the experimental results they obtained supported a mechanistic, teleological, or vitalist interpretation of development was not fully answered at the time. The details of these discussions are not central for my topic, however. What is important for my purposes is that despite the differences in the subject matter, experimental embryology, experimental physiology, and pathology had similar methodological problems pertaining to experimentation, proper methods, and methodologies.

The embryologist Wilhelm Roux thus spoke for many of his contemporaries when he presented experimentation as the "causal method of investigation, κατ' ἐξοχήν"—that is, as the preeminent mode of inquiry through which relations between causes and effects could be established.[75] The causal method alone, not observation or comparison, could reliably identify the causes of effects—or, as Roux put it, "'*Certainty*' in causal deduction *can only come from* experiment." Roux made this point first in German in the programmatic introduction to his journal *Archiv für Entwickelungsmechanik*. This widely read text was quickly translated into English and was presented and discussed at the Marine Biology Laboratory in the summer of 1894 as part of the MBL biology lecture series.[76] The text laid out what proper experimental procedure meant in embryology and what could (and could not) be attained with the "causal method of research."

Roux's ultimate aim was to outline an experiment-based embryology as the alternative to morphological descriptions of embryonic development. Roux compared the epistemic powers of different subfields in the biomedical sciences with respect to the amount of causal information that each could produce. He presented these deliberations as a criterion for demarcation: Only those contributions that "*directly pursue a causal aim*" deserved to be published in his journal.[77] Those that did not had to be published elsewhere.[78] Experimental embryology came out at the top of the hierarchy, but other empirical research was useful, too. Even descriptive anatomy might qualify if the authors drew conclusions useful for the causal understanding of the phenomena of life. Pathology was more instructive than physiology because, like developmental mechanics, it concerned formative functions. If the normal body functions were disturbed by pathological processes, the disturbances would often lead to changes in form—to the growth of scar tissue, for instance. Roux thus called for more descriptions of pathological processes that caused formative processes—their magnitude and duration. Nature could perform experiments too; certainty, he explained, could come "either from '*artificial*' or from *'nature's' experiment*, such as *variation, monstrosity*, or other *pathological phenomena*."[79] Clinicians would see many of those. Actual physical intervention or manipulation was thus not the distinctive feature of experimental practice, but human intervention was the most secure path to knowledge.

Mill's methods, especially the method of difference, remained a reference point in the debates about causes. Roux's starting point was an ideal experimental situation just like Mill's, a situation in which all antecedent conditions had been identified and could be manipulated individually. Indeed, Roux reminded his readers, "[i]n an experiment performed under the most favorable conditions, only *one* of the components known to us is or will be changed, and through the results of this change we apprehend those phenomena which are connected with this component." In such an ideal situation, the causes for an experimental effect could be established with certainty. But Roux raised the familiar concerns as well: In concrete experiments with organic objects, those favorable conditions were typically not fulfilled. Because of the complexity of organisms, the experimenters could never be sure that their interventions produced really only the intended changes, not others. Organisms were so complex that the antecedent conditions could never be fully specified, let alone individually managed. Often an intervention might cause "*accidental* internal or external conditions" or "unintentional *collateral effects* of our own interference" even as we keep thinking we have changed only one component.[80]

For Bernard, comparative experimentation was the way to go. For Roux, the less-than-ideal experimental situation meant that additional measures needed to be taken to secure experimental results. Roux called for independent confirmation of experimental outcomes. Especially at the beginning of a research project, it was important to investigate one and the same phenomenon from a number of different angles, and "only when these different experiments point to the same causal connection should we assume that this is the true one."[81] This procedure of independent confirmation would become a common strategy in the early twentieth century, as we will see in chapter 10.[82]

Invisible Things and Hypotheses

Complexity was not the only challenge to the late nineteenth-century experimenter. The phenomena of interest were often hidden from sight.

Roux emphasized right at the beginning of his discussion on the methodology of developmental mechanics that this was the most difficult methodological problem that experimental embryology faced. The factors assumed to be relevant for the formation and development of embryos could not be directly observed; they could only be "inferred."[83] Just as there was no direct test to reveal the forces driving biological organization, the presence of all those substances that were involved in the immune response (and the immune response itself, for that matter) could not be directly observed either.

Even late nineteenth-century microscopists who prided themselves that they had succeeded in making visible the "invisible enemies" of mankind were concerned about subvisible entities. With the new staining and microphotography techniques that Robert Koch and others had developed, bacteria and blood cells could be observed with microscopes, accentuated by dyes, and counted, but there were still debates about the reliability of these methods and concerns about how to "verify" one's assumptions about the microscopic world. In his presidential address to the *American Microscopical Society*, Albert McCalla reminded his audience that: "Verification, literally, truth-making, proving true, is, or should be, the one aim of the student of science." He added, however, that "this verification means careful and laborious investigation into a thousand minutiae whose after importance cannot always be known, the substantiating a phenomenon observed by chance, by many a set experiment; the framing a hypothesis to account for the facts observed, and testing its truth by a series of observations under many varying conditions."[84]

In immunological experiments, intermediary bodies, complement, and the like were *assumed* to be present or absent in the experimentally generated and manipulated fluids, as predicted by immunological theory.[85] Immunological experiments required procedures of purification and isolation of agents with invisible entities and conjectured processes. Experiments with these substances were highly indirect, because the presence of these entities and processes could only be inferred from a number of background assumptions. We saw that Flexner and Noguchi treated venom and blood components as agents in a complex mechanism. To be able to design experiments for the management of these agents, they had to make a number of background assumptions, using Ehrlich's approach.

The very fact that the experiments dealt with "unobservables"—that is, with entities that could not be directly manipulated and with processes that could not be seen to happen—was not considered problematic in itself. But for many researchers, the problem of how to connect theoretical assumptions to observable phenomena was a point of concern. How would these assumptions be assessed and validated? Members of the medical community around 1900 hotly debated Ehrlich's efforts to immunize against external critique the "ad hoc" introduction of ever more chemical entities to accommodate experimental outcomes that seemed to contradict his theory.

Since the early modern period, researchers had worried about the problem of the "idols of the mind," or, in modern parlance, the theory-ladenness of observation, the danger that we might see only what we hope or expect to see. The worry about idols of the mind is a concern about how best to gather data without interference of preconceived opinions and expectations. If we are dealing with processes and things that are invisible, the worry is even greater. If we are presented with a finding that does not agree with theory-guided predictions or that available hypotheses cannot explain, giving up our theories and hypotheses is not the only option. In principle we have the option to adjust our theoretical framework in such a way that the observation does agree with our main theoretical assumptions. This is because experimenters never confront just one general statement with empirical data. They must rely on a number of auxiliary assumptions, such as assumptions about initial conditions, the correct working of instruments (or their eyes, for that matter),[86] and so on. Which of these adjustments are permissible?

This question is one of the classic problems of twentieth-century philosophy of science. There are countless attempts to deal with it, ranging from Duhem's concerns about underdetermination, Popper's conventionalist falsificationism and Lakatos's methodology of research programs to recent debates about science and values. For philosophers of science who are familiar with these debates, it is striking to see the various practical solutions that experimenters around 1900 developed to address this issue.

Flexner and Noguchi, for instance, were careful to offer independent evidence for their adjustments of Ehrlich's theory when they were trying to protect it from refutation. According to Ehrlich, serum complement was a necessary factor for hemolysis. It was known that complement could be deactivated by heating it. The French immunologist Albert Calmette, however, found that venomized dog's corpuscles were destroyed by previously heated dog serum (that is, serum that presumably no longer contained any complement). To Calmette, this was evidence against Ehrlich's theory, as complement could not have played a role in the process.

Flexner and Noguchi, obtaining the same results, considered two possibilities: Perhaps the red blood corpuscles had not been washed thoroughly enough; some complement might still remain—enough for the venom to act on the globules.[87] Or there might be complement contained within the blood corpuscles—so-called endocomplement—that would be set free by the heating of the mixture.[88] They emphasized that research by Preston Kyes could confirm this assumption. Kyes, another student of Welch, worked in Ehrlich's laboratory. He was collecting evidence for the existence of "endocomplement," which was more thermostable than serum complement.[89] So Ehrlich's theory appeared to be correct after all, if more complicated than originally thought.

Twenty-first-century readers who are familiar with Ian Hacking's discussion of entity realism[90] will be struck by Flexner and Noguchi's experiments to determine whether cytolysis involved just one lytic agent or several specific ones. In these experiments, unobservable entities are manipulated to affect other unobservable entities. Venom (in this case, water moccasin venom) was treated successively with emulsions of different organs from sheep, rat, and dog. In one series of experiments, one sample of venom was treated with sheep testes, a second one with sheep testes and liver, a third one with testes, liver, kidney, and so forth. According to Ehrlich's theory, the tissue cells would bind with the lytic agent in the venom, thereby deactivating them. If only one kind of lytic agent existed, the remaining fluid should be inactive for all kinds of sheep cells. But the sample of venom that was treated with testes did in fact still have a solvent effect on cells from other organs.[91] Flexner and Noguchi did not present their results as conclusive, but they did conclude that the mechanism of cytolysis was due "to a number of solvents, which are distinct one from another."[92] Different solvents must exist, as demonstrated by the ways in which they could be used to manipulate other substances.

Perhaps most striking is Roux's solution to the problem of unobservables. Roux addressed the problem head-on, and the relevant passages in the *Archiv für Entwicklungsmechanik* sound like textbook hypothetico–deductivism. Roux claimed that investigators of the developing embryo had to rely on hypotheses. In this respect, they were in much the same situation as the researchers in the physical and chemical sciences. He insisted that such hypotheses must be evaluated with regard to their predictive success and simplicity. In Roux's words, experimenters in developmental embryology had *"to make as much or even more use of hypotheses, as physicists and chemists are compelled to do* when they cope with the fundamental processes of their respective sciences. And just as in these sciences, we shall have to regard those assumptions as approximating most nearly to the truth which explain the most facts and permit of the successful prediction of new facts; and *ceteris paribus* we shall prefer that explanation which appears to be the "simplest." "[93] Roux added that any such hypothesis was fallible. Roux's statement is noteworthy because of its explicit formulation of requirements for the evaluation of hypotheses about invisible entities and processes. Roux did not provide any references, but these kinds of statements were quite common in German-language books about science dating from the mid- to late nineteenth century. These books, in turn, were inspired by William Whewell's *Philosophy of the Inductive Sciences*.[94] In Whewell's work and the German books and pamphlets on the logic of science, the physical sciences were the focus of attention. Roux changed the focus to the life sciences.[95]

Roux's explicit commitment to hypothetico–deductivism is particularly remarkable in light of Larry Laudan's early work on the so-called method of hypothesis.[96] Laudan identified a turn to hypothetico–deductivism in early nineteenth-century philosophy of science. This turn was the result of a perceived tension between established canons of scientific inference and the features of the theories that scientists were constructing at the time. During the second half of the eighteenth century, theories of electricity, heat, and chemical and physiological processes were formulated that were highly speculative in their claims about the microstructure of matter but that were nevertheless in agreement with observed phenomena. These theories could not be supported by inductive inferences but only via the verification of testable consequences of these theories. Laudan argued that these developments in science were the inspiration for the nineteenth-century philosophers whom we know as the main proponents of the method of hypothesis, most notably Whewell.[97]

Laudan's bold claims about the history of the method of hypothesis were criticized at the time, but they are intriguing. The way in which the scientist Roux again relied on the philosopher's framework to justify his own use of hypotheses should give us pause. If nothing else, the various responses to the problem of hypothetical assumptions show that the question of how to evaluate such hypotheses about unobservables was a real question and point of concern for the practitioners. However, attention to hypotheses could also be motivated by practical concerns about communicating arguments. Late nineteenth-century scientific writing did move toward a more "deductive" composition, but the changes of format did not *necessarily* reflect a profound change in scientists' epistemologies. Rather, it had to do with the efficient organization of the work—and sometimes also with taking a stance.

Composing Deductively

Certain features of the articles on venom research published around 1900 tell us something about the community of investigators. The footnotes and references are evidence that a sizeable international community had begun to form and that this group had close bonds to immunologists and bacteriologists. Many of the more recent works were published by researchers with colonial ties and thus with access to exotic animals and their venoms. The footnotes and acknowledgements also tell us something about the changing institutional situation for biomedical researchers: new academies and institutes offered funding opportunities for investigators. There are several features of Flexner and Noguchi's papers that indicate their connectedness to various institutions as well as to other researchers on both sides of the Atlantic. They make numerous references to the international community of immunologists and bacteriologists. They acknowledge funding from the National Academy of the Sciences, the Carnegie Institution in Washington, DC, and other organizations. The journal articles demonstrate that the social relationships within the group of venom researchers were strengthened through the exchange of goods. There are acknowledgments of gifts of snake venom samples, antivenin, and exotic animals from other researchers. The authors are grateful to "Professor Calmette, of Lille" for the supply of cobra venom and antivenin[98] and express their thanks for a supply of mongoose to "Dr. F. P. Gay, who kindly brought them from Jamaica."[99] These features of scientific articles reflect quite significant shifts in the composition of the community of snake venom researchers as well as a more general change in the funding structure of biomedicine.

The expansion of the publishing industry around 1900 also began to affect the organization of scientific papers. That the bibliography in Noguchi's substantial handbook on venoms was eleven pages long and that the majority of the items on Noguchi's list had appeared after 1880 show how much the scientific literature had increased even in a comparatively small field like venom research. Judged by the guidelines for authors that journals and organizations such as the *American Medical Association* put together, the "shapeless lumps" that Billings had complained about in 1881 were now being whipped in shape much more vigorously than before.

None of these organizations recommended a particular format for scientific articles. As late as 1916, the *Botanical Gazette* stated that as regarded effective methods of presentation, selection of material, and style, "each contributor is a law unto himself."[100] Nevertheless, it is clear from the kinds of recommendations that were being offered to scientific authors that expedience and easy access to information had become priorities in scientific writing.

The AMA called, above all, for "neatness"; manuscripts should be typed, double-spaced, and numbered consecutively, they should have proper margins and should end with the word *end*. The AMA commented drily that it would "seem unnecessary to suggest that the pages in manuscript offered for publication should be arranged in the order in which they are to be read. And yet it is not at all uncommon to receive manuscripts the sheets of which are so mixed up that it is a puzzle to straighten them out. The impression produced on the editor is not such as to influence him favorably."[101] The AMA also wanted subheadings for articles so as to give an "indication of the subtopics," as this information would enable the readers to orient themselves more readily in the text they were perusing.

An entire section of the AMA guidelines concerned the proper construction of tables. At this point in time, it had become common for scientific authors to report results of sets of experiments and multiple trials in synoptic form. The guidelines introduce conventions of presentation such as the consecutive numbering of multiple tables in a publication and the vertical, not horizontal, ordering of like data. Constant factors and measurement units should not be presented in the table itself but in the heading or the footnotes, and so on.

In view of the significance that earlier scholars had attached to the "prolix" account of experiments, it is noteworthy that the AMA required its authors to be brief or at least terse. For Fontana, prolixity was a means to demonstrate to the reader how trustworthy the novel insights were that he had obtained from his experiments. The AMA thought that readers would "pass by the prolix writer for the one who gets down at once to the heart of his subject. The author who wishes to be read will do well to be as chary of words as if they cost him money."[102] Again, we see that quick access to information was a prime concern.

The period around 1900 is still a period of transition in the approach to writing scientifically. The very fact that more and more books and papers on scientific writing and composition were published suggests that writers, readers, and editors became increasingly aware of issues of style, organization, and presentation in technical and scientific writing. In this period, we find treatises like William Cairns's *The Forms of Discourse* that discuss technical and scientific writing as a specific genre while remaining firmly rooted in the classic discipline of rhetoric. Cairns, instructor in rhetoric at the University of Wisconsin distinguished in his textbook a number of different genres of articles, including "popular essays," "informal essays," and "book reviews." Discussions specifically about the form of scientific articles and reports are absent from his survey. The only science-related textual genre Cairns discussed is the "short technical essay," by which he meant encyclopedia essays, especially those in the *Encyclopedia Britannica*.[103] His text is a treatise on rhetoric, which is generally concerned with how to write effectively, how to organize a text or speech, and how to persuade an audience.

In the same period, we also find works like T. Clifford Allbutt's *Notes on the Composition of Scientific Papers*. This is an illuminating book in many ways. The author, the Regius Professor of Physic at the University of Cambridge in the United Kingdom and a historian of medicine, was much admired by Flexner's teacher, the pathologist Welch.[104] The English edition of Allbutt's *Notes* was published in 1894. The first American edition appeared in 1904; the second edition followed only one year later. In the preface, Allbutt stated that he had written the book because every year he had to examine close to a hundred M.B. and M.D. theses (!). The content of these theses was often excellent, but most of them were written rather badly, "some very ill indeed."[105] Like Cairns's *Forms of Discourse*, Allbutt's treatise is very general in scope. Written primarily for the medical student, it offers advice for, and examples from, writings in philosophy, literature, history, and several other fields. A quote from Herschel's *Preliminary Discourse* sits comfortably between passages from Edward Gibbon's *The Decline and Fall of the Roman Empire* and from an essay on Keats by Matthew Arnold. One of Allbutt's main concerns is with style, grammar, and word choice. He advised caution with words such as: "*That.* Keep down your 'that's'; for they multiply like lower organisms: e.g. 'He told me that he told you that you were to see that all was in order,' etc."[106]

Allbutt did have some specific advice to the writer of scientific papers. His recommendations did not concern the formal division of the paper in sections, or the use of titles and headings. Instead, he focused on the logic of the argument. He explicitly advised the authors of scientific texts against describing the investigative pathways they had followed. For him, good scientific practice was "inductive" (as opposed to "deductive") science—which meant, for him, research that built on empirical facts and did not proceed from first principles. In his text, he explicitly drew a distinction between the logic of scientific arguments and the sequence of steps in an investigative project, and he recommended that the text be constructed in a deductive fashion: "Speaking generally, it is better to compose a scientific essay, and to construct its limbs, not on the inductive plan on which the research was pursued, but deductively. In investigation we step first upon the bottom facts; then we make short inferences, and test them by more facts; these inferences widen and widen, and in their turns are tested, and so on; such is the course of research: but as demonstration the system is not telling; the student is held too long in suspense."[107]

The "inductive plan" in Allbutt's description is quite remote from a simple enumerative induction; it evokes Whewell's discoverer's induction. But here the main interesting point is Allbutt's suggestion that a report should present a *reconstruction* of the research that was done. Of course, this suggestion was not a new thing in itself. Even in the late eighteenth century, it was common to select and rearrange the order of experiments to construct a more convincing narrative sequence, and scholars like Senebier explicitly recommended such a reconstruction. But for the eighteenth-century scholars, the reconstruction of the investigative pathway did not conflict with the "inductive plan" of research; on the contrary, it brought the plan out more and thus strengthened the overall argument. For Allbutt, by contrast, an account of an investigation should emphatically *not* begin at the beginning. Allbutt wrote: "It is better to begin, then, by setting forth certain more general views; and from these to proceed to closer and closer quarters with the particulars on which our position is to be established. Logic does not make matter, it arranges matter already gathered."[108]

If we take this sentence at face value, we may conclude that Allbutt's advice reflects an increasing appreciation for the "hypothetical-deductive method" that we found in Roux's and other scholars' methodological writings. However, as we read on, another meaning emerges for it:

> We shall not begin with a crude or heavy lump of our matter, yet we shall try to touch the keynote of the subject, and to engage in the argument easily but directly. We have seen that the "beginnings" of great writers are direct; we shall not begin, then, with apologies, with wayward or fanciful approaches, nor with any kind of skirmishing. After these great examples, we shall try to give first

some glimpse into the heart of the matter, to put the reader at our point of view, and then to lead him briskly into the subject. Hence the beginning is not to be written until we have so cast our argument that we can perceive the exact place whence the best glimpse of its purport is to be had. We may encourage those essayists who may fall so shy of the beginning as never to enter upon their work at all, by assuring them that it is not necessary to begin their essay till they have ended it.[109]

In other words, "composing deductively" is important because it lightens the reader's load and also because it draws the reader in. Logic—language, *logos*—"like good manners, owes not its charm only but also its force and penetration to incalculable, imponderable elements."[110] Allbutt's charming exposition tells us how arguments in a piece of scientific writing really work: "The line of the dryest argument overflows logic in all directions, reason turns and doubles on itself; were it possible to photograph it in a flash, its course would appear not as a straight line, but as one of curves and zigzags; thus as it goes it falls under changing lights, and intimate metaphors creep in even unbidden. Moreover, an author cannot but be aware of his audience; he receives its influence into his fancy, and betrays his wariness by glances and stage asides."[111]

The AMA's recommendation was more down to earth. Revising and reconstructing one's work was strongly encouraged, but not because the AMA was partial to "deductive" instead of "inductive" composition. What was at stake was clarity. To present a case in an informative and convincing way, it was often necessary to make lots of revisions to the original case notes—the description of symptoms, diagnosis, and treatment of individual patient. Revisions were necessary because the authors would use their patient files as resources for the composition of their articles. To present an instructive case, it was not sufficient just to copy hastily scribbled case notes directly from the files into the manuscript. Abbreviations had to be eliminated, vague expressions such as "a month ago" had to be specified, and details that had no bearing on the case presented had to be removed.

In the context of Ehrlich's and competing theories about the immune response, the scientists had several "general views" to choose from, and precisely because there was such a controversial debate about the validity of competing theoretical frameworks, expounding a position was not an attempt merely "to give first some glimpse into the heart of the matter" but indeed "to put the reader at our point of view." It was equivalent to taking a side in a controversy and arguing for it. But this did not necessarily mean that the entire text was presented in terms of hypothetico–deductivism. Flexner and Noguchi began their first joint publication by "setting forth certain more general views"—namely, Ehrlich's theory of the immune response. From there they proceeded to the particulars—specific investigations that illustrate, flesh out, and develop aspects of Ehrlich's views. But they did not derive any testable predictions from them in the strict sense of the hypothetico–deductive framework, and they certainly did not consider how "simple" Ehrlich's theory was compared to other views. At least, its complexity did not count against it. Flexner and Noguchi set forth Ehrlich's views to let the reader know right away where they stood.

From this starting point, the text proceeds sequentially. Each thematic section has a telling title (such as "Are the haemolysins identical with leucolysins?"). Each section starts with a point or question, describes the relevant experiments that address the issue at hand, and ends with a conclusion based on the outcomes of the experiments that are reported. There are no sections exclusively devoted to methods and techniques. Instead, relevant aspects of methods and techniques are briefly described within each thematic section, as in Mitchell's writings. Each comes with a subtitle that gives an indication of the specific topic treated in it: experiments done in vivo, then in vitro, followed by a series of experiments to clarify an additional point—because the first two groups of experiments demonstrate that venom does destroy the bactericidal power of blood, does it affect the intermediary bodies or the complement? In Flexner and Noguchi's reports, many experiments have numbers, but not all experiments are numbered, and those that are often bear numbers that are out of sequence. The subsections *culminate* in a conclusion; they do not begin with a statement of the purport of the argument. As far as the organization of the argument goes, such articles are closer to the eighteenth- and early nineteenth-century tradition of writing scientifically and to demonstrating the discoverability of a result than to a strictly and formally hypothesis-driven composition. References to hypotheses were frequent in the late nineteenth century, but these references ranged from explicit methodological commitments to hypothetico–deductivism to pragmatic concerns with clear and well-constructed pieces of effective scientific writing.

CHAPTER TEN

Fragmentation and Modularity

In the first half of the twentieth century, venom studies grew ever more complex. Research on venoms became an integral part of diverse fields ranging from biochemistry to evolutionary theory. This ramification parallels the overall expansion of biological and medical research fields in the twentieth-century life sciences. The internationalization of the biological and medical communities that had begun even before 1900 continued, collaborations and exchanges among scholars grew even more intense, multiple novel research techniques were devised, big machines reorganized biological and medical practice, and (partly as a result of this expansion) the modular scientific article became the standard format for scientific writing. Tracing all these intertwined developments in close enough detail would be the task for a whole new book, even if I limited myself strictly to the theme of snake venom research. Still, I will pick a path through the thicket of early twentieth-century venom studies if only to show, along the way, how some of the historiographical insights gained in the previous chapters might help in tackling the immense topic. We will see how protocols in venom research resembled those from other areas, how methodological statements were shaped by the constitution of scientific disciplines, and what textual genres tell us about the communities of researchers and the organization of work in the life sciences.

Proteins, Enzymes, and Big Machines

Around 1900, research on venom was part and parcel of the new immunology. Venom researchers in the early twentieth century drew on conceptual resources from even more intricate intellectual contexts—namely, protein and enzyme research. For many chemical and biological researchers in the early decades of the twentieth century, enzymes were the fundamental units of life. Other researchers in the life sciences regarded proteins as the constituents of the animal body.[1] Nineteenth-century investigations of enzymes (or "ferments")[2] focused on alcoholic fermentation, putrefaction, and digestion and on the question of whether fermentation was a purely chemical process or involved a living organism. In the early twentieth century, enzymes came to be identified as the driving forces of more complex biochemical processes. Until the late 1920s, however, the chemical composition of enzymes remained a matter of debate. Many researchers claimed that they were proteins, but it was not until the 1930s, after John Northrop, working at the Rockefeller Institute, had demonstrated the protein nature of the digestive enzyme pepsin, that the protein nature of enzymes became widely accepted.[3]

Nineteenth-century investigations of protein or "albuminous bodies" focused on organic materials such as milk, egg, and blood. At that time, albuminous substance was mainly identified by its behavior when treated with heat or certain chemicals—the similarity of the coagulation of egg white, the curdling of milk by acid, and the clotting of blood suggested that the same substance was involved in all cases.[4]

Powerful instruments such as the ultracentrifuge and electrophoresis began to transform medical and biological research during the 1920s and 1930s. Low-power electricity-driven centrifuges had been used earlier to aid the identification, isolation, and characterization of the components of organic substances, but they were of limited capacity.[5] In the mid-1920s, the Swedish researcher Theodor Svedberg devised the high-precision oil-driven analytic ultracentrifuge.[6] By the 1930s, the instrument could measure—at very high speed—the sedimentation velocity of materials or—at lower speed—the position of materials at the equilibrium between centrifugal force and diffusion.[7] In the 1930s, Svedberg's student Arne Tiselius developed an apparatus for electrophoresis, another technique for the isolation of substances.[8] In electrophoresis, electricity was passed through a tube containing mixtures of organic substances. Some particles then moved to the positive electrodes, others to the negative. The principle of electrophoresis had been known for some time. Incidentally, paper electrophoresis or "electro-capillarization" [*Elektro-Kapillarisation*] was first described by two venom researchers.[9]

The ultracentrifuge and the apparatuses for electrophoresis and chromatography quickly became core technologies for the purification and structural analysis of proteins. The ultracentrifuge helped answer questions about the weight of the large protein molecules. Because the instruments themselves and their operation were so costly, the few institutions that could afford them dominated the field of protein research: notably the Rockefeller Institute and Caltech in the United States and Svedberg's home institute, the Institute of Physical Chemistry, at Uppsala, Sweden. Svedberg's group was an especially powerful force in protein research because at first only it had an analytic ultracentrifuge at its disposal. During the late 1920s and early 1930s, the group measured the molecular weights of more than thirty proteins.[10]

Venoms: Proteins, or Enzymes?

During the first half of the twentieth century, the community of venom researchers that had begun to form around 1900 consolidated into a close-knit group, complete with professional journals, conferences, and proceedings. There was an early attempt to start a journal specifically on venoms, the *Bulletin of the Antivenin Institute of America*, but this periodical was short-lived. Founded in 1927, it lasted just about five years. In the mid-1950s, the first "venom conferences" were held, and several proceedings and collected volumes on venoms were published. Contributors to the 1956 collection *Venoms*, which contained the papers presented in 1954 at the first international conference on venoms, came from five continents. In the early 1960s, the International Society on Toxinology [*sic*]—the study of naturally occurring animal venoms and poisons—and the journal *Toxicon* were founded, with encouragement from the WHO.[11] *Toxicon*, which publishes studies of poisons produced by living organisms, still exists today. In his short history of the society on toxinology and *Toxicon*, American physician and toxicologist Findlay Russell explicitly emphasized the international composition of the venom community: it involved researchers from the United States, Austria, Germany, the USSR, Yugoslavia, Egypt, Israel, India, "the Orient," Central and South America, and Australia.[12]

Between the late 1930s and the mid-1950s, the German-born biochemist Karl Heinrich Slotta, then working in Brazil,[13] was one of the main suppliers of snake venom to researchers in Europe and the Americas. Between the 1930s and the 1950s, numerous papers on snake venom—even those published by European research groups—acknowledged Slotta's gifts of venom. In fact, during this period, much venom research centered on one particular kind of snake, the Brazilian rattlesnake (*crotalus terrificus terrificus*, now *C. durissus terrificus*) simply because these snakes were so abundant that Slotta could easily supply other teams with the material resources for their projects.

Venom research had always been at least partly motivated by the desire to develop antidotes for human and also for nonhuman snakebite victims (such as cattle). As evidenced by the title of the first specialized journal—*Bulletin of the Antivenin Institute of America*—and by the WHO involvement in the founding of the *Society of Toxinology* in the 1960s, the hope of finding antidotes continued to inspire venom research. Nevertheless, many of the investigations that were pursued in the first half of the twentieth century were foundational—they were driven by the desire to clarify the composition and biological action of venoms, not necessarily by the desire to find cures. The *Bulletin of the Antivenin Institute* carried articles about the physiological action and chemical composition of venom and about the purification of active principles.

In the early decades of the twentieth century, venom research was informed by protein and enzyme studies. Just as Flexner and Noguchi transferred Ehrlich's analytic framework and the techniques of immunology to venom studies, venom researchers consciously drew on new analytic techniques and approaches from research on enzymes, proteins, and hormones to understand the composition and biological action of venom. For quite some time, it was a matter of debate whether the toxic factors contained in snake venoms should be regarded as enzymes or as proteins. Along with the intertwined discussions about proteins and enzymes, the conceptualization of venom changed quite dramatically over time, as a glance at some handbooks and reviews from the middle decades of the century reveals. In 1928, the Brazilian researcher Afranio Amaral reviewed the latest insights in a chapter, "Venoms and Antivenins," in the handbook *The Newer Knowledge of Bacteriology and Immunology*. The very fact that a chapter on venom was included in this handbook shows that venom studies were still closely linked to early twentieth-century immunology. According to Amaral, venoms had to be classified with proteins due to their most important reactions and physico–chemical affinities. He stated: "Very little is known concerning the chemical composition and the real nature of the toxic principles of the venom [. . .]. In general, crude venom consists of: *(a)* proteins (albumin, globulin); *(b)* proteoses and peptones; *(c)* mucin and mucin-like substances; *(d)* ferments; *(e)* fat;

(f) detritus (cells, etc.); *(g)* salts (e.g. calcium chloride and calcium, mag-
nesium, and ammonium phosphate)."[14] Amaral listed several "antigenic
principles" that had so far been recognized in venoms, including proteol-
ysin, hemorrhagin, antibactericidin, hemagglutinin, and lecithinase. He
considered the last of these, lecithinase, as the most important because it
might be responsible for the hemolysis taking place with red blood cells. In
line with earlier research, he surmised that hemolysis might be the result
of the action of a ferment on the serum, by which a hemolytic substance,
"lysocithin," is formed.

About ten years later, one of the novel biochemistry journals, the *Annual Review of Biochemistry*, carried a review article entitled "Animal
Poisons."[15] The author of the review, the Australian medical researcher
Charles Kellaway stated: "Snake venoms contain at least two, and in some
cases possibly more, toxic principles which are protein or of protein nature, some being enzymes." He added, however: "The classification of
the toxic actions of venoms and the relation of these to separate active
principles is still uncertain."[16] In the early 1950s, there was still disagreement about whether the main components of venoms were enzymes or
proteins, whether all enzymes in venoms were toxic, or whether all toxins
were enzymes. The handbook *The Enzymes*, edited by biochemist James
Sumner (of Nobel fame) and the Swedish biochemist Karl Myrbäck,
contains a chapter entitled "Enzymes as Essential Components of Bacterial and Animal Toxins." The author, E. Albert Zeller, reconstructed
the history of twentieth-century venom research as a trajectory from the
early works by Flexner, Noguchi, Kyes, and the German medical chemist
Karl Lüdecke on the mechanism of venom-induced hemolysis to the later
insight that the hemolytic factor in snake venom was actually an enzyme.[17]
According to Zeller, the crystalline rattlesnake venom protein "crotoxin"
that had been isolated in the late 1930s simply *was* the enzyme phospholipase A.[18] Nevertheless, he emphasized that "not all the enzymes in [ani-
mal] poisons have clear-cut roles in the mechanism of poisoning, and that
not all toxins have obvious enzyme functions."[19]

The crystallization of the active protein from rattlesnake venom that
Zeller highlighted in his chapter was accomplished in 1938. Two biochemists, Slotta and his brother-in-law Heinz Fraenkel-Conrat,[20] reported the
successful crystallization of a neurotoxic protein component from rattlesnake venom in a brief article in the journal *Nature*, stating that this substance was "the first proteinic toxin which has so far been crystallized. It
contains the whole neurotoxic and the whole haemolytic activity of the

venom. These two properties have hitherto been attributed to two different substances, one an enzyme, the other a toxin."[21] The two named the
substance "crotoxin." The isolation of crotoxin in crystalline form was one
of the key moments in twentieth-century snake venom research.[22]

Coping with Diversity

To the historian of twentieth-century life sciences, the endeavor to characterize the main toxic component of rattlesnake venom is so interesting
because it combines in a nutshell all the distinctive features of biochemical investigation in the first half of the twentieth century: the initial prox-
imity to immunology, the conceptual shifts in the understanding of biochemical mechanisms, the increasing reliance on machines and the influx
of funding from drug companies, and—last, but not least—the multina-
tional makeup of the community of researchers. Debates about crotoxin
involved scholars from the Americas, Europe, Australia, and Asia, and time
and again, Slotta, still based in Brazil, was acknowledged as the provider
of venom samples.

Around 1900, Paul Ehrlich and others developed procedures for the
quantitative determination of the biological activity of a substance, whereby
biological "units"—guinea pigs of a certain weight—served as standard
reference in the calculations. When Slotta began his research at Butantan, he and his coworkers used very similar protocols. They characterized venom by so-called efficacy values [*Wirksamkeits-Werte*], which were
determined as the minimum lethal dose for a white mouse of a certain
weight (15–18 g). They determined the toxin value [*Giftwert*] (a measure
of its neurotoxic activity relative to a biological unit, the mouse), the lecithinase value (the number of lecithinase units in 1 mg of venom), and
the coagulation value (the number of units of coagulase in 1 mg of dry
venom). But with the aid of Fraenkel-Conrat, the investigation of the
biological activity of venom quickly became reconceptualized in terms of
protein chemistry.[23] The two researchers imported experimental designs
from contemporary research on insulin. They noted that venom, unlike
many other proteins, had high sulfur content. Insulin, another physiologically highly active protein, had the same feature. Other investigators had
just found that the disulfide bonds (-S-S-) in insulin could be split by the
amino acid cysteine and that breaking those disulfide bonds in insulin
molecules deactivated the hormone. Building on these findings, Slotta

and Fraenkel-Conrat examined whether breaking the disulfide bonds in venom with cysteine rendered venom inactive, and they found that the *Giftwert* decreased very rapidly.[24] But then the two researchers succeeded in crystallizing an active protein from rattlesnake venom, and subsequent papers from Butantan all deal with the composition and quantitative analysis of this substance.

For the historian of methods discourse, crotoxin research is intriguing as an example of how the researchers were assessing the reliability of their instruments and research techniques and how they established whether the experimental results they obtained were informative. In the last two chapters, we saw that in the late nineteenth century, numerous methodological challenges for biomedical experimentation came to the fore: the establishment of standards of comparison, the control of experiments, the reliable identification of invisible causes, and the question of how to relate theoretical assumptions with experimental outcomes. Debates about the nature and biological action of crotoxin continued for decades, not least because several of the methodological issues that had been raised toward the end of the nineteenth century were still unresolved. In fact, in some respects, the challenges had become even greater. Unlike Flexner and Noguchi, who worked within a relatively stable theoretical framework that they (along with many others) took for granted, the crotoxin researchers found themselves on shifting grounds: theoretical approaches multiplied, concepts changed meanings, and new analytic techniques were being developed. As more and more research techniques and instruments for biochemical analysis were devised, crotoxin could be isolated and investigated in different ways, but this did not lead to a better understanding of its structure and function—at least not right away.

Toward the end of the nineteenth century, Wilhelm Roux put forward in a prominent place clear criteria for the use of hypotheses in explaining the fundamental yet hidden processes of life. Echoing other late nineteenth-century commentators on science, Roux insisted that those hypotheses were closer to the truth that explained the most facts, successfully predicted new facts, and were simpler than others. To philosophically informed twenty-first-century readers, it might seem natural to expect that by the 1920s, this conception of hypothesis assessment (so familiar to philosophy of science today), would have been widely used in the sciences themselves. But this was not the case. Quite often, the notion of hypothesis was used in a much more liberal, everyday sense of "assumption." Any assumption about biochemical processes and entities that available findings suggested was called a "hypothesis," regardless of how many phenomena it could explain and of whether it led to novel predictions.[25] Notably, however, the other strategy that Roux had recommended—the strategy of seeking confirmations for one's findings in multiple ways—did become part of the efforts to secure findings about subvisible biochemical substances and mechanisms.

Late nineteenth-century investigators were concerned both about the complexity of organic bodies and about the invisibility of the things and processes that were of interest to biologists. Roux emphasized that as a novel research field opened up, it was prudent to investigate one and the same thing or process from a number of different angles and to perform different kinds of experiments on it. If these lines of research converged on the same finding, one could be sure that the outcome was sound. The early twentieth-century response to the problem of how to investigate subvisible mechanisms and things was essentially the same as Roux's: multiple lines of research were pursued, using multiple techniques. But the problem of how reliably to investigate subvisible things was exacerbated by the emergence in the early twentieth century of more and more elaborate instruments and technologies for the investigation of ever more remote things, all the way down to the molecular level. The biologically active units on which these researchers experimented were no longer standardized organisms but rather fractions of organic substances. The workings of the elaborate techniques and instruments for investigating the realm of the subvisible were often not completely understood. It was not always clear whether the isolation and separation techniques for organic materials did in fact completely separate one substance from others, whether it really isolated biochemically active units, how the biochemical actions of these units (if they were indeed units) could be correctly determined, and whether these actions as they were determined *in vitro* were the same as the actions they performed in the living body.

Recent philosophical debates about the relative merits of the methodological strategies of "reliable process reasoning"[26] and "robust detection"[27] have suggested that the advantages of obtaining confirmatory evidence from different lines of research over obtaining positive evidence from just one research procedure are not so obvious: If one particular technique works reliably and produces results that are stable, repeatable, meaningful, and agree with theoretical expectations, what would one gain by seeking confirmations from other, less well understood techniques?[28] If researchers seek confirmation for their results from another line of research, how could they be sure that the second technique was not employed in the interpretation of the first?[29] And if none of the available techniques works reliably, then how could one be sure that the apparent convergence of research outcomes is not because all of them produce the same artifact?[30]

Philosophical analyses of these problems have concentrated on the structure, epistemic credentials, and epistemic force of arguments drawing on multiple lines of research in ideal epistemic situations.[31] Debates about the homogeneity of crotoxin, by contrast, offer excellent materials to examine how "real-life methodology" dealt with such a challenge.[32] The goal of real-life methodology is not to establish whether or not the argumentation of early twentieth-century biochemists fell short of the ideal of robust detection or whether the conditions of reliable process reasoning were, or were not, fulfilled. Rather, the goal is to examine what strategies were proposed to solve the problem of using less-than-secure technologies and preliminary theories to investigate phenomena that were hidden from sight, to what extent it *was* a problem for the investigators, and if it was, what kinds of solutions to it were deemed acceptable.

First of all, it is important to attend to a distinction. Sometimes the researchers employed several lines of research to understand a thing or process, because each of the techniques was by itself too narrowly focused and too specific to provide comprehensive understanding of the issue at stake.[33] In the case of the identification of disulfide bonds in amino acids, for instance, multiple techniques were used, because each of the methods had specific merits and specific limitations. In the quantitative determination of the distribution of sulfur across the amino acids contained in crotoxin, for example, the so-called Folin method could detect disulfide bonds in cystine as well as in other amino acids, whereas the Sullivan method could detect disulfide bonds only in cystine. If both methods detected the same amount of disulfide bonds (as was indeed the case), one could conclude that only cystine disulfide bonds were present in crotoxin.[34] Using multiple analytic procedures provided more complete information about the entity than a single technique did. The findings obtained with the different techniques were complementary, and together they revealed information about sulfur components in crotoxin.

The underlying idea that different lines of research might complement one another was explicitly proposed and defended. The first International Conference on Venoms, which took place at the annual meeting of the AAAS at Berkeley in December 1954, yielded a collected volume, aptly titled *Venoms*, that was published in 1956. In the introductory paragraphs to the chapter on comparative biochemical analyses of snake venom components, the authors state: "The classical biochemical, physiopathological, and immunological methods used in investigations of the complex subject of venoms were supplemented some ten years ago by perfected means of approach, such as the quantitative electrophoretic analysis introduced by Tiselius and its application by Polson and collaborators [. . .] to snake venom. The original Tiselius method and the modification of the latter, constituted by filter paper electrophoresis, permit separation and isolation of the respective fractions of venoms, as well as study of their specific characters and actions *in vivo* and *in vitro*. Similarly, introduction of filter paper chromatography [. . .] permitted demonstration, by the spraying method, of the complex biochemical constitution of venoms, and identification of a large number of amino acids present in venom proteins. *However, no one of these various methods is sufficient in itself to supply all the information on chemical constitution, biological activity, and antigenic properties of venoms.*"[35] At the end of the chapter, the authors reiterate their methodological point: "Each investigative method provides results of different but complementary nature. Comparison and correlation of such results constitute the only rational approach to a relatively accurate picture of the complex composition of venoms, of their physiopathological activities and of their modes of interactions with antibodies produced in immunized animals."[36] Taken together, multiple lines of research provided more complete information about a process or object that is hidden from sight.

Intuitively, this is a convincing use of different lines of research. The issue of the uniformity of crotoxin is different because several research techniques were employed to gather epistemic support for *one and the same* finding, and indeed, different lines of research suggested the uniformity of crotoxin. The interesting thing is that the convergence of data from different investigations was ultimately not decisive. The announcement in *Nature* of the successful crystallization of crotoxin[37] came with a brief description of the way in which the venom protein could be isolated and crystallized and with a paragraph describing the quantitative determination of the sulfur content of crotoxin. The crystal was obtained with traditional methods (heat coagulation, precipitation, and fractionation).[38]

At that time, the criterion for the homogeneity of a substance that was considered decisive was the behavior of the substance in the ultracentrifuge. Many years later, Fraenkel-Conrat recalled: "In those days (1937) the homogeneity and molecular weight of a protein was best established by sending it to The Svedberg at Uppsala for ultracentrifugal analysis, which was done. Crotoxin behaved as a single protein."[39] Theodor Sved-berg and his coworker Nils Gralén promptly published a short paper in the *Biochemical Journal* describing the results they had obtained by the ultracentrifugal methods developed in their laboratory. They had analyzed crotoxin by two different procedures: by measuring the sedimentation ve-locity (the rate of settling of the molecules) and by measuring the sedimen-tation equilibrium (the competition between settling and diffusion). Both sets of measurements showed that the substance was homogeneous, and the measurements thus supported Slotta and Fraenkel-Conrat's results.[40] In the acknowledgements, the two researchers thanked Slotta for suggesting the problem and for kindly supplying 120 mg crystallized crotoxin.[41]

The very fact that the homogeneity of crotoxin had not only been dem-onstrated in different ways but also with sophisticated "big" technology became an important piece of evidence when Slotta and Fraenkel-Conrat's findings were challenged, which happened quickly. Although crotoxin was soon widely accepted as a major component of snake venom, several investigators debated whether it was really uniform or whether it was a composite, whether it was an enzyme, and how its biological activity could be characterized. A couple of years after the first announcement of the identification of crotoxin, another article in *Nature* reported that the he-molytic and the neurotoxic substances could get partially separated into two different proteins.[42]

This result is intuitively more plausible because one would not have to assume that one homogenous substance was a producer of two biological mechanisms. But Slotta and Fraenkel-Conrat were unfazed. They imme-diately responded, again in *Nature*, listing the "main points of evidence"—derived from several lines of research—that suggested the uniformity of crotoxin. They argued that they had obtained the same data from dif-ferent samples at different times; five recrystallizations did not alter the hemolytic and neurotoxic activities of the protein, and the solubility curve of crotoxin at different ammonium sulfate concentrations showed the straight line typical for pure substances. Svedberg and Gralén's measure-ments with the ultracentrifuge indicated the homogeneous structure of crotoxin. Therefore, they saw no reason to consider findings from only one experiment—and a "low-tech" experiment at that: The experiment had been performed with "the crude venom secrete," and Slotta stated, rather brusquely, that he saw "no reason to discuss it."[43]

At this point the ultracentrifuge technology was perceived as being su-perior to more traditional precipitation techniques. The fact that Fraenkel-Conrat soon turned to electrophoretic experiments to study the structure

and behavior of crotoxin suggests, however, that the issue was not completely settled even in the minds of the two discoverers. In the early 1940s,
Fraenkel-Conrat (now at the University of California at Berkeley) and his
coworker Choh Hao Li undertook the electrophoresis of crotoxin using the
apparatus devised by Svedberg's student Tiselius. I have already noted that
a technique of paper electrophoresis or "electro-capillarization" was developed at Butantan in the late 1930s,[44] but Slotta's colleagues Klobusitzky and
König initially used this technique to analyze crude venom of the Brazilian
pit viper Bothrops jararaca and the coagulating component of that venom.
The Berkeley group applied the technique to crotoxin and, just like the
investigation with the ultracentrifuge, the electrophoretic behavior of crotoxin suggested that crotoxin was a homogeneous substance.[45] Again, it was
Slotta who had provided the venom.[46]

I will not attempt to reconstruct in detail the complex debates about
crotoxin during the following decade. Suffice it to say that Slotta and
Fraenkel-Conrat's attempts to strengthen their results with evidence obtained from ultracentrifugation and electrophoresis were not enough to
settle the debates. Even though these investigations yielded quantitative
data that did point to the same fact—the homogeneity of crotoxin—the
convergence of the research outcomes did not bring closure. In the early
1950s, Slotta himself acknowledged the mounting evidence for the composite nature of crotoxin. In a German-language article published in the
Swiss journal *Experientia* (currently *Cellular and Molecular Life Sciences*),
he gave an overview of the chemical studies of venom since the late 1930s.
He was quite cautious in his statements, noting that "today we believe we
know that all active components [of venoms] are of protein nature," and
added that it was still a point of debate whether each of the physiological
effects of snake venom was due to a chemically uniform protein.[47] He now
acknowledged the experiment by the two Indian researchers Gosh and
De of which he had been so critical in 1939. Although the experiment
was rather primitive, it was food for thought, and other evidence pointed
to the presence of subunits in crotoxin. His own attempt at separation by
electrophoresis did not yield any definite outcome. Chromatography also
produced inconclusive results. A clear picture did not emerge.[48]

Thus we see that the strategy of pursuing different lines of research
was adopted even after the initial discovery had been announced in *Nature*.[49] Because none of the strategies that were employed was perceived as
completely reliable, and because many theoretical questions about toxins,
enzymes, and proteins were wide open, the apparent convergence of the
data was not decisive enough to settle the issue of the chemical nature of

venom proteins. In 1956, Slotta's last Brazilian article on rattlesnakes was published. Slotta presented the paper at the First International Conference on Venoms in 1954, and it appeared in print in the collection *Venoms*. This contribution again documents the shifting conceptual frameworks for the interpretation of the composition and biological activity of venom. In their early work, Slotta and Fraenkel-Conrat and their coworkers had considered a "proteinic toxin" that had both neurotoxic and hemolytic properties, whereby the toxicity of rattlesnake venom was due to enzymic effects. In 1954, Slotta stated that it had now become "very doubtful" that the action or interaction of enzymes could explain the toxic effect of a venom. But he did not have a good alternative that he could propose. His reference to the "very wise"—and, one might add, well-worn—adage by Paracelsus ("the dose makes the poison") appears rather provisional.[50]

Slotta's contribution documents quite nicely how confusing the situation was at the time. Slotta reported a wealth of results from various crotoxin researchers, but nothing seemed conclusive. There were indications that crotoxin could be split, but the activities of the two parts had not been determined. Separation by means of electrophoresis had been tried, "without, however, any definite results." Some separations yielded nontoxic, inactive fractions. Some separations yielded fractions that were nontoxic but still showed hemolytic activity. Slotta admitted: "It is difficult to decide whether the split products are the inactivated toxins or whether, by these experiments, the crotoxin molecule was only deprived of its toxic activity without losing its hemolytic properties."[51] Other researchers who attempted to separate venom components also commented on the possibility that their findings were distorted by the investigative tools. The German crotoxin researchers Wilhelm Neumann and Ernst Habermann, for instance, noted that the speed of migration of different venom fractions might depend less on the chemical composition than on the presence of factors that were introduced by the experimenters.[52]

Slotta concluded his essay by saying that the political and economic situation in Brazil was no longer conducive to scientific work. He would thus have to abandon his venom research, and he hoped that his North American colleagues would continue where he had to leave it off.[53]

Research on crotoxin did indeed continue, and not just in the United States. The German group around Neumann[54] published a swath of papers (most of them in German) on their attempts at separating crotoxin components with the help of chromatography. In 1955, Neumann's son Wilhelm Paul Neumann reported successful separation of crotoxin by means of chromatography. The analysis produced two fractions; one was highly toxic, the other a nontoxic enzyme (phospholipase A).[55] W. P. Neumann and his colleagues named the toxic fraction "crotactin." They insisted that crystallization alone was not a criterion for the homogeneity of a protein substance.[56]

But this announcement, too, turned out to be premature. Researches on the composition of crotoxin continued, and in the early 1970s, two research groups, Robert A. Hendon, working with Fraenkel-Conrat, and the German group around Habermann distinguished two biologically active components of crotoxin.[57] They agreed that one component was acidic and the other basic and that both were nontoxic. They also agreed that the acidic component showed little or no hemolytic activity, whereby the basic component clearly was hemolytic. Both groups found that recombination of the subunits restored the toxicity of the crotoxin. Nevertheless, the debates remain confusing even in hindsight, because those research groups introduced different terminologies for these subunits and sometimes even invented entirely new terms. Hendon and Fraenkel-Conrat called the two components "crotoxin A" and "crotoxin B." Rübsamen et al. identified the basic, barely toxic component as the enzyme phospholipase A. They called the acidic, nontoxic component "crotapotin" because it was a substance "from CROTAlus venom, which POTentiates the toxicity and INhibits the enzymatic activity of phospholipase A."[58] The debates are confusing also because these new findings were subsequently read into the older literature. For instance, in 1972, Slotta and Fraenkel-Conrat were celebrated for having identified "a toxic component with phospholipase A activity."[59] Their original communication in *Nature*, however, discusses the "neurotoxic" and "hemolytic" activity of crotoxin, in line with the then common conceptions in immunology.

As collaborations and interactions among venom researchers increased, a new conception of crotoxin emerged according to which crotoxin was a complex of different components. One of the components, the basic phospholiphase, had weak toxicity. The other component, the acidic, enzymatically inactive subunit, was nontoxic. Together, they were highly toxic. The toxic effects were due to a synergetic effect of the components, whereby the nontoxic subunit potentiated the toxicity of the weakly toxic subunit.[60] Around 1980, the once uniform crotoxin had literally become the crotoxin *complex*, "a molecular complex consisting of two markedly distinct and dissimilar proteins interacting to express the full biological activity long associated with crystalline crotoxin."[61]

The investigations of crotoxin became even more intricate when different forms of the enzyme phospholipase were distinguished in different
animal venoms as well as in bacteria and body tissues and fluids (phospholipase A and B, then also C and D), and the different mechanisms of their
actions on phospholipid molecules were characterized. Later reviews of
the literature on crotoxin research emphasized the dramatic changes of
the interpretation of venom action, the turn from the view that enzymatic
activities were mainly responsible for venom action to the notion that enzyme action was very weak and back to the notion that venom enzymes
(specifically PhA2) do have potent toxic effects.[62]

Much more could be said about the developments, but even from my
brief sketch it should have become clear how the methodological worries
and discussions of venom researchers were fueled by various concerns about
complexity—not only by the complex organization of biological substances but also by the intricacy and opacity of instruments and techniques. In his review of research on snake venom PhA2, published in
the handbook *Snake Venoms*, Philip Rosenberg explicitly acknowledged
that shifting interpretations of the nature and action of venoms were due
to shifting techniques and procedures. In the introduction to his review,
Rosenberg announced that it was not his purpose "to present an uncritical encyclopedic listing of all studies in which pharmacologic effects of
snake venoms have been attributed to Ph." This would only add a "stamp
of approval to much data which is impossible to interpret or has been
incorrectly interpreted by the authors."[63] The review includes a synoptic
overview and critical discussions of experimental protocols (Rosenberg
actually used the term *experimental protocol*).[64] Rosenberg's synoptic assessment of experimental protocols was facilitated by the rise of a new
element of scientific writing—the methods section.

The Modular Article

In 1990, the authors of a brief article entitled "The Science of Scientific Writing," published in *American Scientist*, reminded their scientist–readers—
the future authors of scientific articles—that "[r]eaders have relatively
fixed expectations about where in the structure of prose they will encounter particular items of its substance."[65] Such expectations could reason-
ably be formed from the mid-twentieth century onward. Earlier readers
would often have to read the complete work to find the particular item

they were looking for. Readers of one of Flexner and Noguchi's articles, for instance, would have had to peruse the entire document to find out how the different experiments had been made. In the mid-twentieth century, by contrast, article content was structured around the components of the project.

The brief article on electrophoresis of crotoxin by Choh Hao Li and Fraenkel-Conrat, published in the *Journal of the American Chemical Society* in 1942, looks just like a paper from a recent scientific journal. It is very concise and matter-of-fact. On two pages, it contains several tables, graphs, and images of electrophoretic patterns as well as acknowledgements to the University to California, the Rockefeller Foundation, a pharmaceutical company (Parke-Davis, a subsidiary to Pfizer), and Slotta (for the supply of crude venom) and abundant references to other scientific literature. The titles of the sections no longer signal investigative goals but orient the reader toward "particular items of its substance." The paper presents the sections in the order that is now the standard in scientific journals: The first short paragraph states the research question and the key finding. It is followed by a section entitled "Experimental," a section "Results," and a short paragraph "Summary" that repeats the main outcomes. It is this structure that became common for biomedical and biochemical journals during the 1940s. In scientific papers having a modular structure, methods discourse has a circumscribed place in the scientific article: the methods section. The busy researcher, eager to find information quickly, knows exactly where to look for it.

Prior to the 1930s, scientific articles combined narratives of experimental procedures with arguments—that is, with discussions of the significance of empirical findings as well as with occasional comments on methodological issues. Flexner and Noguchi's articles, for example, described series of experiments on a number of related topics, whereby protocols and the occasional methodological statement were integral parts of the narrative. The flow of the narrative drove the argument home.

Around 1900, reflections on the art of (scientific) writing gradually moved away from classical rhetoric. Instruction manuals for scientific authors shifted the attention from the tools of persuasion to the problem of how to guarantee ease of access for busy readers. During the first half of the twentieth century, more and more books and articles appeared that contained instructions specifically for the writing of scientific and other kinds of professional texts. These works addressed questions of how to prepare, organize, and write scientific and technical papers and reports and how to handle definitions, descriptions of procedures and machines, and explanations of processes. The modular structure became the recommended format for scientific articles. Although the authors of these manuals and instructions often emphasized that there was a certain degree of freedom in the arrangement of the parts of scientific or technical publications, they usually distinguished three main components: the introduction (which was supposed to include an overview of available literature), the description of experimental procedures and materials, and the statement of results and conclusions.

In her manual *The Writing of Medical Papers* from 1922, Maud Mellish, then the medical editor of the Mayo Clinic, advised that a research paper be structured as follows: (1) introduction, (2) historical notes, (3) materials and methods, (4) results, and (5) summary and conclusion.[66] Mellish had concrete guidelines for what should be in each section. The section "Materials and Methods" should include "an exact statement of the character and amount of material investigated, of the old and new methods of investigation, and of the operations, devices, and so forth, used." The "Results" section "should embody a detailed discussion of the results of the investigation, operative procedures, or experimentation." Mellish clearly assumed that the scientific article should present an argument (rather than a narrative), and she did make an interesting comment about how the argument should be structured: "While findings which prove the author's working hypothesis may properly be given first place in the argument, other findings of a negative character and those of no apparent significance should also be stated."[67] The last part of the comment is noteworthy because it was unusual at the time. References to negative and insignificant results were rare (and remained so until very recently). But the beginning with a "working hypothesis" is what the modular form of scientific articles especially encouraged.

Around 1930, a new genre of text emerged in these writings on writing: the so-called report. The term is not used in the sense in which I have used it in the present book—*viz.* as a label for an article or treatise that presents the findings of experimental research. The textual genre of "report" from the 1930s comprised chronicles of all kinds of projects, not necessarily limited to science. As the authors of the volume *The Preparation of Reports: Scientific—Engineering—Administrative—Business* explain in the preface to their manual, the word *report* refers "to special and periodic communications employed to convey information through certain literal and graphic elements the character of which is determined by the nature of the facts presented. Such documents are indispensable today in practically all work of a scientific, engineering, financial, or administrative character. Because of the complexity of business and professional life in the twentieth century, those charged with the responsibility of large undertakings are seldom able to investigate personally the matters with which they must deal. Consequently reports from consultants or from subordinates frequently form the basis of highly important decisions, often involving heavy commitments."[68]

It is clear from the title of this manual and from the explication of the term *report* that the authors had in mind review articles and summary accounts not just of scientific or technical content but also of bureaucratic operations of supervision and control. The research report or scientific report is just one kind of report among others. Writing reports became a necessity because of the overall complexity of business and professional life in the twentieth century. The instructions for the writing of research reports also call for a modular organization according to the following scheme: (1) introduction (comprising the aim of the study, a survey of the literature, materials, and methods); (2) procedure (experiments and results); (3) conclusion.[69] The authors added that research reports prepared for publication in scientific and technical periodicals "are usually organized in three main divisions: 1. Introduction. 2. Experimental Procedure. 3. Results and Conclusions."[70]

Current scientific articles are usually organized in this modular form—often called the IMRD form—whereby the modules present introductory remarks, descriptions of methods, statements of results, and discussions of the findings. In the mid-twentieth century, this modular structure became the norm for the organization of articles in scientific journals.

As one might expect, research articles in scientific journals took on modular structure only gradually. During this time, structural differences among different genres of scientific articles became even more pronounced depending on their outlet and the purposes for their publication. The articles we find published in the journal *Nature*, which was becoming a major voice in the international scientific community of scientists[71] are typically at most one page in length; they report major novel findings but provide only the bare minimum of information about methods, instruments, and techniques as well as the occasional picture. The articles published in specialized professional journals such as the *Journal of the American Chemical Society* or the German equivalent, the *Berichte der deutschen Chemischen Gesellschaft*, include more detailed descriptions of instruments, techniques, procedures, materials, and methods as well as more in-depth discussions of findings and their implications. In these publications, protocols are usually clearly separated from the other sections of the text, even though they might not be called "methods sections" and the sections might not follow IMRD order. Articles usually contain distinct sections— introductory sections, sections describing instruments and procedures, and sections presenting findings, as well as discussions and interpretations of results, plus a summary section restating the main points. The sections covering methods and materials are placed in varying parts of the article; sometimes they appear early on; often they are presented at the end, just before the summary of the main claims or findings. Many of the German-language articles were organized in this fashion—for instance, the series of communications from Butantan in *Berichte der deutschen Chemischen Gesellschaft* in the late 1930s.

As more and more sophisticated techniques for the investigation of subvisible phenomena were developed, methods sections became more technical and formalized, and methods were typically distinguished by the names of those who had first developed them. Cho Hao Li and Fraenkel-Conrat's section entitled "Experimental" simply states that crystallized crotoxin had been prepared "according to the method described previously." Electrophoretic experiments "were carried out in the Tiselius apparatus [and] were recorded by the method described by Longsworth [. . .] The conductance was measured with the usual Wheatstone bridge type of circuit and a Washburn conductivity cell at 1.5°. The mobility was calculated from the descending boundary as recommended by Longsworth and MacInnes and was determined in the manner described in a previous paper."[72] This bland list of standardized procedures and methods could hardly be any more different from Fontana's graphic prose.

About Methods

This book began as an extended comment on Fontana's careful descriptions and justification of his methods, his approach, and his strategies for making experimental findings more secure. Thinking about Fontana's work has brought into focus an aspect of science that has often been neglected in history as well as in philosophy of science: working scientists' methods discourse, its meanings, the roles it plays in reports of experiments, the textual forms in which it is presented, the link to contemporaneous philosophy of science (or lack thereof), and the historical development of it all.

Methods discourse comes in layers. There are the protocols—scientists' (or experimental philosophers') accounts of the steps of an experiment or observation, as well as of the materials, equipment, and techniques that were used. Methods discourse also encompasses methodological views—scientists' (or experimental philosophers') conceptualization of procedures to assess and secure empirical results. In addition, methods discourse involves broader commitments to experimentalism—specifically to the imperative that scientific ideas must be confronted with, or based on, empirical findings.

Protocols, methodological criteria for successful experimentation, and even the commitment to experimentalism have histories. The procedures and techniques for experimentation changed, different methodological criteria were privileged at different times, entirely new criteria for securing and validating experimental results emerged, and what it meant to be committed to experimentation changed, too. Because these changes did not always occur at the same time—nor did they happen at the same pace—the distinction among the layers is a convenient analytic tool for the history of methods discourse.

Scientists may take descriptive or critical perspectives on methods discourse. Experimenters' pronouncements and commitments relating to methods are sometimes, but not always, accompanied by explicit reflections on, and justifications and defenses of, methods, techniques, and methodologies. Broader commitments to experimentalism are at times explicitly defended, specific methodological strategies are singled out and bolstered with supportive arguments, and even the proper ways of writing about methods-related issues might become a theme for critical reflection.

Fontana's treatise is such a remarkable document because it exemplifies so well both the layering of methods discourse and the distinction between descriptive and critical perspectives on methodological issues. Fontana described at length the plan, experimental setup, and procedure for each of his numerous experimental projects. In his narrative, he expounded again and again the criteria for successful experimentation, and he included in his book a methodological essay in which he described, explained, and defended his methodological principles and his experimentalism. Fontana's work exemplifies how methods discourse is shaped and influenced by a range of factors. Not only the social organization of the scientific community and its place in society but also prevalent conceptions of nature, of proper reasoning, and of scientific writing affect the content and presentation of methods discourse. The organization of the community of naturalists, the recalcitrance of experimental objects, conventions for writing analytically, and conceptions of nature and body impinged on the content of his methodological views, as well as on the ways in which methods discourse was integrated in his writings.

The distinctions among the different layers of and perspectives on methods discourse are not always perfectly clear-cut, neither in Fontana's treatise nor in other experimenters' reports of their activities and endeavors. Nevertheless, these distinctions are helpful tools for a fine-grained analysis of methods discourse in different fields and historical periods as well as for comparisons across fields and periods. Protocols are often bound to specific investigative contexts. Methodological views are typically transcontextual, spanning longer periods and more fields of investigation. Broader commitments—in particular the general commitment to experimentalism—are often of long duration, although, as we have seen, there were a number of different reasons why and various audiences to whom such a commitment was expressed.

Paying attention to the different layers methods discourse shows that protocols, descriptions and justifications of methodological criteria, and commitments to experimentation are incorporated in different portions of experimental reports, whether in introductions, in methodological essays, as integral parts of descriptions of experimental projects, or in highly technical "methods sections." Not only the content but also the stylistic features and organization of methods discourse change over time. Methods discourse does not constitute a specific genre of text. In fact, experimental reports themselves already come as different textual genres—letters, book-length narratives of experimental projects, journal articles, articles in peer-reviewed professional journals, and articles in popular magazines. The organization of such texts is historically variable, too. Modular scientific articles with distinct methods sections became common only from the 1930s onward. Today, new communication tools, especially electronic publication, even allow outsourcing methods sections to separate files.

Of course, little insight would be gained if we simply compared seventeenth-century and present-day methods discourse—say, the seventeenth-century treatment of repetitions and the twentieth-century notion of independent support or the plan of an early modern animal experiment with the design of biochemical analyses of protein molecules, or else the seventeenth-century narrative of an experimental project with the twentieth-century modular article. These things are simply too different for comparisons to be very illuminating. But there is a historical trajectory in methods discourse that is worth uncovering—namely, changes in content, in the ways in which methods discourse is incorporated in experimental reports to establish proper procedure, and in the ways in which the experimenters used statements and reflections about methods to confer epistemic force on the results presented. It is this trajectory that this book has sought to capture, and the distinctions among layers of methods discourse and among descriptive and critical perspectives on methodological issues are means to this end.

Procedures

Protocols comprise the description of the experimental design and setup, procedures, and interventions to obtain an effect. Protocols encapsulate the correct way of doing things within and even beyond a particular laboratory context. The protocol is the most variable ingredient in methods discourse because it is often (but by no means always) closely tied to concrete experimental projects. The life span of experimental designs might be comparatively short; protocols might be adapted quickly or might be re-placed as new or improved instruments, research techniques, and materials become available. Nevertheless, writing the history of methods discourse with an eye to protocols can lead to unexpected insights about the consti-tution and situation of scientific disciplines when it turns out that research-ers working in quite different fields adopted quite similar protocols. The migration of protocols across different fields is especially noticeable in the second half of the nineteenth century, and it binds fields together in ways that the mapping of personal interactions among researchers or the tracking of explicit citations might not capture. For the history of venom research, paying attention to protocols is particularly enlightening. One of the key features of venom research is that it intersects with so many dif-ferent fields in the life sciences, and this is reflected in venom researchers' incorporation of protocols from medical chemistry, physiology, pathology, bacteriology, immunology, and so forth.[1]

Writing the history of methods discourse with an eye to protocols can be informative also because it can tell us how much certain techniques and procedures were standardized, how technologies became established and eventually black-boxed, how resources were exchanged among re-search groups, and so on.[2] For instance, especially from the nineteenth century onward, we find names attached to procedures, techniques, and tools (the Sternberg bulb, the Tiselius apparatus), and we have seen how Slotta's ample resources and generous gifts of venom made an entire gen-eration of venom researchers focus on the Brazilian rattlesnake.

Typically, experimental reports contain at least some descriptions of procedures. We know from older histories of experimentation that the style of presentation of protocols might vary dramatically—between the seventeenth and the early twentieth century, it changed from prolix and vivid narrations and depictions of all the conditions and procedures in-volved in the experiment to terse, parsimonious descriptions, increasingly reduced to mere lists of instruments, processes, and acronyms that only the expert reader can understand. Perhaps the most detailed descriptions of instruments, techniques, equipment, and so on were published in the early modern period, but even in the seventeenth century, by no means were all descriptions prolix. In the second half of the nineteenth century, as experimental reports proliferated, easy access to information became a prime concern, and prolixity turned from a virtue to a vice.

Explicit reflections on protocols are rare, however, except in periods of controversy about appropriate procedures or as part of a retrospective in a review article. As we have seen, authors of reviews sometimes discuss in hindsight procedural errors that were deemed particularly damaging to an unfolding field.

Methodological Issues

Methodological strategies or criteria typically transcend local contexts and remain in effect for extended periods. Repetitions or repeatability are obvious examples for strategies or criteria that have long governed experimental practice. Early modern experimenters insisted that experiments must be repeated several times; to this day, repeatability remains one of the hallmarks of successful experimentation. But even though certain methodological concerns have long remained in place, they have not continuously been foregrounded in scientific discourse. Early modern experimenters were concerned about repetitions; some explicitly required multiple trials. Fontana took this requirement to the extreme. He stated that thousands of repetitions might be necessary to gain confidence in a result, and again and again he pointed out that he had repeated a trial numerous times. Even today, a singular experimental trial that cannot be performed again would not count for much, but this condition is just too basic, too taken for granted, to be discussed in a research article. The *Stanford Encyclopedia of Philosophy* entry on experiments in biology does not mention repetitions. The list of key epistemic strategies for validating experimental outcomes in the entry on experiments in physics also does not include repetitions. In the text, authors Allan Franklin and Slobodan Perovic do briefly discuss repetitions, but only in connection with a specific controversy in high-energy physics— namely, the debates about the "Fifth Force."[3] (High-energy physics is a special case, because repetitions of experiments pose particular challenges.) Replication (in the sense of repeating other people's results) is a different issue. The question of whether it is possible to replicate particular research outcomes is often intensely debated, but predominantly in contexts where certain findings are challenged.

Fontana put repetition first on his list of principles for successful experimentation, and he took his preoccupation with repetitions to the extreme. Still, he and other late eighteenth-century investigators placed the emphasis on variations: through countless "diversifications" of setups, initial conditions, tools, and experimental objects, they tried to establish cause–effect relations and to identify potentially confounding factors.

Charting the history of scientists' methods discourse has led to a better understanding of the changing relations between the experimenters' own methodological reflections and the history of professional philosophy of science. In the course of the nineteenth century, systematic analyses of scientific methods and logic and working scientists' reflections on their practices gradually diverged, and more formal treatments of methods were presented in books on the logic of science (or natural philosophy). In the mid-nineteenth century, the systematic search for and identification of causes through variations and comparative experimentation were explicitly conceptualized in works such as Herschel's *Preliminary Discourse of the Study of Natural Philosophy*, Mill's *System of Logic*, and Comte's *Cours de philosophie positive*. These systematic analyses of the structure of methodological issues were often in tension with the more pragmatic concerns and approaches of working scientists. It is in this context that working experimenters explicitly rejected the idea that many earlier experimenters took for granted—namely, that medical experimentation could profit from an orientation toward the physical sciences. Mead and other early eighteenth-century medical men wanted "Newtonian medicine"—the medicine of numbers, math, and geometry. By contrast, many (but not all!) mid- to late nineteenth-century experimenters emphasized that biomedical experimentation was profoundly different from physical experimentation, that methodological ideas from physics could not be transferred to the life sciences, and that the ideal experimental situation could never be obtained in biomedical investigations. They assumed that life, living organisms, and those organisms' environments were too complex to be fully understood by the experimenter and that rigorous experimentation as required by Mill's methods of experimental reasoning was impossible.

Mill himself acknowledged that his method was idealized and must be amended for practical purposes even in the physical and chemical sciences, as experimental conditions could not always be completely specified in actual experimental situations. Mill was most concerned with the ideal experimental situation and the inferences that could be drawn from such ideal experiments, as well as the epistemic force of these inferences. Many practitioners of experimentation in the second half of the nineteenth century were more concerned with the actual challenges they encountered every day in the laboratory. Some even introduced terminological distinctions to mark the difference—Bernard's notion of comparative experimentation, for instance, encapsulates a retreat from Mill's ideal experimental situation and an explicit acknowledgement that in practice, the experimental conditions for determining truly causal relations could only be approximated.

In these discussions, the difference between the practice of identifying causes in complex, complicated real-life experimentation and the structure of causal reasoning from ideal experimental situations came to the fore. It is, above all, a question of emphasis. The experimenters in the life sciences sought to develop strategies for dealing with complexity—and yet they did try their best to constrain and determine the experimental situation so as to approximate ideal of a completely determined physical experiment. The authors of methods treatises, by contrast, concentrated on the structure and epistemic force of experimental strategies—and yet they did acknowledge that in actual investigations, the ideal was unattainable.

In current historiography and philosophy of science, this tension between systematic, idealized philosophical reconstructions and real-life methodologies has turned into a tug-of-war between different brands of historically informed philosophy. The emergence of the concept of control is a wonderful (and to the analytically inclined philosopher quite maddening) illustration of this tension. In hindsight, it appears plausible to interpret the structure of control experiments with the conceptual tools of Mill's methods of inquiry. But the actual scientific discussions around 1900 about controls, controlling experiments, and control-experiments are so multifaceted and conceptually elusive that they escape easy categorization along these lines. Careful analysis of the different uses of control concepts shows, moreover, that the various terms related to experimental control capture quite different practical concerns—including concerns with the stabilization of experimental conditions, with causality, and with uniformity and comparability across local contexts.

Equally striking are the endeavors to establish the existence and agency of things and processes that were not directly observable. The investigators had to find ways of identifying substances and of assessing their agency, all while trying to stabilize very complex experimental environments. The factors assumed to be relevant for the formation and development of embryos or the pathogens bringing about diseases could only be inferred; intermediary bodies, complement, and even the products of their agency were assumed to be present or absent in experimentally generated and manipulated fluids. These challenges linked the work of investigators in seemingly remote fields, such as experimental embryology, pathology, physiology, and immunology, and they continued to occupy researchers in early twentieth-century life sciences.

The very fact that experiments and observations dealt with unobservables—that is, with entities that could not be directly manipulated and with processes that could not be seen to happen—was not considered problematic in itself. But for many investigators, the problem of how to connect theoretical assumptions to observable phenomena was a point of concern. How could these assumptions be assessed? How could hypotheses about hidden entities and processes be validated? The question of the role of hypotheses in scientific research had been a part of methodological debates at least since the days of Newton. The exchanges between Mill and Whewell about induction brought the issue to the fore. Again, the most systematic treatments of the problem at the time can be found in books of the logic of science. For us, what is at stake in these discussions is the underdetermination of theories and hypotheses about unobservables. Past experimenters themselves were less explicit in their reflections, although we do find occasional suggestions along the line of Whewell's ideas about the testing of hypotheses—namely, that the degrees of predictive success and simplicity of the hypothesis be considered in hypothesis evaluation. Other than that, the responses remained practical and pragmatic: As long as the evidence fit with the hypothesis or theory, as long as the hypothesis explained the phenomena, it was considered acceptable.

For late nineteenth- and early twentieth-century investigators, the question of how to deal with theoretical and instrumental uncertainties became the most pressing methodological issue. Of course, repetitions, variations of experimental conditions, and systematic comparisons were still a relevant part of experimental practice, but the emphasis in methods discourse had shifted. The investigators were confronted with two problems: First, the elaborate techniques and instruments for investigating subvisible phenomena that had become available were not completely understood. Second, the results obtained with those techniques and instruments were too narrow and piecemeal to yield comprehensive understanding of a phenomenon of interest. Again, we find in the strategies of the experimenters a pragmatic, "real-life" counterpart to philosophical discussions about underdetermination—namely, the integration of multiple research techniques and approaches.

Protocols are described at least briefly in virtually every experimental report. Methodological views, by contrast, are not always explicitly stated. If they are mentioned, they are often not discussed in detail. Explicit methodological discussions might indicate novelty. Fontana's methodology of experiment, for instance, reflects not only his attempts to disarm his opponents but also the emergence of a new conception of circumstances of experimentation that demanded a novel approach to experimentation. Most of the time, however, methodological views and strategies come to the fore in controversies and critical debates. Such controversies might concern concrete empirical results or entire approaches, such as the best way to pursue developmental embryology or the soundness of the metaphysical principles underlying experimental medicine. Researchers who had an axe to grind might even include a methodological essay in their work, as Fontana did. Or, like Roux, researchers might frame an entire discipline-specific journal to good effect with a long programmatic introduction saturated with methodological pronouncements. Yet others might publish whole books on the methodology of experiment, as Bernard did in his attempt to intervene in contemporaneous debates about the usefulness of statistics and about the plausibility of vitalism vis-à-vis determinism in physiology.

Today, we often find that the more concrete methodological debates about proper procedure migrate to outlets other than the experimental report. On the one hand, professional philosophers of science discuss in more or less abstract terms the structure and epistemic force of specific experimental strategies and proper reasoning processes—measurement robustness, severe testing, the argument from coincidence, and so forth. On the other hand, scientists themselves might be driven to discuss methodological problems in general science journals or even in popular media—because established professional practices are under scrutiny or because they are held accountable for spending taxpayers' monies, or both.

Commitments to Experimentalism

General commitments to the view that experimentation is the hallmark of science transcend even long-term methodological commitments such as the commitment to repetitions, to the variation of experimental conditions, or to "causal research." Commitments to experimentalism have been expressed ever since the rise of the "new" experimental method in the seventeenth century. In the broadest sense, "proper procedure" for the study of nature just means "experimental inquiry" per se. With the possible exception of Redi's and Charas's contemporary Bourdelot, all investigators who populate this book endorsed experimentation and were experimentalists in this broad sense, and many of them explicitly committed themselves to experimentalism.

Even the very commitment to experimentalism has a history, how-ever: It meant different things and served different purposes for differ-ent people. Sometimes it evoked Bacon's *Novum Organum*, sometimes Newtonianism (which also comes in different flavors). Experimentation was favorably contrasted with the practice of following or trusting ancient authorities; at other times experimentation was favorably contrasted with "mere" (e.g., noninterventive) observation of events and processes in na-ture. The commitment to experimentation could also be made to align oneself with a particular professional community—such as, for instance, those medical men who endorsed scientific medicine.

The audiences for the commitment to experimentalism and the ways in which this commitment was expressed changed accordingly. In the early modern period, scholars who endorsed experimentalism addressed other scholars who engaged in the study of nature, religious authorities, or aris-tocratic patrons. It thus made sense for them to treat the broader com-mitment to experimentalism as an integral part of the letters they wrote to their sponsors and patrons. In the nineteenth century, experimenters such as Mitchell addressed their arguments for experimentalism to uni-versity trustees, perhaps also to antivivisectionism advocates, or else to fellow members of the medical community who doubted that experimen-tal methods could ever advance therapeutics. Here the commitment to experimentalism became a battle cry in disputes among different profes-sional groups and between professional scientists and the larger public. At this time, commitments to experimentalism had become largely separated from experimental reports. They had drifted to introductory sections, and sometimes they were entirely removed from scientific articles and ap-peared in public addresses and commencement speeches. Only rarely do we find commitments to experimentalism within experimental reports in publications after 1850. If we do, we can be sure that specific battles were fought within the relevant community of experimenters. Explicit discus-sions of and apologies for experimentalism migrated from experimental reports, textbooks, and programmatic writings for scientists to public ad-dresses for broader, often nonscientist audiences.

Writing Scientifically

Writings on scientific writing are valuable resources for tracing the dy-namics of methods discourse, because they tell us what was deemed to be an effective means of presentation and persuasion at a certain time. Of course, we cannot assume that experimenters actually read all these books about writing scientifically and followed the guidelines. Some of the medical students at Oxford might well have glanced at Allbutt's treatise on the composition of scientific papers if they had any time to spare before submitting their theses, but they probably generally didn't. If at all, past experimenters will have followed editorial and journal guidelines when they were submitting their texts for publication. But for the historian of methods discourse, the manuals and guidelines on writing scientifically are valuable sources of information because they reflect the different conceptions of good argumentation and effective organization of a scientific text that were held at specific points in time.

The preceding chapters have shown not only that there are different genres of experimental reports but also that there are different genres of "writing on scientific writing"—the eighteenth-century books on scientific methods, the manuals on professional writing that developed out of books on classical rhetoric in the second half of the nineteenth century, and the specific guidelines for writing professionally that appeared around 1900.

These differences are significant, and they point to a transformation in methods discourse that has not attracted much attention. Until the mid-nineteenth century, writings on writing scientifically invoke contemporaneous ideas about both proper experimental procedure and proper reasoning practices. The very first chapter of this book showed that for Boyle, the distinction between experience and speculation was the complement of the distinction between narrative and argument in an experimental report. Late eighteenth-century guidelines bring theoretical notions and empirical facts closer together, but the organization of the account is still expected to elucidate experimental procedures—the guidelines call for the demonstration of the discoverability of a theoretical idea. The organization of Fontana's bulky tome exemplifies this approach.

In contrast, from the late nineteenth century onward, the sequence of experimental steps and the steps of the argument were separated, and argument and methods discourse no longer had to correspond to the actual practices of experimentation. One of the main driving forces for the turn to so-called deductive composition was the notion that "composing deductively" could facilitate the researcher's understanding of the main points of a scientific paper.

Paying attention to the history of methods discourse also sheds new light on Charles Bazerman's suggestion that scientific writing underwent a transformation from "establishing credibility of the *author*" to "establishing credibility of the *procedure*." In his account, seventeenth- and early eighteenth-century experimenter–authors aimed to establish their own credibility when they were presenting their results. They did so by citing illustrious witnesses or by assuring their readers that they had performed experiments with care and caution. Starting in the late eighteenth century, scientist–authors began to offer detailed reports of each individual experiment to support their views. In such reports, the burden of persuasion falls on "the representation: to establish proper procedure (that is, the experiment is done as any scientist might have done it), to specify all the conditions and procedures (that is, replication instructions), and to indicate how the experimental procedure answers potential objections."[4]

The distinction between establishing the credibility of authors and establishing the credibility of procedures is a helpful analytic tool for examining experimental reports, but my study has shown that the notion of an overall transformation from author-centered to procedure-centered writing is misleading. Early modern authors, too, made efforts to establish the credibility of procedures when they wrote about experiments. Redi did try to establish his trustworthiness by referring to witnesses, as did several of his critics, but he also presented, and to some extent discussed and defended, "proper procedure" to establish credibility of his experiments and results. He admonished one of his adversaries who had not performed enough repetitions of experimental trials, for example—not to question his opponent's credibility, however, but to question the reliability of the trial outcome. If we compare Redi's letters with late eighteenth-century writings on experiments, what we find is not so much a transformation but rather a shift of emphasis from person to procedure in the sense that later writings contain a much larger proportion of methods discourse and only a few references to witnesses and their social positions. Establishing the credibility of the experimental procedure was always a prime concern.[5]

Transformations of Methods Discourse

We are left with the question of how the dynamics of methods discourse can be explained. Why were some aspects of methods discourse preserved over longer time periods but others not? What forces drove the shifts, the innovations, and the transformations? Historians and sociologists of experimental science have shown that the social organization of the scientific community shapes certain features of methods discourse (such as eponymy and the modularity of scientific papers). But the social does not fully explain the content of the methodological reflections that are incorporated in scientific texts. Other factors that constitute methods discourse are metaphysical conceptions of nature, life, disease, and causality, new theoretical ideas and investigative techniques, and the very practice of experimentation itself.

Early modern experimenters required that experiments be repeated because they were concerned with the possible effects of accidental events and contingencies that, unknown to the experimenter, might interfere with the experiment. Fontana, by contrast, assumed that the circumstances surrounding the experiment could be identified, and he required all experimenters to identify them—and thus must have assumed that circumstances were not random occurrences but rather regular phenomena amenable to systematic study and general characterization.

Even the diligent application of methodological principles in experimentation might in turn lead to methodological innovation. In the late seventeenth century, the identity of experimental outcomes was the ideal that guided methodological reflections and the evaluation of experimental trials. In the course of the eighteenth century, the commitment to numerous repetitions combined with the increased emphasis on measurement became a driving force for the development of new methodological concepts. Many repetitions of measurements yielded discrepant data sets, which in turn motivated changes of protocols. Sometimes these changes led to more uniform results. At other times, however, greater uniformity of the results could not be achieved. Around 1800, we notice an erosion of the ideal of uniformity as an unintended consequence of the methodological goal of numerous repetitions, and new approaches to data management emerged as well.

Rapid development of new theories and instruments could also drive methodological innovation. Around 1900, the problem of uncertain theories and techniques justified the strategy to seek independent confirmation for one's findings through a different experimental approach. Experimenters remarked that some of the newly emerging and intricate technologies were not yet completely understood and that others were too specialized and too narrow in scope to provide comprehensive information about a particular subject.

Methodological discussions and reflections might get to a point where stock-taking appears necessary. Experimental strategies and methods are explicitly justified and formalized in systematic treatises on logic and method. Sometimes, such books were written by authors who were well read in the sciences (and their history) but who were not (at least not predominantly) experimental researchers themselves—authors such as Bacon, Mill, Whewell, or Comte. These treatises were often the culmination of a period of methodological innovation; they were produced in hindsight, intended as an intervention to advance a certain position or to offer a synopsis and achieve closure in a period of intense debate. Those treatises might in turn become sources of reference for experimental researchers. These researchers might present themselves as Baconians or Newtonians, like Mead and Fontana, or they might engage critically with the programmatic texts, as the nineteenth-century critics of Mill did.

Critics and analysts of science like Mill developed their philosophical views through an engagement with scientific debates and practices, as have many philosophers of science in the late twentieth and early twenty-first centuries. At the same time, working scientists engage with and scrutinize philosophical thought. Mill's explication of the methods of experimental inquiry, for instance, drew on early nineteenth-century science, as did Comte's discussion of experiments in physics and biology. Bernard's methodological conception of "comparative experimentation" emerged from a reflection on his own earlier experiments as well as from a critique of Comte and Mill. Roux's discussion of the causal method of research and of hypotheses combined a critique of Mill with an invocation of Whewell.

In the preface to the third edition of *Against Method*, Paul Feyerabend reassured his most zealous critics that " 'anything goes' is not a 'principle' I hold—I do not think that 'principles' can be used and fruitfully discussed outside the concrete research situation they are supposed to affect—but the terrified exclamation of a rationalist who takes a closer look at history."[6] An even closer look at history shows that there is a sense in which strategies and tools of experimentation transcend the concrete research situation. There were trends and strands, shifts and innovations on all levels of methods discourse—even advancement of sorts. Many strategies and tools of experimentation became more refined, results became more quantitative and precise, and experiments became more informative about subvisible worlds. Overall, however, the long-term history of methods discourse as it emerges from this book's analysis is not simply a steady progression toward more sophisticated or more improved methodologies of experimentation. Methodological advancement means increasing awareness of the obstacles and limitations of experimentation:

the unknown but suspected contingencies, the countless circumstances, the variations among living beings, the complexity of organic bodies, and the uncertainties related to techniques and instruments for the study of sub-visible phenomena. Methodological advancement means increasing efforts to develop strategies for managing and perhaps overcoming these challenges. Methodological advancement also includes the realization that the means through which we make sense of the world might forever remain precarious.

Acknowledgments

This project began to take shape during a wonderful year as a Mellon Fellow at the Institute for Advanced Study (Princeton). I also benefited from the Walter Hines Page Fellowship at the National Humanities Center (Research Triangle Park). I am very grateful to both institutions, and especially to their amazing librarians, for their support of my work. Karen Merikangas Darling at the University of Chicago Press has been a splendid editor, and I am enormously thankful for her constant support over so many years.

Early versions of some of the passages in this book were published in the following articles: "Trying Again and Again: Multiple Repetitions in Early Modern Reports of Experiments on Snake Bites," *Early Science and Medicine* 15 (2010): 567–617; "Scientists' Methods Accounts: S. Weir Mitchell's Research on the Venom of Poisonous Snakes," in *Integrating History and Philosophy of Science: Problems and Prospects*, edited by Tad Schmaltz and Seymour Mauskopf, (Dordrecht: Springer, 2011): 141–61; "The Significance of Re-Doing Experiments: A Contribution to Historically Informed Methodology," *Erkenntnis* 75 (2011): 325–47; and "What Does History Matter to Philosophy of Science? The Concept of Replication and the Methodology of Experiments," *Journal of the Philosophy of History* 5 (2012): 513–32.

I have presented portions of my work at a number of conferences and colloquia on both sides of the Atlantic. I thank the various audiences for their helpful and sometimes challenging feedback. I am especially grateful to the following people, who provided valuable advice as well as comments, critique, encouragement, suggestions, support, and workouts: Gar Allen, Meagan Allen, Theodore Arabatzis, Domenico Bertoloni Meli, Geoffrey Brown,

Jed Buchwald, Richard Burian, Jordi Cat, Hasok Chang, Klodian Çoko, Uljana Feest, Peter Finn (above all!), Yves Gingras, Sander Gliboff, Amit Hagar, Gary Hatfield, Don Howard, Jane Maienschein, William Newman, Heinrich von Staden, Friedrich Steinle, and Robert Wright. Warmest thanks to you all.

Notes

Introduction

1. Felice Fontana, *Treatise on the Venom of the Viper, on the American Poisons, and on the Cherry-Laurel, and Some Other Vegetable Poisons. To Which Are Annexed Observations of the Primitive Structure of the Animal Body, Different Experiments on the Reproduction of the Nerves, and a Description of a New Canal of the Eye*, trans. Joseph Skinner, vol. II (London: J. Murray, 1787), 73.

2. Allan Franklin's work on epistemological strategies of experimentation is perhaps the most comprehensive; see Allan Franklin, *The Neglect of Experiment* (Cambridge, UK: Cambridge University Press, 1986), chapter 6; and, with minor variations, Allan Franklin and Colin Howson, "It Probably Is a Valid Experimental Result: A Bayesian Approach to the Epistemology of Experiment," *Studies in History and Philosophy of Science* 19 (1988): 419–27; Allan Franklin, "The Epistemology of Experiment," in *The Uses of Experiment: Studies in the Natural Sciences*, ed. David Gooding, Trevor Pinch, and Simon Schaffer (Cambridge, UK: Cambridge University Press, 1989), 437–60; Allan Franklin, *Experiment, Right or Wrong* (Cambridge, UK: Cambridge University Press, 1990); Allan Franklin and Colin Howson, "Comment on 'the Structure of a Scientific Paper' by Frederick Suppe," *Philosophy of Science* 65 (1998): 411–16; and Allan Franklin, "Experiment in Physics," in *The Stanford Encyclopedia of Philosophy (Spring 2010 Edition)*, ed. Edward N. Zalta (2010), http://plato.stanford.edu/archives/spr2010/entries/physics-experiment/, as well as the latest edition, Allan Franklin and Slobodan Perovic, "Experiment in Physics," in *The Stanford Encyclopedia of Philosophy*, ed. Edward N. Zalta (2015), http://plato.stanford.edu/archives/sum2015/entries/physics-experiment/. The set includes the following strategies: Experimental checks and calibration; reproducing artifacts that are known in advance to be present; elimination of plausible sources of error and alternative explanations of the result; using the results themselves to argue for their validity; using an independently well-corroborated theory of the phenomena to explain the results; using an apparatus

based on a well-corroborated theory, and using statistical arguments. Franklin's set of strategies can be understood as a collection of procedures that might be—and that have been—carried out to gather empirical support for an experimental result. None of these procedures is necessary for the validation of experience, nor are these procedures meant to be together sufficient to guarantee a valid experimental result (Franklin, "The Epistemology of Experiment," 459). William Bechtel's epistemology of evidence applies more specifically to biology. It comprises three criteria for assessing whether instruments and techniques reliably provide information about the phenomena of interest: the new research techniques repeatedly yield a determinate structure, the results obtained with the new technology agree with results generated by other established techniques, and the results obtained with the new technology agree with accepted theoretical accounts; see William Bechtel, *Discovering Cell Mechanisms: The Creation of Modern Cell Biology* (Cambridge, UK: Cambridge University Press, 2006), 120, 127.

3. Franklin's point is that validation procedures for experiments are rational insofar as they invoke some subset of the strategies that feature in this set. (The rationality of these strategies can be independently demonstrated, as they can be embedded in a Bayesian approach.) To make this point, however, an analysis of the history, significance, or precise meaning of the epistemological strategies *for past scientists* is not required. In fact, there is the tacit implication that the validation strategies Franklin has identified have always been a part of methodological reasoning about experiments (at least since the early modern period). The same is true for Deborah Mayo's work on the experimental strategy of "severe testing." Mayo illustrates her error-statistical approach to experimentation with historical examples, such as Perrin's work on Brownian motion. But her approach is ahistorical in the sense that she seeks to show that all the strategies that experimenters should apply, and indeed that they have been applied, to assess the validity of experimental outcomes can be rendered as severe tests of the specific hypotheses involved in experimentation; see Deborah G. Mayo, *Error and the Growth of Experimental Knowledge* (Chicago: University of Chicago Press, 1996); see also Deborah G. Mayo and Aris Spanos, *Error and Inference: Recent Exchanges on Experimental Reasoning, Reliability, and the Objectivity and Rationality of Science* (Cambridge, UK: Cambridge University Press, 2010). Bechtel ties his epistemology of evidence to a specific period—the early twentieth century—but *begins* by presenting three intuitively convincing epistemological criteria for the assessment of new research instruments or techniques and then shows that his three criteria were indeed used by cytologists in the 1940s and 1950s in their work with the electron microscope and the ultracentrifuge.

4. The recent special issue of the *Journal of the History of Biology* on animal experimentation in the early modern period, for instance, drew together a number of detailed case studies of vivisection experiments but was not concerned with methodological discussions; see Domenico Bertoloni Meli and Anita Guerrini,

"Special Issue on Vivisection," *Journal of the History of Biology* 46 (2013). Recent work on exploratory experimentation in physics and biology has also focused on the practices of experimentation, not on the investigators' methodological discussions—see, e.g., Friedrich Steinle, *Explorative Experimente: Ampere, Faraday und die Ursprünge der Elektrodynamik* (Stuttgart, Germany: Franz Steiner Verlag, 2005); Friedrich Steinle, "Experiments in History and Philosophy of Science," *Perspectives on Science* 10 (2002): 408–32; Maureen O'Malley, "Exploratory Experimentation and Scientific Practice: Metagenomics and the Proteorhodopsin Case," *History and Philosophy of the Life Sciences* 29 (2007): 337–60; Kevin Elliott, "Varieties of Exploratory Experimentation in Nanotoxicology," *History and Philosophy of the Life Sciences* 29 (2007): 313–60—as has recent work on "big data"; see, e.g., Sabina Leonelli, "Special Issue on Data-Driven Research in the Biological and Biomedical Sciences," *Studies in History and Philosophy of Biological and Biomedical Sciences* 43 (2012).

5. See, among others, John Harley Warner, "Ideals of Science and Their Discontents in Late Nineteenth-Century American Medicine," *Isis* 82 (1991): 454–78; John Harley Warner, "The Fall and Rise of Professional Mystery: Epistemology, Authority, and the Emergence of Laboratory Medicine in Nineteenth-Century America," in *The Laboratory Revolution in Medicine*, ed. Andrew Cunningham and Perry Williams (Cambridge, UK: Cambridge University Press, 1992), 310–41; Nicholas Jardine, "The Laboratory Revolution in Medicine as Rhetorical and Aesthetic Accomplishment," in *The Laboratory Revolution in Medicine*, ed. Cunningham and Williams, 304–23; Harry M. Marks, *The Progress of Experiment: Science and Therapeutic Reform in the United States, 1900–1990* (Cambridge, UK: Cambridge University Press, 1997); and Michael Worboys, "Practice and the Science of Medicine in the Nineteenth Century," *Isis* 102 (2011): 109–15.

6. See, e.g., Martha Baldwin, "The Snakestone Experiments: An Early Modern Medical Debate," *Isis* 86 (1995): 394–418; André Ménez, *The Subtle Beast: Snakes, from Myth to Medicine* (London: Taylor and Francis, 2003); Ian A. Burney, *Poison, Detection, and the Victorian Imagination* (Manchester, UK: Manchester University Press, 2006); and Joel Levy, *Poison: A Social History* (Stroud, UK: The History Press, 2011).

7. Excellent examples are Findlay E. Russell, *Snake Venom Poisoning* (Philadelphia, PA: J. B. Lippincott Company, 1980); and Harry Greene, *Snakes: The Evolution of Mystery in Nature* (Berkeley: University of California Press, 1997).

8. Laurence Klauber, "Foreword," *Toxicon* 1 (1962): 1.

9. The situation was more or less the same in early modern anatomy, as Anita Guerrini's recent study of anatomical practice in Louis XIV's Paris has illustrated. She documents a practice of dissection that "is, to current sensibilities, a distasteful and even shocking phenomenon. There was no anesthesia or reliable means of reducing pain, and experimenters seem, to modern eyes, stunningly cavalier in their use of animals"; Anita Guerrini, *The Courtiers' Anatomists: Animals and Humans in*

Louis XIV's Paris (Chicago: University of Chicago Press, 2015), 3. Several investigators populating the present book seem to have had a similar attitude to the countless animals they poisoned with venom—and the maltreatment of countless snakes for the sake of science was not even considered worthy of comment.

10. For the history of ethical discussions about animal experiments, see Andreas-Holger Maehle, *Kritik und Verteidigung des Tierversuchs* (Stuttgart, Germany: Franz Steiner Verlag, 1990); for nineteenth-century debates about vivisection and the anti-vivisection movement, see Anita Guerrini, *Experimenting with Humans and Animals: From Galen to Animal Rights* (Baltimore, MD: Johns Hopkins University Press, 2003).

11. One of the major criticisms raised against philosophers and many historians of science in the 1960s and 1970s was that they focused their attention on "finished scientific achievements" (Kuhn) as they were presented in scientific publications. In the introduction to *The Structure of Scientific Revolutions,* Thomas Kuhn famously complained that a concept of science drawn from published sources, especially from textbooks, "is no more likely to fit the enterprise that produced them than an image of a national culture drawn from a tourist brochure or a language text"; Thomas S. Kuhn, *The Structure of Scientific Revolutions*, 2nd ed. (Chicago: University of Chicago Press, 1970), 1.

12. The classics are, of course, Bruno Latour and Steve Woolgar, *Laboratory Life: The Social Construction of Scientific Facts*, 2nd ed. (Beverly Hills, CA: Sage, 1986); and Bruno Latour, *Science in Action: How to Follow Scientists and Engineers through Society* (Cambridge, MA: Harvard University Press, 1987).

13. The term *literary technology* goes back to Shapin and Schaffer's influential studies of Robert Boyle's experiments: Steven Shapin, "Pump and Circumstance: Robert Boyle's Literary Technology," *Social Studies of Science* 14 (1984): 481–520; and Steven Shapin and Simon Schaffer, *Leviathan and the Air-Pump: Hobbes, Boyle, and the Experimental Life* (Princeton, NJ: Princeton University Press, 1985). Other well-known analyses of the translation of activities and events into the argument of a scientific paper are Karin Knorr-Cetina, *The Manufacture of Knowledge* (Oxford: Pergamon Press, 1981); Latour and Woolgar, *Laboratory Life: The Social Construction of Scientific Facts*; and Latour, *Science in Action*.

14. Shapin and Schaffer, *Leviathan and the Air-Pump*; Frederic L. Holmes, "Argument and Narrative in Scientific Writing," in *The Literary Structure of Scientific Argument: Historical Studies*, ed. Peter Dear (Philadelphia: University of Pennsylvania Press, 1991), 164–81; and Paula Findlen, "Controlling the Experiment: Rhetoric, Court Patronage and the Experimental Method of Francesco Redi," *History of Science* 31 (1993): 35–64.

15. See, e.g., the contributions to Richard Yeo and John A. Schuster, eds., *The Politics and Rhetorics of Scientific Method* (Dordrecht Reidel, 1986).

16. To avoid misunderstandings, I should stress that although the present work examines scientific argumentation, it is not a study of the rhetoric of science. The notion of rhetoric has had a complex and confusing history in science studies; for

an overview of the older literature, see Charles Bazerman, "Studies of Scientific Writing—E Pluribus Unum?" *4S Review* 3 (1985): 13–20. One reason for this confusion is that some scholars associate the notion of "rhetorical strategies" with textual features that are added to a piece of scientific writing to "sell" that content to a more or less gullible audience. Scholars in the field of rhetoric of science, by contrast, operate within the classical concept of rhetoric as a set of appeals to audiences. They analyze scientific texts with respect to pattern of reasoning (*logos*), the way in which the texts establish credibility for the author (*ethos*), and the manner in which the texts engage the audience (*pathos*). See, e.g., Lawrence Prelli, *A Rhetoric of Science: Inventing Scientific Discourse* (Columbia: University of South Carolina Press, 1989); and Marcello Pera and William R. Shea, eds., *Persuading Science: The Art of Scientific Rhetoric* (Canton, MA: Science History Publications, 1991). This study has a narrower focus, being concerned specifically with methods-related accounts as part of the reasoning in a scientific text.

17. Charles Bazerman, "Modern Evolution of the Experimental Report in Physics: Spectroscopic Articles in Physical Review, 1893–1980," *Social Studies of Science* 14 (1984): 165.

18. Knorr-Cetina, *The Manufacture of Knowledge*, 114, 115.

19. In the introduction to their collected volume *Histories of Scientific Observation*, Lorraine Daston and Elizabeth Lunbeck distinguish between observation as a practice, as a word, and as an "epistemic category"—that is, a concept that had become an object of reflection and analysis for philosophers and practitioners; see Lorraine Daston and Elizabeth Lunbeck, eds., *Histories of Scientific Observation* (Chicago: University of Chicago Press, 2011), 12. For them, the eighteenth century is the time when practice, word, and concept converged. Methods discourse comprises several such categories, but my point is that the two aspects of descriptions of and reflections on proper procedure should be kept separate. It remains to be seen how the practitioners brought them together (or not) in an experimental report.

20. Analyses of scientific literature have shown that in times of scientific controversy, methods sections expand and thus become informative resources not only for scientists but also for analysts of science; see, e.g., Greg Myers, *Writing Biology: Texts in the Social Construction of Scientific Knowledge* (Madison: University of Wisconsin Press, 1990), 121–25. So far it has not been sufficiently appreciated that the same applies to methodological discussions about good experimental practice. Methodological concerns that are raised in scientific writings indicate what was not taken for granted in the practice of research at that time.

21. The distinction among "long-term," "middle-term," and "short-term" methodological commitments or constraints is loosely based on Peter Galison's work on experimental cultures in high-energy physics. Galison distinguished three layers—he called them levels—of commitments that together shape experimental practices: long-term metaphysical commitments that transcend the rise and fall of particular beliefs about the nature of matter (Peter Galison, *How Experiments End* [Chicago: University of Chicago Press, 1987], 246), middle-term programmatic goals

(249), and short-term models and local theories (sections 5.2–4). The layering of methods discourse that I present is similar to Galison's framework by also distinguishing between more and less enduring commitments and constraints. I focus, however, on methodological issues, not on the theoretical background for experiments. In my framework, the programmatic commitment to experimentalism is a long-term commitment that transcends the rise and fall of particular metaphysical beliefs about nature.

Chapter One

1. See, e.g., Heinrich von Staden, *Herophilus: The Art of Medicine in Early Alexandria* (Cambridge, UK: Cambridge University Press, 1989); William R. Newman, *Atoms and Alchemy: Chymistry and the Experimental Origins of the Scientific Revolution* (Chicago: University of Chicago Press, 2006); and Robert Adam Leigh, *"On Theriac to Piso," Attributed to Galen: A Critical Edition with Translation and Commentary* (PhD diss., University of Exeter, 2013).

2. W. E. Knowles Middleton, *The Experimenters: A Study of the Accademia Del Cimento* (Baltimore, MD: Johns Hopkins University Press, 1971), 92. As Jay Tribby notes, at the Tuscan court, the difference between "experiment" and experience" was being collapsed. The contemporaneous dictionary of Tuscan language that was being put together at the Accademia della Crusca cross-references the terms; see Jay Tribby, "Cooking (with) Clio and Cleo: Eloquence and Experiment in Seventeenth-Century Florence," *Journal of the History of Ideas* 52 (1991): 425. In general, there does not seem to have been a systematic distinction among *experience, experiment*, and *observation* (or among the equivalent terms in Latin, French, or Italian). See also Paula Findlen, *Possessing Nature: Museums, Collecting, and Scientific Culture in Early Modern Italy* (Berkeley: University of California Press, 1994), 203–5.

3. See Roger Hahn, *The Anatomy of a Scientific Institution: The Paris Academy of Sciences, 1666–1803* (Berkeley: University of California Press, 1971), 5–19; quote on 15–16.

4. See Luciano Boschiero, *Experiment and Natural Philosophy in Seventeenth-Century Tuscany* (New York: Springer, 2007), chapter 4.

5. Middleton, *The Experimenters*, 88.

6. Ibid., 90.

7. Ibid., 91.

8. Tribby, "Cooking (with) Clio and Cleo," 418–20, 439. See also Paula Findlen's analysis of the illustration from Redi's *Generation of Insects*: Findlen, "Controlling the Experiment," 35–39.

9. I am grateful to Luciano Boschiero and Martin Eisner for their instructive discussions of the relevant passage of Dante's text.

10. Thomas Sprat, *The History of the Royal-Society of London, for the Improving of Natural Knowledge* (London: Printed by *T. R.* for *J. Martyn* at the *Bell* without *Temple-bar*, and *J. Allestry* at the *Rose* and *Crown* in *Duck-lane*, Printers to the *Royal Society*, 1667), 327.

11. In her brief overview of the "Empire of Observation, 1600–1800," Lorraine Daston points out that early modern astronomers were concerned not only with observing more things but also with a "desire to test—not just add to—and improve upon past observations, a process that paradoxically led them first to vaunt their own advances and later to cultivate an ever more scrupulous awareness of possible sources of error"; see Lorraine Daston, "The Empire of Observation, 1600–1800," in *Histories of Scientific Observation*, ed. Lorraine Daston and Elizabeth Lunbeck (Chicago: University of Chicago Press, 2011), 93.

12. Guerrini, *The Courtiers' Anatomists*, chapter 3, 99–105.

13. Galileo Galilei, *Dialogues Concerning Two New Sciences* (New York: Macmillan, 1914), 179.

14. Boyle had a long and productive life, and his written works covered a wide range of topics, from experimental studies to metaphysics and theological reflections; see Michael Hunter, ed., *Robert Boyle Reconsidered* (Cambridge, UK: Cambridge University Press, 1994). Here I am mostly concerned with Boyle's early experimental writings.

15. I will return to Bacon's discussion of "new methods" in later chapters.

16. Robert Boyle, "The Second Essay, of Unsucceeding Experiments," in *The Works of Robert Boyle*, ed. Michael Hunter and Edward B. Davies (London: Pickering & Chatto, 1999), 61.

17. "Physiology" is meant in the widest sense of the study of *physis*—it includes anatomy, botany, and chemistry.

18. Boyle, "The First Essay, of the Unsuccessfulness of Experiments," in *The Works of Robert Boyle*, ed. Hunter and Davies, 38.

19. "The Second Essay, of Unsucceeding Experiments," 57; see Rose-Mary Sargent, *The Diffident Naturalist: Robert Boyle and the Philosophy of Experiment* (Chicago: University of Chicago Press, 1995), chapter 3, for a more detailed account of this essay.

20. Boyle, "The Second Essay, of Unsucceeding Experiments," 57.

21. Ibid., 69; "The First Essay, of the Unsuccessfulness of Experiments," 37.

22. Boyle, "The Second Essay, of Unsucceeding Experiments," 77.

23. See ibid., 82.

24. Peter Dear, "Narratives, Anecdotes, and Experiments: Turning Experience into Science in the Seventeenth Century," in *The Literary Structure of Scientific Argument: Historical Studies*, ed. Peter Dear (Philadelphia: University of Pennsylvania Press, 1991), 138.

25. Critics of Shapin and Schaffer's analysis have argued that Boyle did not consciously and purposely craft these literary tools to persuade his readers—and

especially his opponents—that his experiments were trustworthy; see, e.g., Larry Principe, "Virtuous Romance and Romantic Virtuoso: The Shaping of Robert Boyle's Literary Style," *Journal of the History of Ideas* 56 (1995): 377–97. I will not pursue this debate here, as it is not my main goal to add to Boyle scholarship. Rather, I seek to put methodological pronouncements such as Boyle's in broader historical context and perspective.

26. Boyle was quite explicit about the danger of false beliefs, saying that Hippocrates's erroneous belief that "teeming women" should not be let blood would give rise to false treatment; see Robert Boyle, "Essay I. Containing Some Particulars Tending to Shew the Usefulness of Natural Philosophy to the Physiological Part of Physick," in *The Works of Robert Boyle*, ed. Hunter and Davies, 296.

27. The experimenter might choose, for example, "to open the Body or Limbs, to make Ligatures strong or weak on the vessells, or other inward parts, as occasion shall require, to leave them there as long as he pleaseth, to prick, or apply sharp liquors to any nervous or membranous part, and whenever he thinks convenient, to dissect the Animall again, to observe what change his Experiment hath produc'd there: such a Liberty, I say, which is not to be taken in humane bodies, may in some case either confirme or confute the Theories proposd, and so put an end to divers Pathologicall Controversies, and perhaps too occasion the Discovery of the true genuine causes of Phenomena disputed of, or of others really as abstruse"; Boyle, "Essay II," 323–24.

28. Ibid., 326.

29. "Observations out of Mr B. Essay of Turning Poisons into Medicins," in *The Works of Robert Boyle*, ed. Hunter and Davis, 252.

30. Ibid., 251.

31. On van Helmont's conception of disease, see Walter Pagel, "Van Helmont's Concept of Disease—to Be or Not to Be? The Influence of Paracelsus," *Bulletin of the History of Medicine* 46 (1972): 419–54. On the transmission of Helmontian theories in seventeenth-century England, see Antonio Clericuzio, "From Van Helmont to Boyle: A Study of the Transmission of Helmontian Chemical and Medical Theories in Seventeenth-Century England," *British Journal for the History of Science* 26 (1993): 303–34.

32. See Holmes, "Argument and Narrative in Scientific Writing," 165–66.

33. Holmes, "Argument and Narrative in Scientific Writing," 166; see also Frederic L. Holmes, "Scientific Writing and Scientific Discovery," *Isis* 78 (1987), 220–35.

34. Holmes, "Argument and Narrative in Scientific Writing," 164.

35. Ibid., 171.

Chapter Two

1. For secondary sources on the Accademia del Cimento, see, among others, Mario Biagioli, "Etiquette, Interdependence, and Sociability in Seventeenth-Century

Science," *Critical Inquiry* 22 (1996): 193–238;; Marco Beretta, "At the Source of Western Science: The Organization of Experimentalism at the Accademia Del Cimento (1657–1667)," *Notes and Records of the Royal Society of London* 54 (2000): 131–51; and Boschiero, *Experiment and Natural Philosophy in Seventeenth-Century Tuscany.*

2. Luciano Boschiero maintains that it is misleading to pay too much attention to the natural philosophers' talk about experimentation because it does not reflect the real points of concern in the debates; see Luciano Boschiero, "Natural Philosophizing inside the Late Seventeenth-Century Tuscan Court," *British Journal for the History of Science* 35 (2002): 388. Building on earlier work by Paolo Galluzi, Boschiero argues that the kind of experiments performed at the Cimento reflect the natural philosophical interests of the academicians. Their work should thus be understood as a continuation of previous natural philosophical concerns. I agree with the latter argument, but I think that Boschiero goes too far in his dismissal of what he calls "experimental method rhetoric." That the academicians used experiments as vehicles to support or refute broader natural philosophies and to demonstrate the respective merits of Aristotelian and mechanical philosophy does not imply that they were unconcerned about the conditions of successful experimentation.

3. See Tribby, "Cooking (with) Clio and Cleo"; Findlen, "Controlling the Experiment" and *Possessing Nature*; and Baldwin, "The Snakestone Experiments."

4. On theriac, see G. Watson, *Theriac and Mithridatium: A Study of Therapeutics* (London: Wellcome Historical Medical Library, 1966); and, more recently, Christiane Nockels Fabbri, "Treating Medieval Plague: The Wonderful Virtues of Theriac," *Early Science and Medicine* 12 (2007): 247–83.

5. Redi was also involved in a controversy with the Jesuit Athanasius Kircher on the efficacy of snake stones as a remedy for snakebite; see Baldwin, "The Snakestone Experiments."

6. Tribby, "Cooking (with) Clio and Cleo"; and Findlen, "Controlling the Experiment."

7. Findlen, "Controlling the Experiment," 40.

8. Peter K. Knoefel, *Francesco Redi on Vipers* (Leiden: Brill, 1988), 6.

9. Ibid., 7.

10. Ibid., 3, 23.

11. Redi frequently referred to, or used, passages from Dante; see, e.g., Francesco Redi, *Experiments on the Generation of Insects* (Chicago: Open Court, 1909), 20. Knoefel's illuminating and very helpful annotations to his translation of Redi's letters on viper venom point to a number of passages in which Redi quoted or paraphrased Dante.

12. Knoefel, *Francesco Redi on Vipers*, 3.

13. Ibid., 8–9; see Tribby, "Cooking (with) Clio and Cleo," 430–32.

14. Tribby, "Cooking (with) Clio and Cleo," 438, 439.

15. Leigh, *"On Theriac to Piso,"* 59. In Galen's text, the purpose of the second trial was to test the quality of the antidote—if the person who took it and was

not purged by the purgative, the theriac must be of high quality. The design of the trial was very similar to the one that Redi described. Redi did not refer to Galen in the context of this passage, but he mentioned Galen in other parts of his letter (along with Avicenna's work on drugs), so he clearly considered Galen's work as an important context for the experiments he described.

16. Knoefel, *Francesco Redi on Vipers*, 6. "Another test" is Knoefel's translation of *altre prove*. I quote the Italian text following the second edition of Redi's *Opere*; see Francesco Redi, "Osservazioni Intorno Alle Vipere," in *Opere di Francesco Redi* (Naples: A spese di R. Gessari, nella stamperia di A. Carfora, 1740–1741); and Francesco Redi, "Lettera Sopra Alcune Opposizioni Fatte Alle Sue Osservazioni Intorno Alle Vipere," in *Opere di Francesco Redi* (Naples: A spese di R. Gessari, nella stamperia di A. Carfora, 1740–1741).

17. Knoefel, *Francesco Redi on Vipers*, 3. The Italian original reads *iterata, e reiterata esperienza*.

18. Baldwin, "The Snakestone Experiments," 404.

19. In the debates about snake venom, the replication of other people's experiments played hardly any role at first. It became somewhat more important as the dispute with Charas unfolded, when—as readers of *Leviathan and the Air-pump* might expect—failed attempts at replication were at issue.

20. Knoefel, *Francesco Redi on Vipers*, 30. The Italian expression is *per esperienze molte vole fatta*.

21. Ibid., 6.

22. Findlen, "Controlling the Experiment," 55.

23. "Mattiuoli affirms that oak-galls produce spiders as well as worms and flies; he also says that all galls, which have not been pierced, contain one of these three kinds of small animals, from the nature of which he deduces are terrible prognostic, saying that flies in the falls indicate war to occur in that year; if worms are found, the harvest will be poor; if spiders, there will be a pestilence. Father Fabri laughs at this prognostic; as for myself, I could easily cite many experiments to confute Mattiuoli, for in the space of three or four years I have opened more than twenty thousand galls, and have never found a single spider inside, but only flies and gnats and worms, according to the season in which I opened them; nor was there meanwhile war or famine in Tuscany. It is true, to be sure, that I sometimes found a little spider in a gall, but always in an open one, into which it had probably crept for hiding"; Redi, *Experiments on the Generation of Insects*, 70–71. The larger context of this passage is Redi's endeavor to find out whether insects could be generated from nonliving matter. For details of this project, see Emily C. Parke, "Flies from Meat and Wasps from Trees: Reevaluating Francesco Redi's Spontaneous Generation Experiments," *Studies in History and Philosophy of Biological and Biomedical Sciences* 45 (2014): 34–42.

24. Findlen, "Controlling the Experiment," 55.

25. Dear, "Narratives, Anecdotes, and Experiments," 138.

26. Knoefel, *Francesco Redi on Vipers*, 8.

27. Ibid., 6.

28. Ibid., 6.

29. Marco Aurelio Severino was the author of the work *Vipera Pythia. Id est de viperae natura, veneno, medicina demonstrationes et experimenta nova* (Padua, Italy: 1651).

30. Knoefel, *Francesco Redi on Vipers*, 9.

31. Ibid., 9–10.

32. Ibid.

33. Ibid., 21.

34. Moyse Charas, *New Experiments Upon Vipers. Containing Also an Exact Description of All the Parts of a Viper, the Seat of Its Poyson, and the Several Effects Thereof, Together with the Exquisite Remedies, That by the Skilful May Be Drawn from Vipers, as Well for the Cure of Their Bitings, as for That of Other Maladies. Originally Written in French. Now Rendred English.* (London: Printed by T. N. for J. Martyn, 1670). The quotes from Charas's work are drawn from this translation.

35. Another disagreement between the two researchers concerned the anatomical structure and location of the salivary glands of the viper. I will not examine this dispute here.

36. Charas, *New Experiments Upon Vipers*, preface, n.p.

37. In his later work, Charas explicitly referred to van Helmont's theory of disease.

38. Charas also spent some time in England, but only after his dispute with Redi. For biographical details, see Patrizia Catellani and Renzo Console, "Moyse Charas, Francesco Redi, the Viper and the Royal Society of London," *Pharmaceutical Historian: Newsletter of the British Society for the History of Pharmacy* (2004): 2–10; for information about his works, see also Marie Phisalix, "Moyse Charas et les vipères au Jardin du Roy," *Archives du Muséum d'Histoire Naturelle* 12 (1935): 469–72.

39. Charas, *New Experiments Upon Vipers*, 106.

40. Ibid., 103.

41. Ibid., 106.

42. Ibid., 94.

43. Ibid., 109–10.

44. Ibid., 107–8.

45. Ibid., 133–34.

46. Ibid., 98.

47. Charas explained in detail: "We judged, that the vexed spirits unable to penetrate into the body, defended by the Theriaque, had wrought upon the outward part, and round about the place bitten, where they had coagulated the bloud, and caused the lividness; whereas the like spirits, having met with no resistance in the other Pigeon, had gained and wrought upon the inner parts, having left and as

'twere despised the place, at which they were entred. We also wondred not, that the Theriaque, which had vigorously repulsed the Spirits introduc'd by the first bite, could not resist the latter but for half an hour, and that at last it was forced to yield, in regard that the number of the enemies was great, and being weakn'd by the conflict, it had but now endured, had not force enough to bear up against the new assault of the latter"; Charas, *New Experiments Upon Vipers*, 98–99.

48. Charas, *New Experiments Upon Vipers*, 115.

49. Ibid., 116.

50. Ibid., 108.

51. Ibid., 117.

52. See the *Novum Organum*, Book II, aphorism 36. Crucial instances "reveal that the fellowship of one of the natures with the nature under investigation is constant and indissoluble, while that of the other is fitful and occasional. This ends the search as the former nature is taken as the cause and the other dismissed and rejected. Thus instances of this kind give the greatest light and the greatest authority; so that a course of interpretation sometimes ends in them and is completed through them. Sometimes crucial instances simply occur, being four among instances long familiar, but for the most part they are new and deliberately and specifically devised and applied; it takes keen and constant diligence to unearth them"; Francis Bacon, *The New Organon* (Cambridge, UK: Cambridge University Press, 2000), 159. Modern readers might think of Mill's method of difference as a formal representation of Charas's procedure, but this strikes me as misleading. I will discuss Mill's method of inquiry in a later part of this book—namely, in the context of nineteenth-century venom research (see especially chapters 7 and 8).

53. Domenico Bertoloni Meli describes a number of key episodes of what he calls "parallel trials" in seventeenth-century science, both from the physical and the life sciences. His examples include the famous Puy-de-Dôme experiment, Marcello Malpighi's experiments on plant growth, and two projects by Redi: the experiments on spontaneous generation and experiments on the properties of styptic water; see Domenico Bertoloni Meli, "A Lofty Mountain, Putrefying Flesh, Styptic Water, and Germinating Seeds," in *The Accademia Del Cimento and Its European Context*, ed. Marco Beretta, Antonio Clericuzio, and Larry Principe (Sagamore Beach, MA: Science History Publications, 2009), 121–34.

54. Quoted in Michael McVaugh, "The 'Experience-Based Medicine' of the Thirteenth Century," in *Evidence and Interpretation in Studies on Early Science and Medicine*, ed. Edith Sylla and William R. Newman (Leiden: Brill, 2009), 112. In his essay on viper venom, Charas did not explicitly refer to Gordon's work, but later he did refer to Gordon in his *Pharmacopoeea*; Moyse Charas, *The Royal Pharmacopoeea, Galenical and Chymical According to the Practice of the Most Eminent and Learned Physitians of France: And Publish'd with Their Several Approbations* (London: Printed for John Starkey at the Miter within Temple-Bar, and Moses Pitt at the Angel in St. Pauls Church-Yard, 1678).

Chapter Three

1. According to René Taton, Bourdelot was an important figure in the Parisian scientific scene who provided not only material assistance to experimenters but also a space in which French and foreign investigators could meet. He encouraged intellectual exchanges between French and Italian scholars; see René Taton, "Bourdelot, Pierre Michon," in *Dictionary of Scientific Biography*, ed. Charles C. Gillispie (New York: Charles Scribner's Sons, 1970–1980).

2. Knoefel, *Francesco Redi on Vipers*, 40.

3. Ibid., 47–48. "Many many" is Knoefel's translation of *molte, e molte*.

4. Ibid., 45. The Italian expression is *provate, e riprovate molte, e molt'altre volte*.

5. See ibid., 41.

6. Ibid.

7. Ibid., 46.

8. Ibid.

9. Ibid., 49.

10. Ibid., 41.

11. Ibid., 45.

12. Cf. ibid., 15, 29.

13. Ibid., 46.

14. My quotes are from the English translation of 1673.

15. Moyse Charas, *A Continuation of the New Experiments Concerning Vipers* (London: John Murray, 1673), 44.

16. Ibid., 44.

17. Ibid.

18. Ibid., 46, emphasis added.

19. Ibid., 57.

20. Ibid., 47–48.

21. Ibid., 48.

22. Ibid., 50.

23. Ibid.

24. Ibid., 51.

25. Ibid., 53.

26. Ibid., 54.

27. Ibid., 63.

28. Ibid., 106–7.

29. Ibid., 66.

30. Ibid., 83.

31. "Osservationi Intorno Alle Vipere, fatte da Franceso Redi, in Firenze," *Journal des Sçavants* 2 (1666): 9–12; "Francesco Redi, Some Observations on Vipers," *Philosophical Transactions of the Royal Society of London* (1665–1666): 160–62.

32. "Redi. Some Observations on Vipers," 160.

33. Ibid., 162.

34. "Review: An Accompt of Some Books," *Philosophical Transactions of the Royal Society of London* 4 (1669): 1091, 1092.

35. On de la Chambre, see Markus Wild, "Marin Cureau De La Chambre on the Natural Cognition of the Vegetative Soul: An Early Modern Theory of Instinct," *Vivarium* 46 (2008): 443–61.

36. Thomas Platt, "An Extract of a Letter Written to the Publisher by Mr. Thomas Platt, from Florence, August 6, 1672. Concerning Some Experiments, There Made Upon Vipers, since Mons. Charas His Reply to the Letter Written by Signor Francesco Redi to Monsieur Bourdelot and Monsieur Morus," *Philosophical Transactions of the Royal Society of London* 7 (1672): 5061.

37. Ibid., 5062.

38. On Vincenzo Viviani, see Boschiero, *Experiment and Natural Philosophy in Seventeenth-Century Tuscany*, chapter 2.

39. Platt, "An Extract of a Letter Written to the Publisher by Mr. Thomas Platt," 5062–63.

40. Ibid., 5063–64.

41. Ibid., 5066.

42. "Review: An Account of Some Books," *Philosophical Transactions of the Royal Society of London* 7 (1672): 4075.

43. Edward Tyson, "Vipera Caudi-Sona Americana, or the Anatomy of a Rattle-Snake, Dissected at the Repository of the Royal Society in January 1682/3," *Philosophical Transactions of the Royal Society of London* 13 (1683): 48.

Chapter Four

1. W. Coleman, "Mechanical Philosophy and Hypothetical Physiology," in *The Annus Mirabilis of Sir Isaac Newton, 1666–1966*, ed. Robert Palter (Cambridge, MA: MIT Press, 1970), 323.

2. Ibid., 325.

3. For biographical information about Mead, see Richard H. Meade, *In the Sunshine of Life; a Biography of Dr. Richard Mead, 1673–1754* (Philadelphia, PA: Dorrance, 1974); and Matthew Matty, *Authentic Memoirs of the Life of Richard Mead, M.D.* (London: Printed for J. Whiston, and B. White in Fleet-Street, 1755).

4. See Anita Guerrini, "Archibald Pitcairne and Newtonian Medicine," *Medical History* 31 (1987): 70–83.

5. Mead's biographer, Arnold Zuckerman, notes that Mead's interest in poisons dated back to his days as a student in Leiden under Paul Hermann; see Arnold Zuckerman, *Dr. Richard Mead (1673–1754): A Biographical Study* (PhD diss., University of Illinois, 1965), 33.

6. Richard Mead, *A Mechanical Account of Poisons* (London: 1702), preface, n.p. The title of the fifth essay is "Of Venomous Exhalations from the Earth, Poisonous Airs and Waters." Mead included only a few remarks on contagion, but they suggest that he thought of contagion as active particles that were somehow carried through the atmosphere.

7. Ibid., preface, n.p.

8. Friedrich Steinle, "Newton, Newtonianism, and the Roles of Experiment," in *The Reception of Isaac Newton in Europe*, ed. Helmut Pulte and Scott Mandelbrote (Continuum Publishing Corporation, forthcoming).

9. Mead, *A Mechanical Account of Poisons*, 9–10.

10. Ibid., 11; figure 19 on the accompanying plate shows the sharp-edged crystal.

11. Ibid., 10.

12. Ibid., 17.

13. Ibid., 22.

14. Ibid., 23.

15. Ibid., 29–30.

16. One of the most enthusiastic advocates of such an all-encompassing notion of fermentation was the Oxford physiologist Thomas Willis; see Allen G. Debus, "Rise and Fall of Chemical Physiology in the Seventeenth Century," *Memorias da Academia das Ciencias de Lisboa, (Classe de Ciencias)* 36 (1996): 40; and Ku-Ming (Kevin) Chang, "Fermentation, Phlogiston and Matter Theory: Chemistry and Natural Philosophy in Georg Ernst Stahl's 'Zymotechnia Fundamentalis,'" *Early Science and Medicine* 7 (2002): 33–35. Highlighting the widespread concern with fermentation in the second half of the seventeenth century, Kevin Chang even speaks of a "fermentational program" (33–37).

17. See Debus, "Rise and Fall of Chemical Physiology in the Seventeenth Century."

18. Mead, *A Mechanical Account of Poisons*, 12.

19. Ibid., 13.

20. Ibid., 15.

21. Johann Bernoulli, "Dissertation on the Mechanics of Effervescence and Fermentation," *Transactions of the American Philosophical Society* 87 (1997): 55.

22. Mead, *A Mechanical Account of Poisons*, 15.

23. Ibid., 15–16.

24. Coleman, "Mechanical Philosophy and Hypothetical Physiology," 327.

25. Anita Guerrini, "Ether Madness: Newtonianism, Religion, and Insanity in Eighteenth-Century England," in *Action and Reaction: Proceedings of a Symposion to Commemorate the Tercenary of Newton's Principia*, ed. Paul Theerman and Adele Seeff (London: Associated University Presses, 1993), 236.

26. With this work, Mead became entangled in a vicious controversy between his close friend John Freind and the Oxford physician John Woodward about the use of purgatives in the treatment of smallpox.

27. Cf. for the following Arnold Zuckerman, "Plague and Contagionism in Eighteenth-Century England: The Role of Richard Mead," *Bulletin of the History of Medicine* 78 (2004): 273–308.

28. Richard Mead, *A Short Discourse Concerning Pestilential Contagion*, 8th ed. (London: 1722), 38.

29. Ibid., 39. This is a fourth kind of Newtonianism—theoretical rather than methodological.

30. Mead, *A Short Discourse Concerning Pestilential Contagion*, 9th ed. (London: 1744), 48.

31. Ibid., 48.

32. Ibid., 50.

33. Mead, *A Mechanical Account of Poisons*, 4th ed. (London: J. Brindley, 1745), iv–v.

34. Ibid., xxxvii.

35. Ibid., 16–18.

36. Ibid., 20.

37. Ibid., 18.

38. Ibid., 19–20.

39. Ibid., 21.

40. Ibid., 22.

41. Matty, *Authentic Memoirs of the Life of Richard Mead, M.D.*, 11–12.

Chapter Five

1. Klauber, "Foreword," 1. As we will see, this description is highly problematic. I mention it here only to illustrate that Fontana's methods were considered striking and worthy of mention even into the mid-twentieth century.

2. For details on Fontana's life, education, and works, see the rich intellectual biography by Peter Knoefel, *Felice Fontana: Life and Works* (Trento, Italy: Società di Studi Trentini di Scienze Storiche, 1984). For more information about Fontana's involvement with the Florentine Museum of Natural History *La Specola* and specifically with the collection of anatomical wax models, see Anna Maerker, *Model Experts: Wax Anatomies and Enlightenment in Florence and Vienna, 1775–1815* (Manchester, UK: Manchester University Press, 2011).

3. Unlike Mead, Fontana did not provide a general account of poisons. The parts of his treatise that deal with poisons other than viper venom are not elaborated. But he did draw analogies between the working of animal and vegetable poisons, and he followed very similar principles in his experiments.

4. For details on the publication see Knoefel, *Felice Fontana: Life and Works*, 267–68.

5. Felice Fontana, *Treatise on the Venom of the Viper*, I, 119–30.

6. In the following, I will refer to the Italian and the French treatise as the two "books" or two "treatises" on viper venom. The "two volumes" are the two separate printed works. In other words, Fontana's second *book* on viper venom is printed partly in Volume I and partly in Volume II.

7. The quotes are from the English translation of 1787.

8. On *Anatomia animata*, see Shirley A. Roe, "Anatomia Animata: The Newtonian Physiology of Albrecht von Haller," in *Transformation and Tradition in the Sciences: Essays in Honor of I. Bernard Cohen*, ed. Everett Mendelsohn (Cambridge, UK: Cambridge University Press, 1984), 273–300. Fontana's very first work was on Haller's notion of irritability.

9. Knoefel notes that Fontana was introduced to Haller's work by his teacher Leopoldo Caldani, who also conducted research on irritability. Caldani corresponded with Haller and sent him reports of Fontana's research; see Knoefel, *Felice Fontana: Life and Works*, 14–15.

10. Fontana, *Treatise on the Venom of the Viper*, I, 1–2. The Italian text uses the words *osservationi* and *esperienze* in a broad sense.

11. Albrecht von Haller, *A Dissertation on the Sensible and Irritable Parts of Animals* (London: Printed for J. Nourse at the Lamb opposite Katherine-street in the Strand, 1755), 2.

12. Fontana, *Treatise on the Venom of the Viper*, I, 74.

13. Ibid., 2.

14. In his later essay on argument and narrative in scientific writing, Holmes's example for "argument" is Denis Dodard's memoir on chemical experiments performed at the Paris Académie des Sciences in the late seventeenth century. Dodard discussed "in logical rather than chronological sequence the various methods that the Academicians had learned to employ, over the years, to identify the products of their distillation analyses"; see Holmes, "Argument and Narrative in Scientific Writing," 168–69.

15. Holmes, "Scientific Writing and Scientific Discovery," 231.

16. See, e.g., Evan Ragland, "Chymistry and Taste in the Seventeenth Century: Franciscus Dele Boë Sylvius as a Chymical Physician between Galenism and Cartesianism," *Ambix* 59 (2012): 1–21.

17. Fontana, *Treatise on the Venom of the Viper*, I, 55 57.

18. Ibid., 56.

19. Ibid., 43.

20. This is a shift of emphasis, not a change of approach. Redi also varied his experiments, but the emphasis in his case was on many repetitions.

21. Fontana, *Treatise on the Venom of the Viper*, I, 48.

22. Ibid., 51, 52.

23. Ibid., 53. Fontana rarely mentioned witnesses in his book. Generally speaking, Fontana offered much more information about how he had established proper procedures than about the guarantors of proper procedure.

24. Ibid., 54.

25. For a detailed account of Haller's experimental work on irritability, see Hubert Steinke, *Irritating Experiments: Haller's Concept and the European Controversy on Irritability and Sensibility, 1750–90* (Amsterdam: Rodopi, 2005).

26. Haller proposed to call "that part of the human body irritable, which becomes shorter upon being touched, very irritable if it contracts upon a slight touch, and the contrary if by a violent touch it contracts but little." By contrast, a "sensible" part of the human body is one that "upon being touched, transmits the impression of it to the soul." Because it was not clear whether animals had a soul (or because he did not want to take sides on this sensitive issue), Haller ultimately based the distinction on the behavior of the animal as it could be observed in experimental contexts. He called "those parts sensible, the Irritation of which occasions evident signs of pain and disquiet in the animal. On the contrary, I call that insensible, which being burnt, tore, pricked or cut till it is quite destroyed, occasions no sign of pain nor convulsion, nor any change in the situation of the body. For it is very well known, that an animal, when it is in pain, endeavours to remove the part that suffers from the cause that hurts it; pulls back the leg if it is hurt, shakes the skin if it is pricked, and gives over evident signs by which we know that it suffers"; Haller, *A Dissertation on the Sensible and Irritable Parts of Animals*, 4–5.

27. Mead considered a similar group of substances.

28. Fontana, *Treatise on the Venom of the Viper*, I, 97.

29. Ibid., 100.

30. Haller, *A Dissertation on the Sensible and Irritable Parts of Animals*, 2, 3.

31. Fontana, *Treatise on the Venom of the Viper*, I, 115.

Chapter Six

1. Knoefel, *Felice Fontana: Life and Works*, 269.

2. Fontana, *Treatise on the Venom of the Viper*, I, 125. I modified the wording of the first of the three principal methods, because the eighteenth-century translation is misleading. The French original reads: "Il est presque impossible qu'en répétant un si grand nombre de fois les expériences, on ne rencontre les cas fortuits, qui peuvent les varier, et que le résultat final de tant d'expériences ne soit certain et constant"; see Felice Fontana, *Traité Sur Le Vénin De La Vipère, Sur Les Poisons Americains, Sur Le Laurier-Cerise, Et Sur Quelques Autres Poisons Végétaux*, vol. I (Florence: 1781), 102. The English translation in the 1787 version reads: "It is almost impossible, in repeating them so many times, that fortuitous cases do not occur to vary them, and that the final result of so many of them is not certain and constant."

3. In the following, I draw on Jan Golinski's account of Fontana's work on eudiometry; see Jan Golinski, *Science as Public Culture: Chemistry and Enlightenment in Britain, 1760–1820* (Cambridge, UK: Cambridge University Press, 1992), 117–128.

4. Felice Fontana, "Account of the Airs Extracted from Different Kinds of Waters; with Thoughts on the Salubrity of Air at Different Places," *Philosophical Transactions of the Royal Society of London* 69 (1779): 445.

5. Ibid.

6. See Golinski, *Science as Public Culture*, 123.

7. See Friedrich Steinle, "Entering New Fields: Exploratory Uses of Experimentation," *Philosophy of Science Supplement* 64 (1997): S65–S74; see also Steinle, "Experiments in History and Philosophy of Science"; and Steinle, *Explorative Experimente*. Also in 1997, Richard Burian introduced the term *exploratory experimentation* in an analysis of Jean Brachet's cytochemical studies; see Richard M. Burian, "Exploratory Experimentation and the Role of Histochemical Techniques in the Work of Jean Brachet, 1938–1952," *History and Philosophy of the Life Sciences* 19 (1997): 27–45. Burian, however, used the term in a different way; see Jutta Schickore, "'Exploratory Experimentation' as a Probe into the Relation between Historiography and Philosophy of Science," *Studies in History and Philosophy of Science* 55 (2016): 20–26.

8. Melvin Earles's study of Fontana's treatise provides a concise outline of Fontana's account; see Melvin P. Earles, "The Experimental Investigations of Viper Venom by Felice Fontana (1730–1805)," *Annals of Science* 16 (1960): 255–68; see also the outline in Knoefel, *Felice Fontana: Life and Works*, 273–82.

9. Fontana experimented on sparrows, pigeons, fowls, guinea pigs, rabbits, cats, dogs, and frogs and on skin, adipose membrane, muscles, breast, belly, intestines, liver, ears, pericranium, bones, dura mater and brain, marrow of the bones, cornea, comb, gills, nose, neck, and tendons. He considered the empirical basis broad enough to make generalizations to all animals.

10. This is Newtonianism modeled after the *Opticks*.

11. Fontana, *Treatise on the Venom of the Viper*, I, 271–72.

12. Ibid., 131.

13. Ibid., 117.

14. In the original texts, Redi used the term *accidenti*, whereas Fontana used the term *circonstances*.

15. Fontana, *Treatise on the Venom of the Viper*, I, 137–38.

16. Ibid., 138.

17. Ibid., 139.

18. In his work on exploratory experimentation, Steinle has pointed out that the question of when "stability" is achieved is ultimately a pragmatic question. We see this quite clearly in Fontana's case. Of course, it would still be possible to say that death "between five and eight minutes" does not indicate a stable experimental arrangement. Apparently, for Fontana, the result was uniform enough to stop experimenting.

19. Fontana, *Treatise on the Venom of the Viper*, I, 277.

20. For practices of dealing with discrepancies before the concept of average became common, see Jed Z. Buchwald, "Discrepant Measurements and Experimental

Knowledge in the Early Modern Era," *Archive for the History of the Exact Sciences* 60 (2006): 565–649.

21. Fontana, *Treatise on the Venom of the Viper*, I, 285.

22. Ibid., 286.

23. See Haller, *A Dissertation on the Sensible and Irritable Parts of Animals*, 8–12; quotes from pages 9 and 10. As Hubert Steinke shows in his book on Haller's studies of irritability and sensibility, the question of whether tendons were not very or not at all sensible, and the implications for surgical practice, were fiercely debated in the late eighteenth century. See especially Hubert Steinke, *Irritating Experiments*, 108, 133–34, 238–40.

24. Fontana, *Treatise on the Venom of the Viper*, I, 226.

25. Fontana left it to the reader to draw this conclusion. He did not mention Bacon at this point, but the way in which he let the discussion culminate in a decisive comparative trial reminds one very much of Bacon's discussion of the fourteenth privileged instance, the crucial instance. As we saw in chapter 2, Bacon stated that crucial instances "give the greatest light and the greatest authority; so that a course of interpretation sometimes ends in them and is completed through them" (Bacon, *The New Organon*, 159).

26. Fontana, *Treatise on the Venom of the Viper*, I, 232–33.

27. Ibid., 169.

28. I am thinking specifically of John Stuart Mill's systematic discussion of the methods of experimental reasoning, which I will discuss in more detail in chapter 7. We already saw that Bacon's concern with crucial instances was primarily about the function of experiments as decisive, not about their structure.

29. The same considerations apply to James Lind's study of scurvy, which is also often hailed as the first "controlled trial." This is equally problematic. Lind aimed to establish the cause of scurvy and experimented with different diets. In his famous experiment, he compared "groups" (in fact, six pairs) of sailors who suffered from scurvy. The aim of the experiment was to establish the effects of certain foods: the patients' diets were identical except that each pair received one additional food or drink: cider, sulfuric acid, vinegar, sea water, orange and lemon, and a concoction of garlic, mustard seed, radish, and Balsam of Peru (a liquid derived from a tree); see James Lind, *Treatise on the Scurvy*, 2nd ed. (London: 1757), 149–53. In contrast, the numerical comparison of how two groups of patients respond to treatments, which was also undertaken at that time, was couched in statistical terms and clearly originated in social statistics. For arithmetic observations in Britain, see, e.g., Ulrich Tröhler, " 'To Improve the Evidence of Medicine': Arithmetic Observation in Clinical Medicine in the Eighteenth and Early Nineteenth Centuries," *History and Philosophy of the Life Sciences* 10 (1988): 31–40; for more general treatments of the role of social statistics for theories of probability, see Theodore M. Porter, *The Rise of Statistical Thinking, 1820–1900* (Princeton, NJ: Princeton University Press, 1986); Ian Hacking, *The Emergence of Probability: A*

Philosophical Study of Early Ideas about Probability, Induction and Statistical Inference (Cambridge, UK: Cambridge University Press, 1984); and Ian Hacking, *The Taming of Chance* (Cambridge, UK: Cambridge University Press, 1990).

30. Fontana, *Treatise on the Venom of the Viper*, I, 290.

31. Ibid., 294.

32. Ibid., 297.

33. Ibid., 305.

34. Another long series of experiments, described in equally close detail, supported Fontana's position, because together, the trials demonstrated that nerves were *not* affected by venom. I will not discuss these experiments here, because Fontana did not raise any new methodological issues.

35. Fontana, *Treatise on the Venom of the Viper*, I, 326.

36. Ibid., 386–87.

37. Ibid., II, 72.

38. For the notion of discoverability, see Thomas Nickles, "Beyond Divorce: Current Status of the Discovery Debate," *Philosophy of Science* 52 (1985): 177–206.

39. See Larry Laudan, "Why Was the Logic of Discovery Abandoned?" in *Scientific Discovery*, ed. Thomas Nickles (Dordrecht: Reidel, 1980), 173–83; and K. Schaffner, *Discovery and Explanation in Biology and Medicine* (Chicago: University of Chicago Press, 1993).

40. On Newton's methodology, see George E. Smith, "The Methodology of the *Principia*," in *The Cambridge Companion to Newton*, ed. George E. Smith and I. Bernard Cohen (Cambridge, UK: Cambridge University Press, 2002).

41. Jean Senebier, *L'art d'observer*, vol. 2 (Genève, Switzerland: Cl. Philiber & Bart. Chirol, 1775), 1–13.

42. Ibid., 11. We have here yet another kind of Newtonianism: Newtonianism modeled after Newton's letter "Containing His New Theory About Light and Colors"; see Isaac Newton, "A Letter of Mr. Isaac Newton, Professor of the Mathematicks in the University of Cambridge; Containing His New Theory About Light and Colors," *Philosophical Transactions of the Royal Society of London* 6 (1671): 3075–87.

43. Senebier, *L'art D'observer*, 2, 131–45.

44. See Knoefel, *Felice Fontana: Life and Works*.

45. Fontana, *Treatise on the Venom of the Viper*, II, 72.

46. Ibid., 72.

47. Ibid., 75.

48. For an analysis of de Luc's writing, see Christoph Hoffmann, "The Ruin of a Book: Jean Andre De Luc's *Recherches sur les modifications de l'atmosphère* (1772)," *Modern Language Notes* 118 (2003): 586–602.

49. Fontana, *Treatise on the Venom of the Viper*, II, 395.

50. Ibid., I, iv.

51. Ibid., xv.

52. Johann Friedrich Gmelin, "27tes Stück. Den 6. Jul 1782. Florenz," *Zugabe zu den Göttingischen Anzeigen von Gelehrten Sachen* (1782): 417-31. For the attribution of authorship, see Frank P. W. Dougherty, "Nervenmorphologie und-Physiologie in den 8oer Jahren des 18. Jahrhunderts," in *Gehirn–Nerven–Seele: Anatomie und Physiologie im Umfeld S. Th. Soemmerrings*, ed. G. Mann and F. Dumont (Stuttgart, Germany: G. Fischer, 1988), 89.

53. In her study of a number of writings on galvanism around 1800, Maria Trumpler distinguishes four basic patterns of experimental variation according to their intended purposes: simplification (the isolation of the phenomenon under study), optimization (finding the ideal experimental conditions), exploration (investigating an ever wider range of objects), and application (transferring the approach to other contexts); see Maria Trumpler, "Verification and Variation: Patterns of Experimentation in Investigations of Galvanism in Germany, 1790-1800," *Philosophy of Science* 64 (1997): S75-S84. This is quite similar to the patterns of variation that we can find in Fontana's writings. Because galvanism was such a widely studied phenomenon, we have here an indication that the concern with variations was widespread at the time.

Chapter Seven

1. On Mitchell's life and work in the context of the wider American culture, see, e.g., D. J. Canale, "Civil War Medicine from the Perspective of S. Weir Mitchell's 'The Case of George Dedlow,'" *Journal of the History of the Neurosciences* 11 (2002): 11-18; Laura Otis, *Membranes: Metaphors of Invasion in Nineteenth-Century Literature, Science and Politics* (Baltimore, MD: Johns Hopkins University Press, 1999); Christopher Goetz, "Jean Martin Charcot and Silas Weir Mitchell," *Neurology* 48 (1997): 1128-32; Nancy Cervetti, "S. Weir Mitchell and His Snakes: Unraveling the 'United Web and Woof of Popular and Scientific Beliefs,'" *Journal of Medical Humanities* 28 (2007): 119-33; and Nancy Cervetti, *S. Weir Mitchell, 1829-1914: Philadelphia's Literary Physician* (University Park: Pennsylvania State University, 2012). On the institutional conditions of Mitchell's career, see W. Bruce Fye, "S. Weir Mitchell, Philadelphia's 'Lost' Physiologist," *Bulletin of the History of Medicine* 57 (1983): 188-202.

2. See Warner, "Ideals of Science and Their Discontents." See also Jardine (in turn drawing on Walter Pagel) for the "plethora of programmes for making medicine scientific": Jardine, "The Laboratory Revolution in Medicine," 305.

3. Ronald Numbers even thinks that "[p]robably no practice in the historiography of medicine has distorted a proper understanding of the relationship between science and medicine more than the uncritical (and perhaps unconscious) use of the term *scientific medicine* to describe what became of medicine in the late nineteenth century"; Ronald L. Numbers, "Science and Medicine," in *Wrestling with*

Nature: From Omens to Science, ed. Peter Harrison, Ronald L. Numbers, and Michael H. Shank (Chicago: University of Chicago Press, 2011), 214–15.

4. See, e.g., Worboys, "Practice and the Science of Medicine in the Nineteenth Century."

5. Mitchell later recalled that the physiology he had learned had "but two dimensions." A physiology lecture was "a more or less well stated resumé of the best foreign books, without experiments or striking illustrations. It was like hearing about a foreign land into which we were forbidden to enter"; S. Weir Mitchell, *Memoir of John Call Dalton, 1825–1889* (National Academy, 1890), 179–80.

6. See, e.g. Anna Robeson Burr, *Weir Mitchell—His Life and Letters* (New York: Duffield & Co, 1929), 28–29.

7. W. Bruce Fye, *The Development of American Physiology: Scientific Medicine in the Nineteenth Century* (Baltimore, MD: Johns Hopkins University Press, 1987), 57.

8. Claude Bernard, *An Introduction to the Study of Experimental Medicine* (New York: Dover Publications, 1957; 1865), 15. Several of those physicians who later played leading roles in America's medical community went to Claude Bernard's laboratory in Paris; see John Harley Warner, *Against the Spirit of System: The French Impulse in Nineteenth-Century American Medicine* (Princeton, NJ: Princeton University Press, 1998). Mitchell's friend Dalton attended Bernard's lectures on experimental physiology.

9. Cervetti, *S. Weir Mitchell*, 34.

10. Ibid., 48. In his autobiography, Mitchell recalled: "I set about examining the poison, physically and chemically, and for several years I gave up all my leisure to this work. That leisure was small or was won when other men were idling or had left town for their summer holidays. In fact, most of my work was done in summer and with great difficulty, on account of lack of money and time"; quoted in Burr, *Weir Mitchell*, 75.

11. Ibid., 97.

12. Fye, *The Development of American Physiology*, 58.

13. William F. Bynum, *Science and the Practice of Medicine in the Nineteenth Century* (Cambridge, UK: Cambridge University Press, 1994).

14. Robert Frank, "American Physiologists in German Laboratories, 1865–1914," in *Physiology in the American Context, 1850–1940*, ed. Gerald Geison (Bethesda, MD: American Physiological Society, 1987), 11–46.

15. For Mitchell's failed attempts to obtain a chair in physiology, see Fye, *The Development of American Physiology*, 66–67, 68–73; and Cervetti, *S. Weir Mitchell*, 70–71, 93–95.

16. Historian Bruce Fye coined this experession; see Fye, "S. Weir Mitchell, Philadelphia's 'Lost' Physiologist."

17. Fye, *The Development of American Physiology*, 76–91.

18. On Mitchell's clinical work and literary writings during the civil war, see, e.g., Canale, "Civil War Medicine." The horrifying short story "The Case of

George Dedlow" is an impressive example of how Mitchell *incorporated* contemporary medical thought in literary projects.

19. S. Weir Mitchell, "Experimental Contributions to the Toxicology of Rattle-Snake Venom," *The New York Medical Journal* (1868): 289–322.

20. Reichert became the chair of physiology at the University of Pennsylvania in the mid-1880s, with Mitchell's support; see Fye, *The Development of American Physiology*, 88–89.

21. S. Weir Mitchell, *On the Treatment of Rattlesnake Bites, with Experimental Criticisms Upon the Various Remedies Now in Use* (Philadelphia, PA: Lippincott & Co., 1861).

22. S. Weir Mitchell, *Researches on the Venom of the Rattlesnake* (Philadelphia, PA: Smithsonian Institution, 1860), iii–iv.

23. Ibid., 1.

24. For details, see Warner, "The Fall and Rise of Professional Mystery."

25. Gerald Geison, "Divided We Stand: Physiologists and Clinicians in the American Context," in *The Therapeutic Revolution: Essays in the Social History of American Medicine*, ed. M. J. Vogel and Charles E. Rosenberg (Philadelphia: University of Pennsylvania Press, 1979), 67–90. In 1871, for instance, the Harvard surgeon Henry J. Bigelow stated the following in an address on medical education: "In an age of science, like the present, there is more danger that the average medical student will be drawn from what is practical, useful, and even essential, by the well-meant enthusiasm of the votaries of less applicable science." Bigelow was not at all enthusiastic about experimental physiology, which "leads away from broad and safer therapeutic views, and toward a local and exclusive action of chemistry and cells,—uncertain grounds for students, for whom the result of large and well-attested medical experience is here the safest teaching, and a habit of mind leading to experiments on patients the most questionable"; quoted after Fye, *The Development of American Physiology*, 107, 108. See also Christopher Lawrence, "Incommunicable Knowledge: Science, Technology and the Clinical Art in Britain 1850–1914," *Journal of Contemporary History* 20 (1985): 503–20.

26. John Call Dalton, *Vivisection; What It Is, and What It Has Accomplished* (New York: Baillière brothers, 1867), 35.

27. Fye, *The Development of American Physiology*, 35–39.

28. Mitchell, "Experimental Contributions to the Toxicology of Rattle-Snake Venom," 290–91.

29. Dalton, *Vivisection*, 36.

30. Mitchell, *On the Treatment of Rattlesnake Bites*, 4.

31. Mitchell, *Researches on the Venom of the Rattlesnake*, iv.

32. Mitchell acknowledged the aid of the Smithsonian Institution in procuring the snakes.

33. Mitchell, *On the Treatment of Rattlesnake Bites*, 39. In the Smithsonian essay, there is no evidence that he attempted to select experimental animals that

were alike—a dog was a dog, a rabbit was a rabbit. In part, this may have been simply a practical issue of availability or cost, because Mitchell did emphasize in his reports that differences in age or weight might be important (for dogs more so than for rabbits or pigeons) and should be noted in the experimental report.

34. John Call Dalton, *Treatise on Human Physiology, Designed for the Use of Students and Practitioners of Medicine* (Philadelphia, PA: Blanchard and Lea, 1859), 31.

35. John Lesch has traced how chemists gradually came to ascribe peculiar chemical properties to the "active principle." Lesch has shown that the interest in physiologically active plant principles was part of an effort to rationalize drug therapy. It was this motivation that eventually led to the formation of a new group of chemicals, the class of plant alkaloids and to a revision of accepted chemical classifications; see John E. Lesch, "Conceptual Change in an Empirical Science: The Discovery of the First Alkaloids," *Historical Studies in the Physical Sciences* 11 (1981): 321–22.

36. John. E. Lesch, *Science and Medicine in France: The Emergence of Experimental Physiology, 1790–1855* (Cambridge, MA: Harvard University Press, 1984), 137.

37. See Lesch, "Conceptual Change in an Empirical Science," 328.

38. William A Hammond and S. Weir Mitchell, "On the Physical and Chemical Characterization of Corroval and Vao, Two Recently Discovered Varieties of Woorara, and on a New Alkaloid Constituting Their Active Principle," *Proceedings of the Academy of Natural Sciences of Philadelphia* (1860): 4-9, 8.

39. Mitchell, *Researches on the Venom of the Rattlesnake*, 95; and Matteu Orfila, *A General System of Toxicology, or, a Treatise on Poisons, Found in the Mineral, Vegetable, and Animal Kingdoms, Considered in Their Relations with Physiology, Pathology, and Medical Jurisprudence* (Philadelphia, PA: M. Carey & Sons, 1817), 10. For Orfila's work, see José Ramón Bertomeu-Sánchez and Agustí Nieto-Galan, eds., *Chemistry, Medicine, and Crime: Mateu J.B. Orfila (1787–1853) and His Times* (Sagamore Beach, MA: Science History Publications, 2006). We have seen that Mead's work also suggested a link between snake venom and exhalations. In Mead's case, however, the goal was to offer a general account of poisons, whereas Mitchell no longer pursued this goal.

40. See, e.g., François Magendie, *Lectures on the Blood and on the Changes Which It Undergoes during Disease* (Philadelphia, PA: Haswell, Barrington, and Haswell, 1939); and Gabriel Andral, *Pathological Haematology: An Essay on the Blood in Disease* (Philadelphia: Lea and Blanchard, 1844): 17, 36–37.

41. See, e.g., Marcello Pera, *The Ambiguous Frog: The Galvani–Volta Controversy on Animal Electricity* (Princeton, NJ: Princeton University Press, 1992); and Trumpler, "Verification and Variation."

42. Mitchell, *Researches on the Venom of the Rattlesnake*, 86.

43. Ibid., 42.

44. Ibid., 35.

45. Ibid., 45. The alcohol caused nothing but "slight stupefaction, which passed off rapidly" (46).

46. Ibid., 78.

47. Ibid., 47.

48. Ibid., emphasis added.

49. John Herschel, *A Preliminary Discourse of the Study of Natural Philosophy* (London: Longman, 1830), 151.

50. Auguste Comte, *The Positive Philosophy of Auguste Comte* (New York: Calvin Blanchard, 1855).

51. John Stuart Mill, *A System of Logic* vol. VII, Collected Works (Indianapolis, Indiana: Liberty Fund, 2006), 394. In fact, Mill himself pointed out that his "method of difference" required an ideal experimental situation that could rarely (if ever) be attained. The chapter following the formal presentation of the method discusses "miscellaneous examples" of application. Mill used these examples to show that strictly speaking, the ideal situation for the application of the method was unattainable. Mill picked examples from recent science that, he thought, illustrated the power of the methods. Mill discussed, among other things, Liebig's work on metal poisoning. Liebig had shown why arsenious acid and salts of certain metals were fatal if introduced into an animal organism: If solutions of these substances were placed in close contact with organic substances, the acid or salts would combine with these substances and prevent the chemical actions that constitute life. Mill pointed out, however, that the method of difference would only determine the difference between the presence or absence of a single substance, not of a single circumstance, "and as every substance has innumerable properties, there is no knowing what number of real differences are involved in what is nominally and apparently only one difference" (410). Nineteenth-century critics of Mill rarely appreciated Mill's own ideas about the limitations of his method.

52. Comte, *The Positive Philosophy of Auguste Comte*, 310.

53. Ibid., 311.

54. According to Kenneth Schaffner, the logic underlying comparative experiments and Mill's method of difference is the same. Also, Mill acknowledged that his method was idealized and had to be amended for practical purposes, as the causal conditions could not always be completely specified. Thus Bernard should not have emphasized the difference between the two types of methodological tools quite so strongly; see Kenneth Schaffner, "Clinical Trials: The Validation of Theory and Therapy," in *Physics, Philosophy, and Psychoanalysis: Essays in Honor of Adolf Grünbaum*, ed. R. S. Cohen and Larry Laudan (Boston: Reidel, 1983). From the perspective of a historian of methods discourse, by contrast, I am inclined to emphasize the difference. For Bernard, the terminological distinction marked the difference between the practice of experimental physiology and the unattainable (certainly for physiologists) ideal of a completely controlled physical experiment.

Moreover, in the subsequent discussions, investigators criticized the ideal, not the comments that Mill made on his method.

55. Bernard insisted that "experimental counterproof must not be mistaken for comparative experimentation. Counterproof has not the slightest reference to sources of error that may be met in observing facts; it assumes that they are all avoided and is concerned only with experimental reasoning; it has in view only judging whether the relation established between a phenomenon and its immediate cause is correct and rational"; Bernard, *An Introduction to the Study of Experimental Medicine*, 126–27.

56. Bernard, *An Introduction to the Study of Experimental Medicine*, 128.

57. Claude Bernard, *Leçons sur les effets des substances toxiques et médicamenteuses* (Paris: J. B. Bailliere et fils, 1857), 312–25.

58. This is one reason why Bernard never accepted statistics in physiology. I will come back to this issue hereafter. See also Porter, *The Rise of Statistical Thinking*, 160–62.

59. Ronald A. Fisher, *The Design of Experiments* (Edinburgh: Oliver and Boyd, 1935), 49.

60. Mitchell, *On the Treatment of Rattlesnake Bites*, 6.

61. He noted, for instance: "Almost all toxicologists who have investigated this subject, have been content to submit animals, as dogs, etc. to be bitten by the serpents themselves. We have seen, however, that when this course is followed, a number of fallacies interfere to prevent the observer from drawing satisfactory conclusions . . . ." He had tried to avoid these "embarrassments" by extracting and injecting venom; see Mitchell, *On the Treatment of Rattlesnake Bites*, 6, 11.

62. See especially Rosser Matthews, *Quantification and the Quest for Medical Certainty* (Princeton, NJ: Princeton University Press, 1995); and William Coleman, "Experimental Physiology and Statistical Inference: The Therapeutic Trial in Nineteenth-Century Germany," in *The Probabilistic Revolution: Ideas in the Sciences*, ed. Lorenz Krüger, Lorraine Daston, and Michael Heidelberger (Cambridge, MA: MIT Press, 1987), 201–28. For more general treatments of statistics in nineteenth-century science and medicine, see Porter, *The Rise of Statistical Thinking, 1820–1900*; and Hacking, *The Taming of Chance*.

63. See Porter, *The Rise of Statistical Thinking*, 151–62; and Coleman, "Experimental Physiology and Statistical Inference," 204.

64. Pierre Charles Alexandre Louis, *Researches on the Effects of Bloodletting in Some Inflammatory Diseases, and on the Influence of Tartarized Antimony and Vesication in Pneumonitis* (Boston: Hilliard, Gray, and Company, 1836), 60. Louis was aware that the groups had to be large. He noted that epidemics offered the ideal conditions for the application of the numerical method, because one could compare the efficacy of treatments on groups of hundreds of people.

65. Bernard, *An Introduction to the Study of Experimental Medicine*, 136.

66. Ibid., 137. Bernard referred to a controversy between François Magendie

and François Achille Longet about their experiments on the stimulation of the anterior root of the spinal nerve—a controversy that he had resolved. See Mirko Grmek, "Bernard, Claude," in *Dictionary of Scientific Biography*, ed. C. Gillespie (New York: Scribner, 1970–1980), 30.

67. Coleman, "Experimental Physiology and Statistical Inference."

68. Mitchell, "Experimental Contributions to the Toxicology of Rattle-Snake Venom," 315.

69. Mitchell, *Researches on the Venom of the Rattlesnake*, 70.

70. Mitchell, *On the Treatment of Rattlesnake Bites*, 9.

71. Ibid., 9.

72. I am not aware of Mitchell's having had any financial interest in the production of antidotes.

73. Alan G. Gross, Joseph E. Harmon, and Michael Reidy, *Communicating Science: The Scientific Article from the 17th Century to the Present* (Oxford: Oxford University Press, 2002), 143–44.

74. Mitchell, *Researches on the Venom of the Rattlesnake*, 44.

75. Ibid., 45.

76. Mitchell, "Experimental Contributions to the Toxicology of Rattle-Snake Venom," 299, 300.

77. This is the chief difference between Mitchell's report and the early modern presentation of experimental projects such as Redi's, which also began with a particular question—the question of whether the yellow liquor was poisonous.

Chapter Eight

1. S. Weir Mitchell and Edward T. Reichert, *Researches Upon the Venoms of Poisonous Serpents* (Washington, DC: Smithsonian Institution, 1886), iii.

2. Kathy Ryan, "APS at 125: A Look Back at the Founding of the American Physiological Society," *Advances in Physiological Education* 37 (2013): 10–14.

3. John Billings, "An Address on Our Medical Literature," *British Medical Journal* (1881): 264.

4. On homeopathy in the American context, see, e.g., John Harley Warner, *The Therapeutic Perspective: Medical Practice, Knowledge, and Identity in America, 1820–1885* (Cambridge, MA: Harvard University Press, 1986).

5. For a general overview, see Marks, *The Progress of Experiment*.

6. Numbers, "Science and Medicine," 212–14, 215–16.

7. Billings, in fact, offered a rather detailed explanation of what it meant for medicine to become "scientific." He stated in his address: "The connections of medicine with the physical sciences are yearly becoming closer, and the methods by which these sciences have been brought to their present condition are those by which progress has been, and is to be, made in therapeutics, as well as in diagnosis,

or in physiological research. These methods turn mainly upon increasing the delicacy and accuracy of measurements; of expressing manifestations of force in terms of another force, or of dimension in space or time. The balance and the galvanometer, the microscope and the pendulum, the camera, the sphygmograph, and the thermometer, are some of the means by which investigators, at the bedside and in the laboratory, are seeking to obtain records which shall be independent of their own sensations or personal equations; which shall be taken and used as expressing, not opinions, but facts; and with every addition to, or improvement in, these means of measurement and record, the field of observation widens, and new and more reliable materials are furnished for the application of logical and mathematical methods"; Billings, "An Address on Our Medical Literature," 267.

8. Horatio C. Wood, *A Treatise on Therapeutics, Comprising Materia Medica and Toxicology* (Philadelphia, PA: Lippincott & Co., 1874), 7.

9. Jardine, "The Laboratory Revolution in Medicine." For a discussion of the historiographical category of "laboratory revolution," see Michael Worboys, "Was There a Bacteriological Revolution in Late Nineteenth-Century Medicine?" *Studies in History and Philosophy of Biology and Biomedical Sciences* 38 (2007): 20–42.

10. Newman, *Atoms and Alchemy*.

11. Michael Faraday, *Chemical Manipulation: Being Instructions to Students in Chemistry, on the Methods of Performing Experiments of Demonstration or of Research, with Accuracy and Success* (London: John Murray, 1827). The manual is more than 600 pages long; its first chapter is titled "The Laboratory." For eighteenth-century experimental research on drugs, see Andreas-Holger Maehle, *Drugs on Trial: Experimental Pharmacology and Therapeutic Innovation in the Eighteenth Century* (Amsterdam: Rodopi, 1999).

12. Ian Hacking, "Disunities in the Sciences," in *The Disunity of Science: Boundaries, Contexts, and Power*, ed. Peter Galison and David Stump (Stanford, CA: Stanford University Press, 1996), 37–74. Hacking's conception of research styles draws on Alistair Crombie's monumental work on "styles of scientific thinking." Crombie's "styles" mark broad historical developments. Beginning in antiquity, he identifies several styles of scientific thinking, including methods of postulation, the experimental method, modeling, classification, statistical analyses of populations, and studies of development. Drawing on Crombie, Hacking outlines the "laboratory style" as a combination of two of Crombie's styles, experimentation and modeling; see Ian Hacking, "Styles of Scientific Thinking or Reasoning: A New Analytical Tool for Historians and Philosophers of the Sciences," in *Trends in the Historiography of Science*, ed. K. Gavroglu, J. Christianidis, and N. Nicolaidis (Dordrecht, Netherlands: Kluwer Academic Publishers, 1994), 31–48.

13. The most detailed study of such an "urban lab" is Sven Dierig's account of the wondrous machines and apparatuses that du Bois-Reymond's Berlin laboratory provided to the experimenting physicist and physiologist; see Sven Dierig, *Wissenschaft in der Maschinenstadt: Emil Du Bois-Reymond und seine Laboratorien*

in Berlin (Berlin: Wallstein Verlag, 2006). For earlier studies on the instruments and experiments that embodied "organic physics," see, among others, Robert Brain and Norton Wise, "Muscles and Engines: Indicator Diagrams and Helmholtz's Graphical Methods," in *Universalgenie Helmholtz: Rückblick nach 100 Jahren*, ed. Lorenz Krüger (Berlin: Akademie-Verlag, 1994), 124-45; Kathryn M. Olesko and Frederic L. Holmes, "The Images of Precision: Helmholtz and the Graphical Method in Physiology," in *The Values of Precision*, ed. Norton Wise (Princeton, NJ: Princeton University Press, 1995), 198-221.

14. Henry Pickering Bowditch's physiology laboratory at the Harvard Medical School is usually considered to have been the first of these institutions. Johns Hopkins University was founded in 1876, explicitly to emulate the structure of the German research university. The physiology laboratory at the University of Pennsylvania was established in 1878, with Mitchell as the principal speaker at the dedication; see Fye, *The Development of American Physiology*, 83.

15. Numbers, "Science and Medicine," 210-11. The Medical School at Johns Hopkins opened much later, in 1893, but the Johns Hopkins biology laboratory for biomedical research was established shortly after the university was founded. It was founded with a view to the planned opening of the Medical School, and it was to emphasize the basic training of medical students; see Philip J. Pauly, *Controlling Life: Jacques Loeb & the Engineering Ideal in Biology* (Oxford, UK: Oxford University Press, 1987), 134-36. The individual who was initially responsible for teaching physiology and who succeeded in obtaining funding for new laboratory buildings was the chair of the biology department, the British instructor H. Newell Martin, who had received a degree in physiology from the University of Cambridge, England, working with the physiologist Michael Forster, and who had also worked as Thomas Henry Huxley's assistant. For more biographical details on H. Newell Martin, see Fye, *The Development of American Physiology*, chapter 4.

16. H. Newell Martin, "The Study and Teaching of Biology: An Introductory Lecture Delivered at Johns Hopkins University, October 23, 1876," *Popular Science Monthly* 10 (1877), 304-5.

17. He asked his audience: "How can one ignorant of physics have any real appreciation of the statement that the transmission of a nervous impulse is accompanied by a molecular alteration in the structure of a nerve-fibre, one sign of which is a certain very definite and peculiar alteration in its electrical properties; or how can one ignorant of chemistry grasp the fundamental statement that muscular work is in the long-run dependent on the breaking down of complex chemical molecules into simpler and more stable ones?" Ibid., 305-6.

18. On the founding of the APS, see Toby Appel, "Biological and Medical Societies and the Founding of the American Physiological Society," in *Physiology in the American Context, 1850-1940*, ed. Gerald Geison (Bethesda, MD: American Physiological Society, 1987), 155-76.

19. See Fye, *The Development of American Physiology*, 76. Fye thinks, quite

plausibly, that the doctors Wood had in mind were probably H. Newell Martin at Johns Hopkins, Henry P. Bowdich at Harvard, Austin Flint at the Bellevue Hospital Medical College and Dalton at the College of Physicians and Surgeons of New York; ibid., 250, n. 86.

20. This passage is quoted after Cervetti, *S. Weir Mitchell*, 182.

21. Mitchell and Reichert, *Researches Upon the Venoms of Poisonous Serpents*, 10.

22. John Attfield, *Chemistry: General, Medical, and Pharmaceutical*, 10th ed. (Philadelphia, PA: Henry C. Lea's Son & Co., 1883).

23. On Marey's tambour, see James Putnam, "On the Reliability of Marey's Tambour in Experiments Requiring Accurate Notations of Time," *Journal of Physiology* 2 (1879): 209–13. On the graphic method more generally, see Robert Brain, "The Graphic Method: Inscription, Visualization, and Measurement in Nineteenth-Century Science and Culture" (PhD diss., University of California, 1996); John W. Douard, "E.-J. Marey's Visual Rhetoric and the Graphic Decomposition of the Body," *Studies in History and Philosophy of Science* 26 (1995); Soraya de Chadarevian, "Graphical Methods and Discipline: Self-Recording Instruments in Nineteenth-Century Physiology," *Studies in History and Philosophy of Science* 24 (1992); and Robert G. Frank, "The Telltale Heart: Physiological Instruments, Graphic Methods, and Clinical Hopes 1854–1914," in *The Investigative Enterprise: Experimental Physiology in Nineteenth-Century Medicine*, ed. William Coleman and Frederic L. Holmes (Berkeley: University of California Press, 1988), 211–70.

24. Cervetti detects in the second Smithsonian essay the influence of antivivisectionist concerns mainly because Mitchell and Reichert rarely experiment on dogs and occasionally use ether in their experiments; see Cervetti, *S. Weir Mitchell*, 185. I do not find this argument plausible, because the main part of the report is such an extensive record of violent deaths. The impression of cold cruelty that the antivivisectionist would get from looking at the columns of "remarks" is hardly mitigated by the occasional mention of anesthesia. It seems more plausible to me that rabbits were used instead of dogs because the experimenters strove to create equal conditions for their experiments, and obtaining large numbers of dogs of similar size and breed would have been much more difficult and costly.

25. Mitchell and Reichert, *Researches Upon the Venoms of Poisonous Serpents*, 60, 88.

26. Ibid., 87–88.

27. Ibid., 56.

28. Ibid., 56.

29. Ibid., 136.

30. E.g., Edgar March Crookshank, *Manual of Bacteriology* (London: H. K. Lewis, 1887).

31. See, e.g., Leonard Landois, *A Textbook of Human Physiology, Including Histology and Microscopical Anatomy*, 3rd American ed. (Philadelphia, PA: P. Blakiston, Son & Company, 1889), 162.

32. See, e.g., Heinrich Frey, *The Microscope and Microscopical Technology* (New York: William Wood & Co, 1872).

33. For the long history of the concept of contagion; see Vivian Nutton, "The Seeds of Disease: An Explanation of Contagion and Infection from the Greeks to the Renaissance," *Medical History* 27 (1983): 1–34; and Vivian Nutton, "The Reception of Fracastoro's Theory of Contagion: The Seed That Fell among Thorns?" *Osiris* 6 (1990): 196–234.

34. See Christoph Gradmann, "Isolation, Contamination, and Pure Culture: Monomorphism and Polymorphism of Pathogenic Micro-Organisms as Research Problem 1860–1880," *Perspectives on Science* 9 (2001): 147–72.

35. For details, see Gradmann, *Krankheit im Labor: Robert Koch und die medizinische Bakteriologie* (Göttingen, Germany: Wallstein Verlag, 2005), 35–42.

36. In chapter 10, Mitchell acknowledged the contribution of Henry F. Formad, a pathologist and a coroner's physician at Philadelphia.

37. On the history of these postulates, see Christoph Gradmann, "Alles eine Frage der Methode. Zur Historizität der Kochschen Postulate 1840–2000," *Medizinhistorisches Journal* 43 (2008): 121–48; and K. Codell Carter, "Koch's Postulates in Relation to the Work of Jacob Henle and Edwin Klebs," *Medical History* 29 (1985): 353–75. On debates about pathogens and etiology more generally, see K. Codell Carter, *The Rise of Causal Concepts of Disease* (Aldershot, UK: Ashgate, 2003).

38. Mitchell and Reichert, *Researches Upon the Venoms of Poisonous Serpents*, 136–37.

39. A passage by Athenaeus (AD 200), for instance, describes how some convicted criminals had been thrown among the asps; they survived. It turned out that they had been given lemons prior to their punishment. The governor of the city ordered the next day "that a piece of citron be given, exactly as before, to one convict, but not to the other, and the one who ate suffered no injury when bitten by the reptiles, but the other died the moment he was struck." The passage is from the Deipnosophists, 3.84 d–f:2; the reference is given in Eugene S. McCartney, "A Control Experiment in Antiquity," *The Classical Weekly* 36 (1942): 5–6. Stigler even finds an example for a control experiment in the Old Testament: Daniel, Book I, 164 BC; see Stephen Stigler, "Gergonne's 1815 Paper on the Design and Analysis of Polynomial Regression Experiments," *Historia Mathematica* 1 (1974): 431.

40. Raffaele Roncalli Amici, "The History of Italian Parasitology," *Veterinary Parasitology* 98 (2001): 3–30.

41. This is actually just a plan, which surely was never realized: "Let us take out of the Hospitals, out of the Camps, or from elsewhere, 200, or 500 poor People, that have Fevers, Pleurisies, etc. Let us divide them in Halfes, let us cast lots, that one half of them may fall to my share and the other to yours; I will cure them with-

out bloodletting and sensible evacuation; but do you do as ye know [...] we shall see how many Funerals both of us shall have: But let the reward of the contention or wager, be 300 Florens, deposited on both sides: Here your business is decided." The link to probability theory is evident; quoted in Iain Chalmers, "Comparing Like with Like: Some Historical Milestones in the Evolution of Methods to Create Unbiased Comparison Groups in Therapeutic Experiments," *International Journal of Epidemiology* 30 (2001): 1157.

42. Surprisingly, there are few systematic and detailed historical or philosophical studies of the concept outside the contributions to the history of randomized controlled trials in clinical medicine. The history of RCTs in the early twentieth century has been studied in some detail; see Ian Hacking, "Telepathy: Origins of Randomization in Experimental Design" *Isis* 79 (1988): 427–51; see also Peter Keating and Alberto Cambrosio, *Cancer on Trial* (Chicago: University of Chicago Press, 2012). Much work remains to be done for the period prior to the twentieth century.

43. Edwin Garrigues Boring, "The Nature and History of Experimental Control," *American Journal of Psychology* 67 (1954): 589.

44. Boring claims: "the methodological status of control as a check or comparison is to be understood by reference to J. S. Mill's Method of Difference"; ibid., 589. See also Kenneth Schaffner, "Clinical Trials," 202; and Gerd Gigerenzer, "Survival of the Fittest Probabilist: Brunswik, Thurstone, and the Two Disciplines of Psychology," in *The Probabilistic Revolution: Ideas in Science*, ed. Lorenz Krüger, Gerd Gigerenzer, and Mary S. Morgan (Cambridge, MA: MIT Press, 1989), 53. Boring refers to late nineteenth-century encyclopedia definitions to support his point: those definitions paraphrase Mill; see Boring, "The Nature and History of Experimental Control," 574. He notes that the methodological term *control* came to be used in the late nineteenth century, but he does not discuss the emergence of the term.

45. Mitchell and Reichert, *Researches Upon the Venoms of Poisonous Serpents*, 36.

46. Ibid., 136.

47. Ibid., 144.

48. Ibid., 149.

49. Bowditch's monograph is concerned with the analysis, purification, and use of coal gas. The dictionary offers a second definition, which reads: "control-experiment. . . . An experiment made to establish the conditions under which another experiment is made"; William Dwight Whitney, *The Century Dictionary and Cyclopedia, a Work of Universal Reference in All Departments of Knowledge with a New Atlas of the World* vol. II (Century Company, 1897), 1237. The distinction that is suggested by the twofold definition—controlling to confirm causes and controlling to identify confounding factors—is not consistently made in the scientific literature, but, as we will see in chapter 9, both conceptions of controlling experiments were common around 1900.

50. Thomas P. Hughes, *American Genesis: A History of the American Genius for Invention* (New York: Penguin Books, 1989). There are, of course, terminological links between engineering and the life sciences; Jacques Loeb comes to mind. I will come back to the issue of control, and to Loeb, in chapter 9.

51. Karl M. Figlio, "The Historiography of Scientific Medicine: An Invitation to the Human Sciences," *Comparative Studies in Society and History* 19 (1977): 272.

52. Post Office Dept. City of Washington, *Letter of the Postmaster General and Opinion of the Attorney General in Reference to the Power of the Circuit Court for the District of Columbia to Control Executive Officers of the United States in the Performance of Their Official Duties, Also, an Exposition of the Reasons of the Postmaster General for Refusing to Execute a Part of the Award of the Solicitor in Favor of Messrs. Stockton, Stokes, and Others United States* (Washington, DC: Blair & Rives, 1837).

53. The Board, *Report of the Principal of the Central High School to the Committee of the Board of Controllers of the Public Schools, for the Year Ending July, 1842* (Philadelphia, PA: 1843).

54. James Ward, *The True Policy of Organising a System of Railways for India: A Letter to the Right Hon. The President of the Board of Control* (London: Smith, Elder, and Co., 1847).

55. John Lane Gardner, *Military Control* (A. B. Claxton & Company, 1839).

56. Hamilton Smith, *Cotton and the Only Practical Methods Presented to Its Producers of Advancing and Controlling Its Price: Article IV, in July No. 1849, of De Bow's Commercial Review* (Louisville, KY: Prentice and Weissinger, 1850).

57. John Barlow, *On Man's Power over Himself to Prevent or Control Insanity* (London: Pickering, 1849).

58. J. Clerk Maxwell, "On Governors," *Proceedings of the Royal Society of London* 16 (1867): 270.

59. See W. C. Bond, "Description of the Nebula About the Star Θ Orionis," *Memoirs of the American Academy of Arts and Sciences, New Series* 3 (1848): 87–96; Gideon Algernon Mantell, "On the Structure of the Jaws and Teeth of the Iguanodon," *Philosophical Transactions of the Royal Society of London* 138 (1848): 183–202; Humphry Davy, "An Account of Some Experiments and Observations on the Constituent Parts of Certain Astringent Vegetables; and on Their Operation in Tanning," *Abstracts of the Papers Printed in the Philosophical Transactions of the Royal Society of London* 1 (1800–1814): 114–18 (the context is tanning procedures); and Michael Faraday, "Experimental Researches in Electricity. Tenth Series," *Philosophical Transactions of the Royal Society of London* 125 (1835): 263–74.

60. From the 1860s onward, the terms *standard of comparison* and *control* begin to appear together. The embryologist W. H. Ransom, for instance, conducted "control experiments" expressly "[f]or the sake of having a standard of comparison"; see W. H. Ransom "Observations on the Ovum of Osseous Fishes," *Philosophical Transactions of the Royal Society of London*, 157 (1867), 431–501, quote on 483.

(In this case, the standards of comparison are "normal" germinal disks and egg yolks, i.e. those that are not manipulated in experiments.) Another nice instance of the expression "standard of comparison or control"—with a twist—can be found in the second edition of Charles Darwin's book *The Power of Movement in Plants* from 1880. Here the "standard of comparison or control" is the radicle of a plant that grows in an unconstrained way. Its growth is compared to the growth of radicles whose tips had bits of cardboard attached; see Charles Darwin, assisted by Francis Darwin, *The Power of Movement in Plants* (London: John Murray, 1880), 162.

61. S. Weir Mitchell, "The Poison of the Rattlesnake," *The Atlantic Monthly* (1868), 452, 453. Literary scholars have drawn attention to Mitchell's intimate relation with "his" snakes. For a discussion of the metaphors and anthropomorphic language Mitchell used to describe snakes and their behavior, see Cervetti, "S. Weir Mitchell and His Snakes," 54–57.

62. Mitchell, "The Poison of the Rattlesnake," 454, 455.

63. Ibid., 459.

64. The medical men in the 1880s did not analyze the textual genres that were available to the scientific writer. Only at the very end of the nineteenth century did literary scholars and philologists begin to comment on the differences between crafting literary texts and writing other kinds of essays for broader audiences and to offer new categorizations and classifications of textual genres.

65. Billings, "An Address on Our Medical Literature," 262. Billings did stress that these numbers had to be taken with a grain of salt, because they depended on subjective assessments. If someone else repeated the counting, the numbers would turn out a little different.

66. Ibid., 268.

67. Ibid., 267.

68. Ibid.

69. Ibid.

70. Edward T. Reichert, *The Differentiation and Specificity of Corresponding Proteins and Other Vital Substances in Relation to Biological Classification and Organic Evolution: The Crystallography of Hemoglobins* (Washington, DC: Carnegie Institution of Washington, 1909). At the beginning of the twentieth century, investigators assumed that the structure of the crystals they obtained reflected the composition and structure of the molecules that composed the crystal. The comparison of crystals that were obtained from the hemoglobin of different species of animals was thought to provide information about the relations among species; see Alexander McPherson, "A Brief History of Protein Crystal Growth," *Journal of Crystal Growth* 110 (1991): 3. Reichert's extended study from 1909 belongs in this context.

71. Alexander C. Abbott, "Lunching with Reichert," *Scope (University of Pennsylvania Medical Students' Yearbook)* 21 (1924): 65–70.

72. George W. Corner, *Two Centuries of Medicine: A History of the School of Medicine, University of Pennsylvania* (Philadelphia, PA: Lippincott & Co., 1965), 179.

73. Twentieth-century historians have unearthed several exemplary cases for productive concatenations—e.g., Hans-Jörg Rheinberger in his study of protein synthesis, *Toward a History of Epistemic Things* (Stanford, CA: Stanford University Press, 1997); Nicholas Rasmussen in his work on the electron microscope, *Picture Control: The Electron Microscope and the Transformation of Biology in America, 1940–1960* (Stanford, CA: Stanford University Press, 1997); and Carsten Reinhardt in *Shifting and Rearranging: Physical Methods and the Transformation of Modern Chemistry* (Sagamore Beach, MA: Science History Publications, 2006), among others.

Chapter Nine

1. S. Weir Mitchell and Alonzo H. Stewart, "A Contribution to the Study of the Action of the Venom of the Crotalus Adamanteus Upon the Blood," *Transactions of the College of Physicians of Philadelphia* 19 (1897): 105–10.

2. For an instructive overview of Flexner and Noguchi's venom research, see Karen Deane Ross, "Making Medicine Scientific: Simon Flexner and Experimental Medicine at the Rockefeller Institute for Medical Research, 1901–1945" (PhD diss., University of Minnesota, 2006), 29–60.

3. Unlike Mitchell, Noguchi did not list all the traditional sources—notably, he included Redi (the classic) and, of course, Mitchell (the inspiration), but he did not include Fontana.

4. For the development of the fields of bacteriology and immunology, see Bulloch, *The History of Bacteriology*; and Arthur M. Silverstein, *A History of Immunology* (San Diego, CA: Academic Press, 1989). Bulloch's account, a traditional history of ideas, is still useful, because it presents these ideas clearly and concisely, and it contains a wealth of primary sources.

5. Fye, *The Development of American Physiology*, 185.

6. Jane Maienschein, *Defining Biology* (Cambridge, MA: Harvard University Press, 1986).

7. On the early attempts to integrate laboratory and clinical research at the Rockefeller Institute, see Olga Amsterdamska, "Research at the Hospital of the Rockefeller Institute for Medical Research," in *Creating a Tradition of Biomedical Research: Contributions to the History of the Rockefeller University*, ed. Darwin H. Stapleton (New York: Rockefeller University Press, 2004), 111–26.

8. Appel, "Biological and Medical Societies," 168; see also Philip J. Pauly, *Biologists and the Promise of American Life: From Meriwether Lewis to Alfred Kinsey* (Princeton, NJ: Princeton University Press, 2000), 131–32. Pauly has described the organization of the *American Society of Naturalists* as a "confederacy," a larger organization made up of smaller disciplinary groups, such as the American Physiological Society, the American Morphological Society, and the American Psychological Association; see ibid., 132.

9. For the following, see Pauly, *Biologists and the Promise of American Life*, chapter 6.

10. Ibid., 159.

11. This information and the following details are drawn from Jane Maienschein, "Physiology, Biology, and the Advent of Physiological Morphology," in *Physiology in the American Context, 1850–1940*, ed. Gerald Geison (Bethesda, MD: American Physiological Society, 1987), 180–81.

12. Brooks was not an advocate of experimental methods. He valued experience, but as a morphologist, he placed the emphasis on careful observation. For him, fitness or adaptive response was the main subject of biology; naturalists thus had to study living things in their environment; see "William Keith Brooks. A Sketch of His Life by Some of His Former Pupils and Associates," *Journal of Experimental Zoology* 9 (1910): 41.

13. According to his biographer, Fleming, Welch had chosen as his title "professor of pathology" rather than "pathological anatomy" to indicate that the chair should cover the origin and nature of diseases and to give equal importance to physiology and bacteriology; see Donald Fleming, *William H. Welch and the Rise of Modern Medicine* (Little, Brown and Company, 1954), 105.

14. See William H. Welch, "Adaptation in Pathological Processes," *The American Journal of the Medical Sciences* 113 (1897): 631–55. For a broad overview of arguments in support of experimental study in this community, see Jane Maienschein, "Arguments for Experimentation in Biology," *PSA: Proceedings of the 1986 Biennial Meeting of the Philosophy of Science Association* 2 (1987): 180–95.

15. Welch, "Adaptation in Pathological Processes," 635.

16. For Flexner's education and early years, see James Thomas Flexner, *An American Saga: The Story of Helen Thomas and Simon Flexner* (Boston: Little, Brown and Company, 1983); and Ross, "Making Medicine Scientific."

17. In 1937, Flexner recalled that the Louisville Medical School was "a school in which the lecture was everything." There was no laboratory instruction; the anatomy laboratory was the only place for practical instruction; see Simon Flexner, "A Half Century of American Medicine," *Science* 85 (1937): 506.

18. Welch had studied at Strasbourg as well as at Breslau with the pathologist Julius Cohnheim, who was Rudolf Virchow's and Kühne's student. Welch's assistant in 1891–93 was George Henry Nuttall, the bacteriologist. Nuttall had worked at Göttingen, in Paul Ehrlich's laboratory, and at Oswald Schmiedeberg's pharmacological laboratory at Strasbourg.

19. Flexner, *An American Saga*.

20. Flexner reviewed this work in his address to the Pathological Society of Philadelphia in 1894, his first major public speech; see Simon Flexner, "Original Address: The Pathologic Changes Caused by Certain So-Called Toxalbumins," *Medical News* 65 (1894): 116–24. According to Flexner, in *An American Saga*, 235, this address established Flexner's reputation and was a key factor in his receiving an offer from the University of Pennsylvania.

21. See Simon Flexner, "Infection and Intoxication," in *Biological Lectures Delivered at the Marine Biological Laboratory of Wood's Holl in the Summer Session of 1885* (Boston: Ginn & Company, 1886); Simon Flexner, "The Regeneration of the Nervous System of Planaria Torva and the Anatomy of the Nervous System of Doubleheaded Forms," *Journal of Morphology* 14 (1897): 337–46; and Peyton Rous, "Simon Flexner. 1863–1947," *Obituary Notices of Fellows of the Royal Society* (1947): 414.

22. As director of the Rockefeller Institute, Flexner became an extremely influential figure for American biomedicine; see J. Rogers Hollingsworth, "Institutionalizing Excellence in Biomedical Research: The Case of the Rockefeller University," in *Creating a Tradition of Biomedical Research: Contributions to the History of the Rockefeller University*, ed. Darwin H. Stapleton (New York: Rockefeller University Press, 2004), 25–27.

23. Almost all biographies of Flexner and Noguchi include some version of the story of how Noguchi managed to obtain a position at Philadelphia; see, e.g., Flexner, *An American Saga*, 246–47; Atsushi Kita, *Dr. Noguchi's Journey: A Life of Medical Search and Discovery* (Tokyo: Kodansha International, 2003), 131–35; and Isabel R. Plesset, *Noguchi and His Patrons* (Cranbury, NJ: Associated University Presses, 1980), 72–74.

24. Hideyo Noguchi, *Snake Venoms: An Investigation of Venomous Snakes with Special Reference to the Phenomena of Their Venoms* (Washington, DC: The Carnegie Institution of Washington, 1909).

25. William H. Welch, "On Some of the Humane Aspects of Medical Science," in *Papers and Addresses—Medical Education—History and Miscellaneous—Vivisection* (Baltimore, MD: Johns Hopkins University Press, 1892), 3.

26. Ibid., 16. Appealing to his audience's national pride, he reported that a "distinguished professor of physiology in a German university" had asked him what became of the young men from [Welch's] country. Quite embarrassingly, Welch had to explain "that the facilities and encouragement from carrying out scientific investigations in the medical institutions of this country are in general very meager, and that one great impetus to such work is almost wholly lacking here—namely, the assurance of even likelihood that good scientific work will pave the way to an academic career"; Welch, "Some of the Advantages of the Union of Medical School and University," in *Papers and Addresses—Medical Education—History and Miscellaneous—Vivisection* (Baltimore, MD: Johns Hopkins University Press, 1892), 32. Welch reiterated all these points, as well as a call for more funding for biomedical research, in several speeches and addresses.

27. See Cunningham and Williams, *The Laboratory Revolution in Medicine*; and Bynum, *Science and the Practice of Medicine in the Nineteenth Century*, chapter 4. For references and for a critique of the concept of "bacteriological revolution," see Worboys, "Was There a Bacteriological Revolution in Late Nineteenth-Century Medicine?"

28. Welch explicitly acknowledged the recent progress of bacteriology: "We have learned to recognize, in certain minute organisms, the specific causes of many of the most devastating diseases. These discoveries, which belong alike to the department of pathology, and to that of hygiene, have already had an important influence upon the management of disease, particularly in surgical practice. But it is in the prevention of disease, especially of epidemic diseases, that this increasing knowledge of their causation is destined to do the most good." Welch, "On Some of the Humane Aspects of Medical Science," 4.

29. "The therapeutic use of anti-toxine, though still in its infancy, shows by the unimpeachable records of hospital practice that the physician has now within his grasp the means of successfully treating one of our most dreaded diseases. The anxiety, almost amounting to despair, with which a physician formerly approached a serious case of diphtheria, has given place to a feeling of well grounded hope of a favorable result. Who can estimate the burden of terror and distress thus removed from the anxious watchers by the bedside, and who will dare to say that the boon has been dearly purchased by the lives of some thousands of guinea pigs?" H. P. Bowditch, "The Advancement of Medicine by Research," *Science* 4 (1896): 99.

30. Abraham Flexner, *Medical Education in the United States and Canada: A Report to the Carnegie Foundation for the Advancement of Teaching* (New York: The Carnegie Foundation for the Advancement of Teaching, 1910), chapter IV. The most momentous organizational reforms of American medicine began with the philanthropic activities of the Rockefellers and others, which helped finance large-scale research facilities at medical schools and independent institutes. See Marks, *The Progress of Experiment*; and Darwin H. Stapleton, ed., *Creating a Tradition of Biomedical Research: Contributions to the History of the Rockefeller University* (New York: Rockefeller University Press, 2004).

31. Noguchi, too, was a keen experimentalist, but he was not a public figure like Flexner was, so there were few occasions (if any) for him to deliver programmatic speeches and addresses.

32. Simon Flexner, "Triumphs of Experimental Medicine," *Scientific Monthly* 36 (1933): 514. Flexner never (at least to my knowledge) discussed in detail the relation between biology and physiology or pathology. Only on the occasion of his Friday evening lecture "Infection and Intoxication" at Woods Hole, he opened with the declaration: "The science of biology in its widest sense comprises the study of life in all its forms and activities, both normal and abnormal" ("Infection and Intoxication"), but we may assume that it was a concession to his audience at Woods Hole. If Flexner mentioned biology in his other speeches, he mentioned it along with physics and chemistry as the foundation of scientific medicine—see, for example, "A Half Century of American Medicine," 508. His teacher Welch, by the way, used the very same phrase in his address delivered on the occasion of the opening of the Toronto biology laboratory in 1889, defining biology as "the study of life in all its forms and activities, both normal and abnormal"; William H.

Welch, "Pathology in Its Relation to General Biology," *Johns Hopkins Hospital Bulletin* 1 (1890): 25.

33. Arthur M. Silverstein, "The Heuristic Value of Experimental Systems: The Case of Immune Hemolysis," *Journal of the History of Biology* 27 (1994): 441.

34. For a brief overview of the various contributions to immunological research, see Bulloch, *The History of Bacteriology*. See also Silverstein's *A History of Immunology* for a detailed and more recent analysis.

35. See, in particular, Paul Ehrlich's work on ricin and abrin, "Experimentelle Untersuchungen über Immunität: I. Über Ricin" and "Experimentelle Untersuchungen über Immunität: II. Über Abrin," in *The Collected Papers of Paul Ehrlich: Immunology and Cancer Research*, ed. F. Himmelweit (London: Pergamon Press, 1957). To point to the broad range of defense actions, Ehrlich later introduced a more general term for all the substances that the animal body would produce— *antibodies*. In this broader perspective, "antitoxins" become a subset of antibodies.

36. Around 1900, there were actually two competing explanations being debated: Ehrlich's chemical and Jules Bordet's colloidal theory of the immune mechanism; see Eileen Crist and Alfred I. Tauber, "Debating Humoral Immunity and Epistemology: The Rivalry of the Immunochemists Jules Bordet and Paul Ehrlich," *Journal of the History of Biology* 30 (1997): 321–56. This debate continued for many years and involved several investigators on both sides. In the following, I concentrate on Ehrlich's perspective because it was the one that informed Flexner and Noguchi's work.

37. Paul Ehrlich and J. Morgenroth, "Contributions to the Theory of Lysin Action," in *The Collected Papers of Paul Ehrlich*, ed. Himmelweit, 151–52.

38. See Ross, "Making Medicine Scientific," 39–50, for a detailed account of these investigations.

39. *Mechanism* has become a technical term in the philosophical analysis of experimental biology; see Peter Machamer, Lindley Darden, and Carl F. Craver, "Thinking about Mechanisms," *Philosophy of Science* 67 (2000): 1–25; and Carl F. Craver and Lindley Darden, *In Search of Mechanisms: Discoveries across the Life Sciences* (Chicago: University of Chicago Press, 2013). In philosophical discussions, mechanisms are what biological researchers aim to explore. Prominent examples include the mechanism of DNA replication and the urea cycle. Mechanisms are characterized as "*entities and activities organized such that they are productive of regular changes from start or set-up to finish or termination condition*"; ibid., 15. They are different from "mere" machines in that they are active—they move, change, synthesize, or transmit things. And they can be further characterized— for instance, through descriptions of their start and finish conditions, of how their various different stages give rise to further stages, and of their spatiotemporal organization; ibid., 16–17. Needless to say, nineteenth-century investigators did not have such an intricate and conceptually well-developed understanding of the concept of mechanism, but they did use the term. *Mechanism* was used rather broadly to refer to the action of causal agents on body parts or organic substances, but it

was not explicitly introduced in the discussion. For each of these mechanisms, the experimenters aimed to specify the causal agents or "difference makers" (Darden & Craver) involved at the start and to determine the actual difference that they made to the process they were investigating.

40. Simon Flexner and Hideyo Noguchi, "Snake Venom in Relation to Haemolysis, Bacteriolysis, and Toxicity," *The Journal of Experimental Medicine* 6 (1902): 278.

41. Late nineteenth-century work on snake venom gave an entirely new twist to immunological research. The mechanisms involved in hemolysis, bacteriolysis, agglutination, and so forth were thought to be crucial not only as means of defense but also as means of attack on the body of a victim. In defense, the animal body used these means to destroy invasive agents such as germs. In offense, the attacker—in this case, the snake or, rather, the venom—exploited these means to destroy the victim's body. Venom exploited, in a sense, the destructive power that the animal body possessed against intruders and turned it against the animal body itself. In 1902, even before Flexner and Noguchi's second article on venom research had come out, Welch praised their work (along with that of another of his students, Preston Kyes) for precisely this reason. According to Welch, these researchers had been able to show that "venom serves merely to bring into the necessary relations with constituents of the body cells poisons we already harbour or may generate, but which are harmless without the intervention of intermediary bodies. These poisons within us are powerful weapons, which when seized by hostile hands may be turned with deadly effect against our own cells, but which are also our main defence against parasitic invaders"; William H. Welch, "The Huxley Lecture on Recent Studies of Immunity, with Special Reference to Their Bearing on Pathology," *British Medical Journal* 2 (1902): 1110.

42. The washing of blood corpuscles and blood clots had been used as a technique in research on blood throughout the nineteenth century; see, e.g., Andral, who according to his *Pathological Haematology* used a "saltwater solution of sulphate of soda" to separate the globules from the fibrin (34). Around 1900, the technique involved collecting fresh blood in a test tube containing a saline solution, gently centrifuging the blood until the particles separate, and removing the upper layer, repeating the procedure several times.

43. Flexner and Noguchi, "Snake Venom in Relation to Haemolysis, Bacteriolysis, and Toxicity," 287.

44. Albert Henry Buck, *Reference Handbook of the Medical Sciences Embracing the Entire Range of Scientific and Practical Medicine and Allied Science* (New York: William Wood and Company, 1915), 171.

45. Recall Mitchell and Reichert's investigations of physiological functions: the investigators mentioned that different venoms might differ in their physiological actions, but they did not address this systematically.

46. Simon Flexner and Hideyo Noguchi, "The Constitution of Snake Venom and Snake Sera," *University of Pennsylvania Medical Bulletin* (1902): 349.

47. See Theodore M. Porter, *Trust in Numbers: The Pursuit of Objectivity in*

Science and Public Life (Princeton, NJ: Princeton University Press, 1995). Porter has emphasized that processes of standardization—of establishing a particular standard—were by no means smooth and were indeed often beset with difficulty. Martha Lampland and Susan Leigh Star also remind us that the most interesting and most challenging historical question about standards and established procedures is the question of how particular standards could come into being; see Martha Lampland and Susan Leigh Star, *Standards and Their Stories: How Quantifying, Classifying and Formalizing Practices Shape Everyday Life* (Ithaca, NY: Cornell University Press, 2009). The history of standardization within the exact sciences is quite well understood; see David Cahan, *An Institute for an Empire: The Psysikalisch-Technische Reichsanstalt, 1871–1918* (Cambridge, UK: Cambridge University Press, 2004); and Kathryn Olesko, *Physics as a Calling: Discipline and Practice in the Königsberg Seminar for Physics* (Ithaca, NY: Cornell University Press, 1991). Outside the exact sciences, historical research has concentrated on early twentieth-century standardization of humans, laboratory animals, and therapeutic agents; see Lampland and Star, *Standards and Their Stories*; see Karen Rader, *Making Mice: Standardizing Animals for American Biomedical Research* (Princeton, NJ: Princeton University Press, 2004); and Christoph Gradmann and Jonathan Simon, eds., *Evaluating and Standardizing Therapeutic Agents, 1890–1950* (New York: Palgrave Macmillan, 2010).

48. See especially the contributions to Gradmann and Simon, *Evaluating and Standardizing Therapeutic Agents.*

49. A. Wassermann and K. Takaki, "Ueber tetanusantitoxische Eigenschaften des normalen Centralnervensystems," *Berliner Klinische Wochenschrift* (1898): 5–6. Wassermann and Takaki's paper served as a model for many similar investigations of the power of resistance of body tissue; see, e.g., W. Kempner and E. Schepilewsky, "Ueber antitoxische Substanzen gegenüber dem Botulismusgift" *Zeitschrift für Hygiene und Infektionskrankheiten* 27 (1898): 213–22.

50. For details, see Axel C. Hüntelmann, "Pharmaceutical Markets in the German Empire: Profits between Risk, Altruism and Regulation," *Historical Social Research* 36 (2011): 182–201.

51. For details, see William D. Forster, *A History of Medical Bacteriology and Immunology* (1970), 115–19. See also Jonathan Liebenau, "Paul Ehrlich as a Commercial Scientist and Research Administrator," *Medical History* 34 (1990): 65–78; and Timothy Lenoir, "A Magic Bullet: Research for Profit and the Growth of Knowledge in Germany around 1900," *Minerva* 26 (1988) for the commercialization of Ehrlich's research, as well as Cay-Rüdiger Prüll, "Paul Ehrlichs Standardization of Serum: *Wertbestimmung* and Its Meaning for Twentieth-Century Biomedicine," in *Evaluating and Standardizing Therapeutic Agents*, ed. Gradmann and Simon, 13–30; and Axel C. Hüntelmann, "Evaluation as a Practical Technique of Administration: The Regulation and Standardization of Diphtheria Serum," in *Evaluating and Standardizing Therapeutic Agents*, ed. Gradmann and Simon, 31–51.

52. "Report of the Lancet Special Commission on the Relative Strengths of Diphtheria Antitoxic Serums," *Lancet* (1896): 194.

53. Paul Ehrlich, "Croonian Lecture: On Immunity with Special Reference to Cell Life," *Proceedings of the Royal Society of London* 66 (1900): 426.

54. To be precise, Flexner and Noguchi's guinea-pigs weighed "from 250 to 300 grammes"; Flexner and Noguchi, "The Constitution of Snake Venom and Snake Sera," 291.

55. In his later work with Reichert, Mitchell referred to the differences between different snakes, but this difference did not turn into a task for experimentation. The two researchers assumed that different venoms might act differently, but (most likely for pragmatic reasons) they treated the venoms from different kinds of snakes as similar for the purposes of their inquiry. Flexner and Noguchi, by contrast, systematically studied the differences among different kinds of venom.

56. Ibid., 353.

57. Welch, "The Huxley Lecture on Recent Studies of Immunity."

58. Flexner, "Tendencies in Pathology," 133.

59. Charles O. Whitman, "A Biological Farm. For the Experimental Investigation of Heredity, Variation and Evolution and for the Study of Life-Histories, Habits, Instincts and Intelligence," *Biological Bulletin* 3 (1902): 217.

60. Henry Sewall, "Experiments on the Preventive Inoculation of Rattlesnake Venom," *Journal of Physiology* 8 (1887): 206.

61. E. Behring and S. Kitasato, "Ueber das Zustandekommen der Diphtherie-Immunität und der Tetanus-Immunität bei Thieren. Deutsche Medizinsche Wochenschrift 16:1113–1114," in *Milestones in Microbiology*, ed. Thomas D. Brock (Washington, DC: ASM Press, 1998), 139.

62. See Wassermann and Takaki, "Ueber tetanusantitoxische Eigenschaften des normalen Centralnervensystems," 5.

63. Jacques Loeb, *Untersuchungen Zur physiologischen Morphologie der Thiere* (Würzburg, Germany: Hertz, 1891), 27. Loeb saw himself as an "engineer" in biology; see Pauly, *Controlling Life*. The engineering conception of "controlling life," however, is quite remote from the methodological discourse about how controls can help make experimental results more reliable. For Loeb, an engineer was someone who could control (i.e., manipulate at will) the movements, actions, and the development of living forms and even create new forms of life. (Loeb called this phenomenon "heteromorphosis," the replacement of an organ by another physiologically and morphologically different; see Loeb, *Untersuchungen zur physiologischen Morphologie der Thiere*, 10.) Physiological morphology had a "synthetical or constructive" aim: "to form new combinations from the elements of living nature, just as the physicist and chemist form new combinations from the elements of non-living nature"; Jacques Loeb, "On Some Facts and Principles of Physiological Morphology," in *Biological Lectures Delivered at the Marine Biological Laboratory of Wood's Holl in the Summer Session of 1893* (Boston: Ginn &

Company, 1894), 61. The point was literally the fabrication—the engineering—of *new* things, not "the analysis of the existent"; see Pauly, *Controlling Life*, 8.

64. Whitney, *The Century Dictionary and Cyclopedia* II, 1237.

65. This project was inspired by George Nuttall's study of the blood's bactericidal properties. Nuttall's substantial 1888 paper on "bacteria-hostile" [*bacterienfeindlich*] agency, the defense mechanism of the animal body is a case in point. In it, Nuttall described experiments showing that the defibrinated blood of rabbits, mice, pigeons, sheep, and dogs had the power of destroying anthrax bacilli. Samples of fresh blood were taken from different kinds of animals, and a certain number of anthrax bacteria were added to each of the samples. The bacteria were added at specific times and counted at specific intervals. The decrease in the number of bacteria indicated that the blood could kill bacteria—and how quickly it could do so. Nuttall also showed that heating the blood destroyed its bactericidal power; see George Henry Falkiner Nuttall, "Experimente über die bacterienfeindlichen Einflüsse des thierischen Körpers," *Zeitschrift für Hygiene* 4 (1888): 353–93; see also Bulloch, *The History of Bacteriology* (London: Oxford University Press, 1938), 257–58.

66. Only in a later paper did Flexner and Noguchi spell out why this was an important finding: because the bactericidal power of the serum was diminished, the tissue of the bite victim was much more susceptible to bacteria invasion, and putrefaction of the tissue could occur; see Simon Flexner and Hideyo Noguchi, "Upon the Production and Properties of Anti-Crotalus Venin," *Journal of Medical Research* VI (1904): 364.

67. Flexner and Noguchi, "Snake Venom in Relation to Haemolysis, Bacteriolysis, and Toxicity," 295–6.

68. Ibid., 295–98.

69. In his study on the rise of the causal concept of disease, K. Codell Carter describes the "etiological research programme" in nineteenth-century medicine as a broad movement of which bacteriology was but one part; see Carter, *The Rise of Causal Concepts of Disease*, 7. This broad movement is, in turn, part of an even broader methodological endeavor to explore the possibilities and limits of experimental searches for causes.

70. See Dietrich von Engelhardt, "Kausalität und Konditionalität in der modernen Medizin," in *Pathogenese; Grundzüge und Perspektiven einer theoretischen Pathologie* (Berlin: Springer, 1985), 32–58.

71. See Carter, *The Rise of Causal Concepts of Disease*.

72. Philip Pauly makes this point for experimental physiology; see Pauly, *Controlling Life*, 17.

73. See Jane Maienschein, "The Origins of Entwicklungsmechanik," in *A Conceptual History of Modern Embryology*, ed. Scott Gilbert (Baltimore, MD: Johns Hopkins University Press, 1994), 43–61; Garland Allen, "Mechanism, Vitalism and Organicism in Late Nineteenth and Twentieth-Century Biology: The Importance of Historical Context," *Studies in History and Philosophy of Biological and Biomedical Sciences* 35 (2005): 261–83; Frederick B. Churchill, "From Machine-

Theory to Entelechy: Two Studies in Developmental Teleology," *Journal of the History of Biology* 2 (1969): 165–85; and Pauly, *Controlling Life*.

74. Loeb insisted that it was only the "engineering approach" that made experimental approaches to life possible and meaningful in the first place. For Loeb, that he could control life like the engineer proved those biologists wrong who found in animated nature "a new category of causes, such as are said continually to produce before our eyes great effects, without it being possible for an engineer ever to make use of these causes in the physical world"; Jacques Loeb, *Studies in General Physiology* (Chicago: University of Chicago Press, 1905), 107. His own investigations of the phenomena of heliotropism, geotropism, heteromorphosis, and related phenomena showed that there was no special category of causes in animated nature. Physical and chemical causes such as light, gravity, and chemical agents were responsible for movement and regeneration.

75. Wilhelm Roux, "The Problems, Methods, and Scope of Developmental Mechanics," in *Defining Biology*, ed. Jane Maienschein, 117.

76. The American biologist William Morton Wheeler translated the introduction; see "Einleitung," *Archiv für Entwickelungsmechanik* 1 (1894): 1–42. This was during the year before Flexner spent the summer with Loeb at Woods Hole. Jane Maienschein included this introduction in her collection *Defining Biology*.

77. Roux, "The Problems, Methods, and Scope of Developmental Mechanics," 144.

78. On Roux's editorial politics, see S. J. Counce, "Archives for Developmental Mechanics. W. Roux, Editor (1894–1924)," *Roux's Archive of Developmental Biology* 204 (1994): 79–92.

79. Roux, "The Problems, Methods, and Scope of Developmental Mechanics," 120.

80. Ibid., 120.

81. Ibid., 121.

82. To the twenty-first century reader, Roux's suggestion is evocative of recent debates about robust detection. I will return to this discussion.

83. Roux, "The Problems, Methods, and Scope of Developmental Mechanics," 117.

84. Albert McCalla, "President's Address: The Verification of Microscopic Observation," *Proceedings of the American Society of Microscopists* 5 (1883): 4. For more details on the debates in microscopy, see Jutta Schickore, "Test Objects," *History of Science* 47 (2009): 117–45; and Jutta Schickore, " '. . . as Many and as Diverse Methods as Possible Ought to Be Employed . . .'—Methodological Reflections in *General Cytology* in Historical Perspective," in *Visions of Cell Biology: Reflecting on Cowdry's* General Cytology, ed. Karl Matlin, Jane Maienschein, and Manfred Laubichler (Chicago: University of Chicago Press, forthcoming).

85. Crist and Tauber argue that the experimental styles of Bordet and Ehrlich are very different—they describe Bordet's style as Baconian, Ehrlich's style as heavily theory-laden; see Crist and Tauber, "Debating Humoral Immunity and

Epistemology: The Rivalry of the Immunochemists Jules Bordet and Paul Ehrlich." It seems to me, however, that the styles of the experimenters are in fact very
similar. Bordet's experiments, too, are guided by his assumptions about the nature
of the immune response; the difference is in the interpretation of the experimental
objects and outcomes and in the amount of theoretical baggage that the interpretations carried.

86. Of course, in principle, this skeptical response can also be made in observa-
tional contexts—we might say that we were deceived or prone to hallucinations.
But in complex, complicated experimental designs, the possibilities of raising
doubts are considerably more numerous.

87. Flexner and Noguchi, "The Constitution of Snake Venom and Snake Sera,"
345.

88. Ibid., 346.

89. These results were published a couple of years later; see Preston Kyes, "Ueber die Wirkungsweise des Cobragiftes," in *Gesammelte Arbeiten zur Immunitätsforschung*, ed. Paul Ehrlich (Berlin: August Hirschwald, 1904).

90. See Ian Hacking, *Representing and Intervening* (Cambridge, UK: Cambridge
University Press, 1983).

91. Simon Flexner and Hideyo Noguchi, "On the Plurality of Cytolysins in
Snake Venom," *Journal of Pathology* (1905): 115.

92. Ibid., 116.

93. Roux, "The Problems, Methods, and Scope of Developmental Mechanics,"
126.

94. See William Whewell, *The Philosophy of the Inductive Sciences*, ed. Gerd
Buchdahl and Larry Laudan, 2nd rev. ed., 2 vols. (London: Frank Cass, 1967; 1847).

95. Jane Maienschein has drawn attention to Roux's reflections on hypotheses.
She also emphasizes that not all experimental embryologists were equally optimistic about the epistemic force of experimentation and that, generally speaking,
American biologists were more skeptical than their German colleagues about the
certainty of the knowledge that could be derived from experiment; see Maienschein, "Arguments for Experimentation in Biology," 185–89.

96. See especially Larry Laudan, *Science and Hypothesis* (Dordrecht Reidel,
1981).

97. Ibid., 12–15.

98. Calmette, by the way, explicitly put his research in a colonial context. In
the introduction to his lecture on antivenins, delivered to the conjoint Board of
the Royal Colleges of Physicians (London) and Surgeons (England) in 1896, he
emphasized that the development of antivenins was necessary because the "very
considerable" financial loss through the death of cattle as a result of snake bites
had put financial strain on people in the colonies, especially in India. The *British Medical Journal*, which printed the text of the lecture (in English translation)
remarked that Calmette's talk had been "listened to with great interest"; see

Albert Calmette, "The Treatment of Animals Poisoned with Snake Venom by the Injection of Antivenomous Serum," *British Medical Journal* (1896): 399, 400.

99. Flexner and Noguchi, "The Constitution of Snake Venom and Snake Sera," 345, 352.

100. "Preparation of Copy for the *Botanical Gazette*," *Botanical Gazette* 61 (1916): 337.

101. American Medical Association, "Suggestions to Authors" (Chicago: American Medical Association Press, 1912), 6.

102. Ibid., 10.

103. William Cairns, *The Forms of Discourse* (Boston: Ginn & Company, 1896), 198.

104. Fleming, *William H. Welch and the Rise of Modern Medicine*, 197.

105. Clifford Allbutt, *Notes on the Composition of Scientific Papers*, 2nd ed. (New York: Macmillan, 1905), v.

106. Ibid., 57.

107. Ibid., 17.

108. Ibid., 18.

109. Ibid., 22–23.

110. Ibid., 18.

111. Ibid., 19.

Chapter Ten

1. For the role of enzymes for life, see James B. Sumner and Karl Myrbäck, "Introduction," in *The Enzymes: Chemistry and Mechanism of Action*, ed. James B. Sumner and Karl Myrbäck (New York: Academic Press, 1950), 1; and Robert E. Kohler, "The Enzyme Theory and the Origin of Biochemistry," *Isis* 64 (1973): 185. On the role of proteins for life, see Lily Kay, "Laboratory Technology and Biological Knowledge: The Tiselius Electrophoresis Apparatus, 1930–1945," *History and Philosophy of the Life Sciences* 10 (1988): 55; and Lily Kay, *The Molecular Vision of Life* (1993), 104–20. For the intertwined history of protein and enzyme research, see Joseph Fruton, *Proteins, Enzymes, Genes: The Interplay of Chemistry and Biology* (New Haven, CT: Yale University Press, 1999), especially chapters 4 and 5.

2. For the relation between the terms *enzyme* and *ferment*, see Fruton, *Proteins, Enzymes, Genes*; see also Mikuláš Teich, "Ferment or Enzyme, What's in a Name?" *History and Philosophy of the Life Sciences* 3 (1981): 193–215.

3. See Joseph Fruton, "Early Theories of Protein Structure," *Annals of the New York Academy of Sciences* (1979): 1–18; and Fruton, *Proteins, Enzymes, Genes*, 153–54. The term *protein* came into general use in the mid-nineteenth century, and a number of investigators sought to establish the chemical nature of protein constituents. In the first two decades of the twentieth century, only few techniques for

separating and purifying proteins were available. It was known that proteins were easily affected by heat, acid, and alkali. The most common technique was "salting out." In this procedure, the protein is precipitated by adding a salt—sodium chloride, sodium sulfate, or ammonium sulfate. Different proteins required different concentrations of salt to precipitate them, which made it possible to separate different protein fractions from one sample; see John T. Edsall, "The Development of the Physical Chemistry of Proteins, 1898–1940," *Annals of the New York Academy of Sciences* (1979): 54.

4. Cf. Fruton, *Proteins, Enzymes, Genes*, 161–62.

5. Edsall, "The Development of the Physical Chemistry of Proteins, 1898–1940," 56.

6. For details, see Boelie Elzen, "Two Ultracentrifuges: A Comparative Study of the Social Construction of Artefacts," *Social Studies of Science* 16 (1986): 621–62. Svedberg's invention combined a high-speed centrifuge with an optical device that traced the process of sedimentation of materials.

7. Angela Creager, *The Life of a Virus: Tobacco Mosaic Virus as an Experimental Model, 1930–1965* (Chicago: University of Chicago Press, 2002), 82–86.

8. See Kay, "Laboratory Technology and Biological Knowledge."

9. See Dionysius von Klobusitzky and P. König. "Biochemische Studien über die Gifte der Schlangengattung Bothrops. VI. Mitteilung: Kurzer Bericht über verschiedene, in den Jahren 1936–37 gewonnene Versuchsergebnisse, " *Naunyn-Schmiedebergs Archiv für experimentelle Pathologie und Pharmakologie* (1939): 271–75.

10. See Kay, *The Molecular Vision of Life*, 114. By the end of the 1930s, when filter paper came to be used as supporting medium, simpler and less expensive apparatuses for electrophoresis could be developed, and the technique became commercially available; see Louis Rosenfeld, "A Golden Age of Clinical Chemistry: 1948–1960," *Clinical Chemistry* 46 (2000): 1705–14. Less expensive, air-driven ultracentrifuges became widely available only by the 1950s.

11. Cf. Findlay E. Russell, "History of the International Society on Toxinology and Toxicon. I. The Formative Years, 1954–1965," *Toxicon* 25 (1987): 12.

12. Ibid., 3.

13. Slotta initially worked at the renowned biomedical facility Instituto Butantan in São Paolo. The institute was founded in 1901 and still exists today; it is still a major producer of sera and vaccines. In the 1930s, the institute's progressive director favored foundational research. He hired several émigrés from Nazi Germany and made it possible for them to bring the latest chemical equipment; see Robert A. Hendon and Allan L. Bieber, "Presynaptic Toxins from Rattlesnake Venom," in *Rattlesnake Venoms: Their Actions and Treatment*, ed. Anthony Tu (New York: Marcel Dekker, 1982), 213. Slotta, whose wife was Jewish and who fervently opposed the Nazi regime, was under increasing pressure at his home institution, the University of Breslau (now Wrocław, Poland). In 1935, the government of the State

of São Paolo offered him the directorship at the newly established section of chemical and experimental pharmacology at Butantan. Slotta accepted and emigrated with his wife and daughter; Barbara J. Hawgood, "Karl Heinrich Slotta (1895–1987) Biochemist: Snakes, Pregnancy and Coffee," *Toxicon* 39 (2001), 1279. But—at least for a time—he continued to publish in German and in German journals. Slotta obtained an external research grant, which allowed him to employ for a year his brother-in-law, the protein expert Heinz Fraenkel-Conrat; see Tu, *Rattlesnake Venoms*, iii. Slotta recalled in later years that on arrival at Butantan, he was asked to continue his previous work on sex hormones and to investigate the chemistry of coffee. But in 1937, Slotta took up snake venom research, taking advantage of the abundance of research materials, including a daily supply of hundreds of venomous snakes. (The institute was connected to a snake farm.) Slotta's stint at the institute was brief; the chemical and experimental pharmacology section was closed in 1938, when a new regime took over in São Paulo. Fraenkel-Conrat moved back to the United States. Slotta cofounded a biopharmaceutical company, which allowed him to continue his research in Brazil for a number of years; see Hawgood, "Karl Heinrich Slotta," 1280.

14. Afranio do Amaral, "Venoms and Antivenins," in *The Newer Knowledge of Bacteriology and Immunology*, ed. Edwin O. Jordan and I. S. Falk (Chicago: University of Chicago Press, 1928), 1070.

15. According to Robert Kohler, this new field was held together by a general awareness of the importance of enzymes for physiological processes; see Kohler, "The Enzyme Theory and the Origin of Biochemistry."

16. C. H. Kellaway, "Animal Poisons," *Annual Review of Biochemistry* 8 (1939): 545.

17. This insight is expressed by Donald Fairbairn in "The Phospholipase of the Venom of the Cottonmouth Moccasin (Agkistrodon Piscivorus L.)" *Journal of Biological Chemistry* 157 (1945): 633–44.

18. E. A. Zeller, "Enzymes as Essential Components of Bacterial and Animal Toxins," in *The Enzymes*, ed. Sumner and Myrbäck, 991.

19. Ibid., 1000.

20. When Fraenkel-Conrat joined Slotta in Brazil, he had just completed a year at the Rockefeller Institute, studying proteins. Fraenkel-Conrat is of course a much more familiar figure in history of biology than Slotta; on Fraenkel-Conrat's work on TMV, see Creager, *The Life of a Virus*; and Christina Brandt, *Metapher und Experiment: Von der Virusforschung zum genetischen Code* (Göttingen, Germany: Wallstein Verlag, 2004).

21. Karl Heinrich Slotta and Heinz Fraenkel-Conrat, "Two Active Proteins from Rattlesnake Venom," *Nature* 3587 (1938): 213.

22. Later reviewers described this accomplishment as the most momentous development in twentieth-century venom research, not least because it stimulated further research on the isolation of toxins from venoms; see Anthony Tu, "The Mechanism

of Snake Venom Actions—Rattlesnakes and Other Crotalids" in *Neuropoisons;
Their Pathophysiological Action*, ed. Lance L. Simpson (New York: Plenum Press,
1971), 92–93. Looking back at his early work, Slotta himself proudly stated in 1982:
"In a way, this discovery may have provided the impetus for world-wide research
on animal venoms, the founding of the International Society on Toxinology in 1962
and its publication *Toxicon*, and all the great achievements up to the present time:
research on snake venom enzymes, neurotoxins, cardiotoxins, phospholipase A,
L-amino-acid oxidase, and hemolysis and blood coagulation and the search for the
pharmacological effects of all animal venoms"; see Tu, *Rattlesnake Venoms*, iii.

23. Decades later, in 1994, Fraenkel-Conrat recalled that his brother-in-law had
initially assumed that the toxic component of venom was a sterol. Slotta had just
finished another project on an indigenous animal, the cane toad. Together with
a colleague, he had analyzed the secretion of that toad and found, among other
things, steroidal components. But he subsequently found that it was, in fact, a protein: "When the snake venom neurotoxins proved to be proteinaceous, and his
'protein chemist' brother-in-law happened to be visiting, he quickly made arrangements for him (me) to stay for a year, at a living wage"; H. Fraenkel-Conrat, "Early
Days of Protein Chemistry," *The FASEB Journal* 8 (1994): 452.

24. Karl Heinrich Slotta and Heinz Fraenkel-Conrat, "Schlangengifte, II. Mitteilung: Ueber die Bindungsart des Schwefels," *Berichte der deutschen chemischen
Gesellschaft* 71 (1938): 265.

25. See, e.g., ibid., 266–67. This wider notion of hypothesis was also common
around 1900; in fact some commentators on scientific methodology explicitly contrasted their views with this everyday use of the term.

26. Robert G. Hudson, *Seeing Things: The Philosophy of Reliable Observation*
(Oxford: Oxford University Press, 2014).

27. Brett Calcott, "Wimsatt and the Robustness Family: Review of Wimsatt's
Re-Engineering Philosophy for Limited Beings," *Biology and Philosophy* 26 (2011):
281–93; and James Woodward, "Some Varieties of Robustness," *Journal of Economic Methodology* 13 (2006): 219–40.

28. See Nancy Cartwright, "Replicability, Reproducibility and Robustness:
Comments on Harry Collins," *History of Political Economy* 23 (1991): 143–55.

29. William Bechtel, "Aligning Multiple Research Techniques in Cognitive
Neuroscience: Why Is It Important?" *Philosophy of Science* 69 (2002): S48–S58.

30. This was the case in the much-discussed example of the investigation of
bacterial mesosomes; see Nicolas Rasmussen, "Facts, Artifacts, and Mesosomes:
Practicing Epistemology with the Electron Microscope," *Studies in History and
Philosophy of Science* 24, no. 2 (1993): 227–65; Sylvia Culp, "Defending Robust-
ness: The Bacterial Mesosome as a Test Case," *PSA: Proceedings of the Biennial
Meeting of the Philosophy of Science Association* (1994): 46–57; and Robert G.
Hudson, "Mesosomes: A Study in the Nature of Experimental Reasoning," *Phi-
losophy of Science* 66 (1999): 289–309.

31. For a detailed survey of recent philosophical works on multiple means

of determination and the historical roots of this epistemic strategy, see Klodian Çoko, "The Structure and Epistemic Import of Empirical Multiple Determination in Scientific Practice" (PhD diss., Indiana University, 2015).

32. I am borrowing this term from Nancy Cartwright. In a commentary on Harry Collins's work on replication, Cartwright explicitly puts herself at the "epistemological" end of a line connecting real-life methodology and ideal epistemology. She describes her epistemological considerations as "abstract," preoccupying the philosopher "even if they play a negligible role in the real world"; see Cartwright, "Replicability, Reproducibility and Robustness," 143. I do not wish to align my project with Collins's work, but I think that Cartwright's notion of the line with two endpoints is a helpful conception of analytic approaches to scientific methodology.

33. Bechtel has examined an episode from recent cognitive neuroscience that also highlights this strategy of seeking "complementary" information. He shows that the purpose for aligning instrumental techniques (lesion, cell recording, and neuroimaging) is not to obtain converging information about brain functions but to obtain complementary results, which are integrated into a comprehensive account of brain function than could not be sustained if only the findings obtained with one single technique were considered; see Bechtel, "Aligning Multiple Research Techniques in Cognitive Neuroscience: Why Is It Important?"

34. Karl Heinrich Slotta and Walter Forster, "Schlangengifte, IV. Mitteilung: Quantitative Bestimmung der schwefelhaltigen Bausteine," *Berichte der deutschen chemischen Gesellschaft* 71 (1938): 1082.

35. E. Grasset, et al., "Comparative Analysis and Electrophoretic Fractionations of Snake Venoms," in *Venoms*, ed. E. E. Buckley and N. Porges (Washington, DC: American Association of the Advancement of Science, 1956), 153, emphasis added.

36. Ibid., 168.

37. The authors announced the successful identification of two principles. The other protein was found to cause blood coagulation, but it could not be crystallized.

38. Karl Heinrich Slotta and Heinz Fraenkel-Conrat, "Schlangengifte, III. Mitteilung: Reinigung und Krystallisation des Klapperschlangen-Giftes," *Berichte der deutschen chemischen Gesellschaft* 71 (1938): 213.

39. Fraenkel-Conrat, "Early Days of Protein Chemistry," 452.

40. For a detailed description of the methods, see The Svedberg, "The Ultra-Centrifuge and the Study of High-Molecular Compounds," *Nature* 139 (1937): 1051–62.

41. Nils Gralén and The Svedberg, "The Molecular Weight of Crotoxin," *Biochemical Journal* 32 (1938): 1375, 1377.

42. B. N. Gosh and S. S. De, "Proteins of Rattlesnake Venom," *Nature* 143 (1939): 380–81.

43. Karl Heinrich Slotta and Heinz Fraenkel-Conrat, "Crotoxin," Nature 144 (1939): 290–91. The experiment was "low-tech" because the two researchers simply produced a saturated solution of rattlesnake venom with sodium chloride and

cooled it overnight. Although separation could not be achieved, the proportion of hemolytic and neurotoxic activity differed from the crude venom: the percentage ratio of hemolysin appeared to be much higher. The two investigators took this to be evidence for the substance's composite nature and concluded that the hemolytic and neurotoxic activities must be due to two different substances; see Gosh and De, "Proteins of Rattlesnake Venom," 381.

44. Klobusitzky and König, "Biochemische Studien über die Gifte der Schlangengattung Bothrops," 274.

45. Choh Hao Li and Heinz Fraenkel-Conrat, "Electrophoresis of Crotoxin," *Journal of the American Chemical Society* 64 (1942): 1586–88.

46. Ibid., 1586, n. 2a.

47. Karl Heinrich Slotta, "Zur Chemie der Schlangengifte," *Experientia* 9 (1953): 81.

48. Ibid., 86, 87.

49. Emiliano Trizio suggests that this strategy functions as "methodological attractor"; see Emiliano Trizio, "Achieving Robustness to Confirm Controversial Hypotheses: A Case Study in Cell Biology," in *Characterizing the Robustness of Science*, ed. Léna Soler, Emiliano Trizio, Thomas Nickles, and William Wimsatt, (Boston: Springer, 2012), 119–20. For another explicit statement of the methodological strategy, see Jean Brachet's call for a variety of methods in cell physiology: "The biologist who is interested in cell physiology should not be a morphologist, or a physiologist, or a biochemist: he should not only be capable of using physiological and biochemical methods as well as the microscope, but he should utilize them all in attacking his problem. Neither the variety of the methods nor the acquisition of a wide knowledge in very different fields should frighten him. This is the price which has to be paid if cell physiology is to progress. The same price has been paid in the field biochemical genetics of micro-organisms, in which such outstanding advances have been made recently," quoted in Burian, "Exploratory Experimentation and the Role of Histochemical Techniques in the Work of Jean Brachet, 1938–1952," 27.

50. Karl Heinrich Slotta, "Further Experiments on Crotoxin," in *Venoms: Papers Presented at the First International Conference on Venoms December 27–30, 1954 at the Annual Meeting of the American Association of the Advancement of Science, Berkeley, California*, ed. E. E. Buckley and N. Porges (Washington, DC: American Association of the Advancement of Science, 1956), 254.

51. Ibid., 255.

52. Ernst Habermann and Wilhelm Neumann, "Beiträge zur Charakterisierung der wirksamen Komponenten von Schlangengiften," *Archiv für experimentelle Pathologie und Pharmakologie* 223 (1954): 396.

53. Slotta would continue to work and publish on toxins, however. In 1956, Slotta became professor of biochemistry at the School of Medicine at the University of Miami; see Hawgood, "Karl Heinrich Slotta," 1280.

54. Wilhelm Neumann, professor of toxicology and pharmacology at Würzburg University, is an interesting figure. He had been involved in toxicological research on poison gas during WWII. Because Neumann had been a member of the NSDAP and involved in research on chemical weapons, he was suspended from his position at Würzburg University immediately after the war. But during denazification, he was eventually found to be a "Follower" [*Mitläufer*], not an offender, and he was permitted to return to his position as professor of pharmacology and toxicology. He continued to work on poisons but turned his attention to animal poisons—especially bee and snake venom—as well as to environmental poisons and food additives. The snake venom had been obtained from Slotta. For biographical information about Neumann, more details about his involvement with the National Socialist Party, and information about his career during and after the Third Reich, see Stefanie Kalb, *Wilhelm Neumann (1898–1965): Leben und Werk unter besonderer Berücksichtigung seiner Rolle in der Kampfstoff-Forschung* (Univ. diss., Würzburg, 2005).

55. Neumann added that toxic action was determined through injection of the fractions into white mice—"in most cases 6 animals"; Wilhelm Paul Neumann, "Chromatographische Trennung von 'Crotoxin' in zwei verschiedene Wirkstoffe," *Die Naturwissenschaften* (1955): 370.

56. Wilhelm Paul Neumann and Ernst Habermann, "Ueber Crotactin, das Haupttoxin des Giftes der Brasilianischen Klapperschlange (Crotalus terrificus terrificus)," *Biochemische Zeitschrift* 327 (1955): 173.

57. Robert A. Hendon and Heinz Fraenkel-Conrat, "Biological Roles of the Two Components of Crotoxin," *Proceedings of the National Academy of Sciences of the United States of America* 68 (1971): 1560–63; Klaus Rübsamen, Henning Breithaupt, and Ernst Habermann, "Biochemistry and Pharmacology of the Crotoxin Complex. 1. Subfractionation and Recombination of the Crotoxin Complex," *Naunyn-Schmiedebergs Archiv der Pharmakologie* 270 (1971): 274–88.

58. Rübsamen, Breithaupt, and Habermann, "Biochemistry and Pharmacology of the Crotoxin Complex," 274n.

59. Ernst Habermann and K. Rübsamen, "Biochemical and Pharmacological Analysis of the So-Called Crotoxin," in *Toxins of Animal and Plant Origin*, ed. André De Vries and Elazar Kochva (New York: Gordon and Breach, 1973), 333.

60. The synergetic effect "does not appear to depend upon physical complex formation . . ."; see Philip Rosenberg, "Pharmacology of Phospholipase A2 from Snake Venoms," in *Snake Venoms*, ed. Chen-Yuan Lee (Berlin: Springer, 1979), 403–47, quote on 408; see also Sandra C. Sampaio et al., "Crotoxin: Novel Activities for a Classic B-Neurotoxin," *Toxicon* 55 (2010): 1046–47.

61. Hendon and Bieber, "Presynaptic Toxins from Rattlesnake Venom," 218–19; for a recent review, see Sampaio et al., "Crotoxin: Novel Activities for a Classic B-Neurotoxin."

62. Rosenberg, "Pharmacology of Phospholipase A2 from Snake Venoms," 410.

63. Ibid., 403.

64. Ibid., 408.

65. George Gopen and Judith Swan, "The Science of Scientific Writing," *American Scientist* (1990): 551.

66. Maud Mellish, *The Writing of Medical Papers* (Philadelphia, PA: W. B. Saunders, 1922), 73–75.

67. Ibid., 74.

68. Ray P. Baker and Almonte C. Howell, *The Preparation of Reports, Scientific, Engineering, Administrative, Business*, rev. ed. (New York: The Ronald Press Company, 1938), iii.

69. Ibid., 501–2.

70. Ibid., 503.

71. See Melinda Baldwin, *Making Nature: The History of a Scientific Journal* (Chicago: University of Chicago Press, 2015). Baldwin shows that *Nature* has an intriguing history, but her discussion focuses on the problem of demarcation and on the political context of *Nature* in the interwar period, not so much on the changing organization of the *Nature* article.

72. Li and Fraenkel-Conrat, "Electrophoresis of Crotoxin," 1586.

Conclusion

1. Such migration of protocols is also visible in other fields. Late nineteenth-century histologists, for instance, used experimental designs from experimental physiology in their studies of fluids and tissues; see Schickore, " '. . . as Many and as Diverse Methods as Possible Ought to Be Employed . . . .' "

2. Hannah Landecker's book on tissue cultures shows how productive the study of methods sections can be; see Hannah Landecker, *Culturing Life: How Cells Became Technologies* (Cambridge, MA: Harvard University Press, 2007).

3. Franklin and Perovic, "Experiment in Physics," section 3.2.1.

4. Bazerman, *Shaping Written Knowledge*, 140–41.

5. In turn, establishing credibility of the author remains a concern until today—according to Shapin, the perceived need to do so even increased in the course of the twentieth century; see Steven Shapin, *The Scientific Life: A Moral History of a Late Modern Vocation* (Chicago: University of Chicago Press, 2008).

6. Paul Feyerabend, *Against Method: Outline of an Anarchistic Theory of Knowledge*, 3rd ed. (London: Verso, 1993), vii.

Bibliography

Abbott, Alexander C. "Lunching with Reichert." *Scope (University of Pennsylvania Medical Students' Yearbook)* 21 (1924): 65–70.

Allbutt, Clifford. *Notes on the Composition of Scientific Papers.* 2nd ed. New York: Macmillan, 1905.

Allen, Garland. "Mechanism, Vitalism and Organicism in Late Nineteenth and Twentieth-Century Biology: The Importance of Historical Context." *Studies in History and Philosophy of Biological and Biomedical Sciences* 35 (2005): 261–83.

Amaral, Afranio do. "Venoms and Antivenins." In *The Newer Knowledge of Bacteriology and Immunology*, edited by Edwin O. Jordan and I. S. Falk, 1066–77. Chicago: University of Chicago Press, 1928.

American Medical Association. "Suggestions to Authors." Chicago: American Medical Association Press, 1912.

Amici, Raffaele Roncalli. "The History of Italian Parasitology." *Veterinary Parasitology* 98 (2001): 3–30.

Amsterdamska, Olga. "Research at the Hospital of the Rockefeller Institute for Medical Research." In *Creating a Tradition of Biomedical Research: Contributions to the History of the Rockefeller University*, edited by Darwin H. Stapleton, 111–26. New York: Rockefeller University Press, 2004.

Andral, Gabriel. *Pathological Haematology: An Essay on the Blood in Disease.* Philadelphia, PA: Lea and Blanchard, 1844.

Appel, Toby. "Biological and Medical Societies and the Founding of the American Physiological Society." In *Physiology in the American Context, 1850–1940*, edited by Gerald Geison, 155–76. Bethesda, MD: American Physiological Society, 1987.

Attfield, John. *Chemistry: General, Medical, and Pharmaceutical.* 10th ed. Philadelphia, PA: Henry C. Lea's Son & Co., 1883.

Bacon, Francis. *The New Organon.* Cambridge, UK: Cambridge University Press, 2000.

Baker, Ray P., and Almonte C. Howell. *The Preparation of Reports, Scientific, Engineering, Administrative, Business.* Rev. ed. New York: The Ronald Press Company, 1938.

Baldwin, Martha. "The Snakestone Experiments: An Early Modern Medical Debate." *Isis* 86 (1995): 394–418.

Baldwin, Melinda. *Making Nature: The History of a Scientific Journal.* Chicago: University of Chicago Press, 2015.

Barlow, John. *On Man's Power over Himself to Prevent or Control Insanity.* London: Pickering, 1849.

Bazerman, Charles. "Modern Evolution of the Experimental Report in Physics: Spectroscopic Articles in Physical Review, 1893–1980." *Social Studies of Science* 14 (1984): 163–96.

———. *Shaping Written Knowledge.* Madison: University of Wisconsin Press, 1988.

———. "Studies of Scientific Writing—E Pluribus Unum?" *4S Review* 3 (1985): 13–20.

Bechtel, William. "Aligning Multiple Research Techniques in Cognitive Neuroscience: Why Is It Important?" *Philosophy of Science* 69 (2002): S48–S58.

———. *Discovering Cell Mechanisms: The Creation of Modern Cell Biology.* Cambridge, UK: Cambridge University Press, 2006.

Behring, Emil von, and S. Kitasato. "Ueber das Zustandekommen der Diphtherie-Immunität und der Tetanus-Immunität bei Thieren. Deutsche Medizinsche Wochenschrift 16:1113–1114." In *Milestones in Microbiology*, edited by Thomas D. Brock, 139. Washington, DC: ASM Press, 1998.

Beretta, Marco. "At the Source of Western Science: The Organization of Experimentalism at the Accademia Del Cimento (1657–1667)." *Notes and Records of the Royal Society of London* 54 (2000): 131–51.

Bernard, Claude. *An Introduction to the Study of Experimental Medicine.* New York: Dover Publications, 1957. 1865.

———. *Leçons sur les effets des substances toxiques et médicamenteuses.* Paris: J. B. Baillière et fils, 1857.

Bernoulli, Johann. "Dissertation on the Mechanics of Effervescence and Fermentation." *Transactions of the American Philosophical Society* 87 (1997): 35–97.

Bertoloni Meli, Domenico. "A Lofty Mountain, Putrefying Flesh, Styptic Water, and Germinating Seeds." In *The Accademia Del Cimento and Its European Context*, edited by Marco Beretta, Antonio Clericuzio, and Larry Principe. Sagamore Beach, MA: Science History Publications, 2009.

Bertoloni Meli, Domenico, and Anita Guerrini. "Special Issue on Vivisection." *Journal of the History of Biology* 46 (2013).

Bertomeu-Sánchez, José Ramón, and Agustí Nieto-Galan, eds. *Chemistry, Medicine, and Crime: Mateu J.B. Orfila (1787–1853) and His Times.* Sagamore Beach, MA: Science History Publications, 2006.

Biagioli, Mario. "Etiquette, Interdependence, and Sociability in Seventeenth-Century Science." *Critical Inquiry* 22 (1996): 193–238.

Billings, John. "An Address on Our Medical Literature." *British Medical Journal* (1881): 262–68.

Board, The. *Report of the Principal of the Central High School to the Committee of the Board of Controllers of the Public Schools, for the Year Ending July, 1842.* Philadelphia, PA: 1843.

Bond, W. C. "Description of the Nebula About the Star Θ Orionis." *Memoirs of the American Academy of Arts and Sciences, New Series* 3 (1848): 87–96.

Boring, Edwin Garrigues. "The Nature and History of Experimental Control." *American Journal of Psychology* 67 (1954): 573–89.

Boschiero, Luciano. *Experiment and Natural Philosophy in Seventeenth-Century Tuscany.* New York: Springer, 2007.

———. "Natural Philosophizing inside the Late Seventeenth-Century Tuscan Court." *British Journal for the History of Science* 35 (2002): 383–410.

Bowditch, Henry P. "The Advancement of Medicine by Research." *Science* 4 (1896): 85–101.

Boyle, Robert. "Essay I. Containing Some Particulars Tending to Shew the Usefulness of Natural Philosophy to the Physiological Part of Physick." In *The Works of Robert Boyle,* edited by Michael Hunter and Edward B. Davies, 295–311. London: Pickering & Chatto, 1999.

———. "Essay II. Offering Some Particulars Relating to the Pathologicall Part of Physick." In *The Works of Robert Boyle,* edited by Michael Hunter and Edward B. Davies, 312–29. London: Pickering & Chatto, 1999.

———. "The First Essay, of the Unsuccessfulness of Experiments." In *The Works of Robert Boyle,* edited by Michael Hunter and Edward B. Davies, 37–56. London: Pickering & Chatto, 1999.

———. "Observations out of Mr B. Essay of Turning Poisons into Medicins." In *The Works of Robert Boyle,* edited by Michael Hunter and Edward B. Davis. London: Pickering & Chatto, 2000.

———. "The Second Essay, of Unsucceeding Experiments." In *The Works of Robert Boyle,* edited by Michael Hunter and Edward B. Davies, 57–82. London: Pickering & Chatto, 1999.

Brain, Robert. "The Graphic Method: Inscription, Visualization, and Measurement in Nineteenth-Century Science and Culture." PhD diss., University of California, 1996.

Brain, Robert, and Norton Wise. "Muscles and Engines: Indicator Diagrams and Helmholtz's Graphical Methods." In *Universalgenie Helmholtz: Rückblick nach 100 Jahren,* edited by Lorenz Krüger, 124–45. Berlin: Akademie-Verlag, 1994.

Brandt, Christina. *Metapher und Experiment: Von der Virusforschung zum genetischen Code.* Göttingen: Wallstein Verlag, 2004.

Buchwald, Jed Z. "Discrepant Measurements and Experimental Knowledge in the Early Modern Era." *Archive for the History of the Exact Sciences* 60 (2006): 565–649.

Buck, Albert Henry. *Reference Handbook of the Medical Sciences Embracing the Entire Range of Scientific and Practical Medicine and Allied Science.* New York: William Wood and Company, 1915.

Bulloch, William. *The History of Bacteriology.* London: Oxford University Press, 1938.

Burian, Richard M. "Exploratory Experimentation and the Role of Histochemical Techniques in the Work of Jean Brachet, 1938–1952." *History and Philosophy of the Life Sciences* 19 (1997): 27–45.

Burney, Ian A. *Poison, Detection, and the Victorian Imagination.* Manchester:
Manchester University Press, 2006.

Burr, Anna Robeson. *Weir Mitchell—His Life and Letters.* New York: Duffield &
Co, 1929.

Bynum, William F. *Science and the Practice of Medicine in the Nineteenth Century.*
Cambridge, UK: Cambridge University Press, 1994.

Cahan, David. *An Institute for an Empire: The Psysikalisch-Technische Reichsanstalt, 1871–1918.* Cambridge, UK: Cambridge University Press, 2004.

Cairns, William. *The Forms of Discourse.* Boston: Ginn & Company, 1896.

Calcott, Brett. "Wimsatt and the Robustness Family: Review of Wimsatt's Re-
Engineering Philosophy for Limited Beings." *Biology and Philosophy* 26 (2011):
281–93.

Calmette, Albert. "The Treatment of Animals Poisoned with Snake Venom by the
Injection of Antivenomous Serum." *British Medical Journal* (1896): 399–400.

Canale, D. J. "Civil War Medicine from the Perspective of S. Weir Mitchell's 'the
Case of George Dedlow.'" *Journal of the History of the Neurosciences* 11 (2002):
11–18.

Carter, K. Codell. "Koch's Postulates in Relation to the Work of Jacob Henle and
Edwin Klebs." *Medical History* 29 (1985): 353–75.

———. *The Rise of Causal Concepts of Disease.* Aldershot: Ashgate, 2003.

Cartwright, Nancy. "Replicability, Reproducibility and Robustness: Comments on
Harry Collins." *History of Political Economy* 23 (1991): 143–55.

Catellani, Patrizia, and Renzo Console. "Moyse Charas, Francesco Redi, the Viper and the Royal Society of London." *Pharmaceutical Historian: Newsletter of
the British Society for the History of Pharmacy* (2004): 2–10.

Cervetti, Nancy. "S. Weir Mitchell and His Snakes: Unraveling the 'United Web
and Woof of Popular and Scientific Beliefs.'" *Journal of Medical Humanities* 28
(2007): 119–33.

———. *S. Weir Mitchell, 1829–1914: Philadelphia's Literary Physician.* University
Park: Pennsylvania State University, 2012.

Chalmers, Iain. "Comparing Like with Like: Some Historical Milestones in the
Evolution of Methods to Create Unbiased Comparison Groups in Therapeutic
Experiments." *International Journal of Epidemiology* 30 (2001): 1156–64.

Chang, Ku-Ming (Kevin). "Fermentation, Phlogiston and Matter Theory: Chemistry and Natural Philosophy in Georg Ernst Stahl's 'Zymotechnia Fundamentalis.'" *Early Science and Medicine* 7 (2002): 31–64.

Charas, Moyse. *A Continuation of the New Experiments Concerning Vipers.* London: John Murray, 1673. 1671.

———. *New Experiments Upon Vipers. Containing Also an Exact Description of*

All the Parts of a Viper, the Seat of Its Poyson, and the Several Effects Thereof, Together with the Exquisite Remedies, That by the Skilful May Be Drawn from Vipers, as Well for the Cure of Their Bitings, as for That of Other Maladies. Originally Written in French. Now Rendered English. London: Printed by T. N. for J. Martyn, 1670. 1669.

———. *The Royal Pharmacopoeea, Galenical and Chymical According to the Practice of the Most Eminent and Learned Physitians of France: And Publish'd with Their Several Approbations.* London: Printed for John Starkey at the Miter within Temple-Bar, and Moses Pitt at the Angel in St. Pauls Church-Yard, 1678.

Churchill, Frederick B. "From Machine-Theory to Entelechy: Two Studies in Developmental Teleology." *Journal of the History of Biology* 2 (1969): 165–85.

Clericuzio, Antonio. "From Van Helmont to Boyle: A Study of the Transmission of Helmontian Chemical and Medical Theories in Seventeenth-Century England." *British Journal for the History of Science* 26 (1993): 303–34.

Çoko, Klodian. "The Structure and Epistemic Import of Empirical Multiple Determination in Scientific Practice." PhD diss., Indiana University, 2015.

Coleman, William. "Mechanical Philosophy and Hypothetical Physiology." In *The Annus Mirabilis of Sir Isaac Newton, 1666–1966*, edited by Robert Palter, 322–32. Cambridge, MA: MIT Press, 1970.

———. "Experimental Physiology and Statistical Inference: The Therapeutic Trial in Nineteenth-Century Germany." In *The Probabilistic Revolution: Ideas in the Sciences*, edited by Lorenz Krüger, Lorraine Daston, and Michael Heidelberger, 201–28. Cambridge, MA: MIT Press, 1987.

Comte, Auguste. *The Positive Philosophy of Auguste Comte.* New York: Calvin Blanchard, 1855.

Corner, George W. *Two Centuries of Medicine: A History of the School of Medicine, University of Pennsylvania.* Philadelphia, PA: Lippincott & Co., 1965.

Counce, S. J. "Archives for Developmental Mechanics. W. Roux, Editor (1894–1924)." *Roux's Archive of Developmental Biology* 204 (1994): 79–92.

Craver, Carl F., and Lindley Darden. *In Search of Mechanisms: Discoveries across the Life Sciences.* Chicago: University of Chicago Press, 2013.

Creager, Angela. *The Life of a Virus: Tobacco Mosaic Virus as an Experimental Model, 1930–1965.* Chicago: University of Chicago Press, 2002.

Crist, Eileen, and Alfred I. Tauber. "Debating Humoral Immunity and Epistemology: The Rivalry of the Immunochemists Jules Bordet and Paul Ehrlich." *Journal of the History of Biology* 30 (1997): 321–56.

Crookshank, Edgar March. *Manual of Bacteriology.* London: H. K. Lewis, 1887.

Culp, Sylvia. "Defending Robustness: The Bacterial Mesosome as a Test Case." *PSA: Proceedings of the Biennial Meeting of the Philosophy of Science Association* (1994): 46–57.

Cunningham, Andrew. "Transforming Plague: The Laboratory and the Identity of Infectious Diseases." In *The Laboratory Revolution in Medicine*, edited by

Andrew Cunningham and L. Pearce Williams, 209–44. Cambridge, UK: Cambridge University Press, 1992.

Cunningham, Andrew, and Perry Williams, eds. *The Laboratory Revolution in Medicine*. Cambridge, UK: Cambridge University Press, 1992.

Dalton, John Call. *Treatise on Human Physiology, Designed for the Use of Students and Practitioners of Medicine*. Philadelphia, PA: Blanchard and Lea, 1859.

———. *Vivisection; What It Is, and What It Has Accomplished*. New York: Baillière et fils, 1867.

Darwin, Charles. *The Power of Movement in Plants*. Assisted by Francis Darwin. London: John Murray, 1880.

Daston, Lorraine. "The Empire of Observation, 1600–1800." In *Histories of Scientific Observation*, edited by Lorraine Daston and Elizabeth Lunbeck, 81–113. Chicago: University of Chicago Press, 2011.

Daston, Lorraine, and Elizabeth Lunbeck, eds. *Histories of Scientific Observation*. Chicago: University of Chicago Press, 2011.

Davy, Humphry. "An Account of Some Experiments and Observations on the Constituent Parts of Certain Astringent Vegetables; and on Their Operation in Tanning." *Abstracts of the Papers Printed in the Philosophical Transactions of the Royal Society of London* 1 (1800–1814): 114–18.

de Chadarevian, Soraya. "Graphical Methods and Discipline: Self-Recording Instruments in Nineteenth-Century Physiology." *Studies in History and Philosophy of Science* 24 (1992): 267–291.

Dear, Peter. "Narratives, Anecdotes, and Experiments: Turning Experience into Science in the Seventeenth Century." In *The Literary Structure of Scientific Argument: Historical Studies*, edited by Peter Dear, 135–63. Philadelphia: University of Pennsylvania Press, 1991.

Debus, Allen G. "Rise and Fall of Chemical Physiology in the Seventeenth Century." *Memorias da Academia das Ciencias de Lisboa, (Classe de Ciencias)* 36 (1996): 36–60.

Dierig, Sven. *Wissenschaft in der Maschinenstadt: Emil Du Bois-Reymond und seine Laboratorien in Berlin*. Berlin: Wallstein Verlag, 2006.

Douard, John W. "E.-J. Marey's Visual Rhetoric and the Graphic Decomposition of the Body." *Studies in History and Philosophy of Science* 26 (1995): 175–204.

Dougherty, Frank P. W. "Nervenmorphologie und -Physiologie in den 8oer Jahren des 18. Jahrhunderts." In *Gehirn–Nerven–Seele: Anatomie und Physiologie im Umfeld S. Th. Soemmerrings*, edited by G. Mann and F. Dumont, 55–91. Stuttgart: G. Fischer, 1988.

Earles, Melvin P. "The Experimental Investigations of Viper Venom by Felice Fontana (1730–1805)." *Annals of Science* 16 (1960): 255–68.

Edsall, John T. "The Development of the Physical Chemistry of Proteins, 1898–1940." *Annals of the New York Academy of Sciences* (1979): 53–73.

Ehrlich, Paul. "Croonian Lecture: On Immunity with Special Reference to Cell Life." *Proceedings of the Royal Society of London* 66 (1900): 424–48.

————. "Experimentelle Untersuchungen über Immunität: I. Über Ricin." In *The Collected Papers of Paul Ehrlich: Immunology and Cancer Research*, edited by F. Himmelweit, 21–26. London: Pergamon Press, 1957.

————. "Experimentelle Untersuchungen über Immunität: II. Über Abrin." In *The Collected Papers of Paul Ehrlich: Immunology and Cancer Research*, edited by F. Himmelweit, 27–31. London: Pergamon Press, 1957.

Ehrlich, Paul, and J. Morgenroth. "Contributions to the Theory of Lysin Action." In *The Collected Papers of Paul Ehrlich: Immunology and Cancer Research*, edited by F. Himmelweit, 150–55. London: Pergamon Press, 1957.

Elliott, Kevin. "Varieties of Exploratory Experimentation in Nanotoxicology." *History and Philosophy of the Life Sciences* 29 (2007): 313–60.

Elzen, Boelie. "Two Ultracentrifuges: A Comparative Study of the Social Construction of Artefacts." *Social Studies of Science* 16 (1986): 621–62.

Engelhardt, Dietrich von. "Kausalität und Konditionalität in der modernen Medizin." In *Pathogenese: Grundzüge und Perspektiven einer theoretischen Pathologie*, ed. Heinrich Schipperges, 32–58. Berlin: Springer, 1985.

Fairbairn, Donald. "The Phospholipase of the Venom of the Cottonmouth Moccasin (Agkistrodon Piscivorus L.)." *Journal of Biological Chemistry* 157 (1945): 633–44.

Faraday, Michael. *Chemical Manipulation: Being Instructions to Students in Chemistry, on the Methods of Performing Experiments of Demonstration or of Research, with Accuracy and Success.* London: John Murray, 1827.

————. "Experimental Researches in Electricity: Tenth Series." *Philosophical Transactions of the Royal Society of London* 125 (1835): 263–74.

Feyerabend, Paul. *Against Method: Outline of an Anarchistic Theory of Knowledge.* 3rd ed. London: Verso, 1993. 1978.

Figlio, Karl M. "The Historiography of Scientific Medicine: An Invitation to the Human Sciences." *Comparative Studies in Society and History* 19 (1977): 262–86.

Findlen, Paula. "Controlling the Experiment: Rhetoric, Court Patronage and the Experimental Method of Francesco Redi." *History of Science* 31 (1993): 35–64.

————. *Possessing Nature: Museums, Collecting, and Scientific Culture in Early Modern Italy.* Berkeley: University of California Press, 1994.

Fisher, Ronald A. *The Design of Experiments.* Edinburgh: Oliver and Boyd, 1935.

Fleming, Donald. *William H. Welch and the Rise of Modern Medicine.* Boston: Little, Brown and Company, 1954.

Flexner, Abraham. *Medical Education in the United States and Canada: A Report to the Carnegie Foundation for the Advancement of Teaching.* New York: The Carnegie Foundation for the Advancement of Teaching, 1910.

Flexner, James Thomas. *An American Saga: The Story of Helen Thomas and Simon Flexner.* Boston: Little, Brown and Company, 1983.

Flexner, Simon. "A Half Century of American Medicine." *Science* 85 (1937): 505–12.

———. "Infection and Intoxication." In *Biological Lectures Delivered at the Marine Biological Laboratory of Wood's Holl in the Summer Session of 1885*. Boston: Ginn & Company, 1886.

———. "Original Address: The Pathologic Changes Caused by Certain So-Called Toxalbumins." *Medical News* 65 (1894): 116–24.

———. "The Regeneration of the Nervous System of Planaria Torva and the Anatomy of the Nervous System of Doubleheaded Forms." *Journal of Morphology* 14 (1897): 337–46.

———. "Tendencies in Pathology." *Science* 27 (1908): 128–36.

———. "Triumphs of Experimental Medicine." *Scientific Monthly* 36 (1933): 512–15.

Flexner, Simon, and Hideyo Noguchi. "The Constitution of Snake Venom and Snake Sera." *University of Pennsylvania Medical Bulletin* (1902): 345–62.

———. "On the Plurality of Cytolysins in Snake Venom." *Journal of Pathology* (1905): 111–24.

———. "Snake Venom in Relation to Haemolysis, Bacteriolysis, and Toxicity." *The Journal of Experimental Medicine* 6 (1902): 277–301.

———. "Upon the Production and Properties of Anti-Crotalus Venin." *Journal of Medical Research* VI (1904): 363–76.

Fontana, Felice. "Account of the Airs Extracted from Different Kinds of Waters; with Thoughts on the Salubrity of Air at Different Places." *Philosophical Transactions of the Royal Society of London* 69 (1779): 432–53.

———. *Traité sur le vénin de la vipère, sur les poisons americains, sur le laurier-cerise, et sur quelques autres poisons végétaux*. Vol. I. Florence: 1781.

———. *Treatise on the Venom of the Viper, on the American Poisons, and on the Cherry-Laurel, and Some Other Vegetable Poisons. To Which Are Annexed Observations of the Primitive Structure of the Animal Body, Different Experiments on the Reproduction of the Nerves, and a Description of a New Canal of the Eye*. Translated by Joseph Skinner. Vol. I. London: J. Murray, 1787. 1781.

———. *Treatise on the Venom of the Viper, on the American Poisons, and on the Cherry-Laurel, and Some Other Vegetable Poisons. To Which Are Annexed Observations of the Primitive Structure of the Animal Body, Different Experiments on the Reproduction of the Nerves, and a Description of a New Canal of the Eye*. Translated by Joseph Skinner. Vol. II. London: J. Murray, 1787. 1781.

Forster, William D. *A History of Medical Bacteriology and Immunology*. London: Heinemann, 1970.

Fraenkel-Conrat, Heinz. "Early Days of Protein Chemistry." *The FASEB Journal* 8 (1994): 452–53.

"Francesco Redi. Some Observations on Vipers." *Philosophical Transactions of the Royal Society of London* (1665–1666): 160–62.

Frank, Robert G. "American Physiologists in German Laboratories, 1865–1914." In *Physiology in the American Context, 1850–1940*, edited by Gerald Geison, 11–46. Bethesda, MD: American Physiological Society, 1987.

———. "The Telltale Heart: Physiological Instruments, Graphic Methods, and Clinical Hopes 1854–1914." In *The Investigative Enterprise: Experimental Physiology in Nineteenth-Century Medicine*, edited by William Coleman and Frederic L. Holmes, 211–70. Berkeley: University of California Press, 1988.

Franklin, Allan. "The Epistemology of Experiment." In *The Uses of Experiment: Studies in the Natural Sciences*, edited by David Gooding, Trevor Pinch, and Simon Schaffer, 437–60. Cambridge, UK: Cambridge University Press, 1989.

———. "Experiment in Physics." In *The Stanford Encyclopedia of Philosophy (Spring 2010 Edition)*, edited by Edward N. Zalta, 2010. http://plato.stanford.edu/archives/spr2010/entries/physics-experiment/.

———. *Experiment, Right or Wrong*. Cambridge, UK: Cambridge University Press, 1990.

———. *The Neglect of Experiment*. Cambridge, UK: Cambridge University Press, 1986.

Franklin, Allan, and Colin Howson. "Comment on 'the Structure of a Scientific Paper' by Frederick Suppe." *Philosophy of Science* 65 (1998): 411–16.

———. "It Probably Is a Valid Experimental Result: A Bayesian Approach to the Epistemology of Experiment." *Studies in History and Philosophy of Science* 19 (1988): 419–27.

Franklin, Allan, and Slobodan Perovic. "Experiment in Physics." In *The Stanford Encyclopedia of Philosophy*, edited by Edward N. Zalta, http://plato.stanford.edu/archives/sum2015/entries/physics-experiment/. Stanford, CA: Metaphysics Research Lab, 2015.

Frey, Heinrich. *The Microscope and Microscopical Technology*. New York: William Wood & Co, 1872.

Fruton, Joseph. "Early Theories of Protein Structure." *Annals of the New York Academy of Sciences* (1979): 1–18.

———. *Proteins, Enzymes, Genes: The Interplay of Chemistry and Biology*. New Haven, CT: Yale University Press, 1999.

Fye, W. Bruce. *The Development of American Physiology: Scientific Medicine in the Nineteenth Century*. Baltimore, MD: Johns Hopkins University Press, 1987.

———. "S. Weir Mitchell, Philadelphia's 'Lost' Physiologist." *Bulletin of the History of Medicine* 57 (1983): 188–202.

Galilei, Galileo. *Dialogues Concerning Two New Sciences*. New York: Macmillan, 1914.

Galison, Peter. *How Experiments End*. Chicago: University of Chicago Press, 1987.

Gardner, John Lane. *Military Control*. A. B. Claxton & Company, 1839.

Geison, Gerald. "Divided We Stand: Physiologists and Clinicians in the American Context." In *The Therapeutic Revolution: Essays in the Social History of American Medicine*, edited by M. J. Vogel and Charles E. Rosenberg, 67–90. Philadelphia: University of Pennsylvania Press, 1979.

Gigerenzer, Gerd. "Survival of the Fittest Probabilist: Brunswik, Thurstone, and the Two Disciplines of Psychology." In *The Probabilistic Revolution: Ideas in Science*, edited by Lorenz Krüger, Gerd Gigerenzer, and Mary S. Morgan. Cambridge, MA: MIT Press, 1989.

Gmelin, Johann Friedrich. "27tes Stück. Den 6. Jul 1782. Florenz." *Zugabe zu den Göttingischen Anzeigen von gelehrten Sachen* (1782): 417–31.

Goetz, Christopher. "Jean Martin Charcot and Silas Weir Mitchell." *Neurology* 48 (1997): 1128–32.

Golinski, Jan. *Science as Public Culture: Chemistry and Enlightenment in Britain, 1760–1820*. Cambridge, UK: Cambridge University Press, 1992.

Gopen, George, and Judith Swan. "The Science of Scientific Writing." *American Scientist* (1990): 550–58.

Gosh, B. N., and S. S. De. "Proteins of Rattlesnake Venom." *Nature* 143 (1939): 380–81.

Gradmann, Christoph. "Alles eine Frage der Methode. Zur Historizität der Koch-schen Postulate 1840–2000." *Medizinhistorisches Journal* 43 (2008): 121–48.

———. "Isolation, Contamination, and Pure Culture: Monomorphism and Polymorphism of Pathogenic Micro-Organisms as Research Problem 1860–1880." *Perspectives on Science* 9 (2001): 147–72.

———. *Krankheit im Labor: Robert Koch und die medizinische Bakteriologie*. Göttingen: Wallstein Verlag, 2005.

Gradmann, Christoph, and Jonathan Simon, eds. *Evaluating and Standardizing Therapeutic Agents, 1890–1950*. New York: Palgrave Macmillan, 2010.

Gralén, Nils, and The Svedberg. "The Molecular Weight of Crotoxin." *Biochemical Journal* 32 (1938): 1375–77.

Grasset, E., T. Brechbuhler, D. E. Schwartz, and E. Pongratz. "Comparative Analysis and Electrophoretic Fractionations of Snake Venoms." In *Venoms*, edited by E. E. Buckley and N. Porges, 153–69. Washington, DC: American Association of the Advancement of Science, 1956.

Greene, Harry. *Snakes: The Evolution of Mystery in Nature*. Berkeley: University of California Press, 1997.

Grmek, Mirko. "Bernard, Claude." In *Dictionary of Scientific Biography*, edited by C. Gillespie, 24–34. New York: Scribner, 1970–1980.

Gross, Alan G., Joseph E. Harmon, and Michael Reidy. *Communicating Science: The Scientific Article from the 17th Century to the Present*. Oxford: Oxford University Press, 2002.

Guerrini, Anita. "Archibald Pitcairne and Newtonian Medicine." *Medical History* 31 (1987): 70–83.

———. *The Courtiers' Anatomists: Animals and Humans in Louis XIV's Paris*. Chicago: University of Chicago Press, 2015.

———. "Ether Madness: Newtonianism, Religion, and Insanity in Eighteenth-Century England." In *Action and Reaction: Proceedings of a Symposion to Commemorate the Tercenary of Newton's Principia*, edited by Paul Theerman and Adele Seeff, 232–54. London: Associated University Presses, 1993.

————. *Experimenting with Humans and Animals: From Galen to Animal Rights.* Baltimore: Johns Hopkins University Press, 2003.

Habermann, Ernst, and Wilhelm Neumann. "Beiträge zur Charakterisierung der wirksamen Komponenten von Schlangengiften." *Archiv für experimentelle Pathologie und Pharmakologie* 223 (1954): 388–98.

Habermann, Ernst, and K. Rübsamen. "Biochemical and Pharmacological Analysis of the So-Called Crotoxin." In *Toxins of Animal and Plant Origin*, edited by André De Vries and Elazar Kochva, 333–41. New York: Gordon and Breach, 1973.

Hacking, Ian. "Disunities in the Sciences." In *The Disunity of Science: Boundaries, Contexts, and Power*, edited by Peter Galison and David Stump, 37–74. Stanford, CA: Stanford University Press, 1996.

————. *The Emergence of Probability: A Philosophical Study of Early Ideas About Probability, Induction and Statistical Inference.* Cambridge, UK: Cambridge University Press, 1984.

————. *Representing and Intervening.* Cambridge, UK: Cambridge University Press, 1983.

————. "Styles of Scientific Thinking or Reasoning: A New Analytical Tool for Historians and Philosophers of the Sciences." In *Trends in the Historiography of Science*, edited by K. Gavroglu, J. Christianidis, and N. Nicolaidis, 31–48. Dordrecht: Kluwer Academic Publishers, 1994.

————. *The Taming of Chance.* Cambridge, UK: Cambridge University Press, 1990.

————. "Telepathy: Origins of Randomization in Experimental Design." *Isis* 79 (1988): 427–51.

Hahn, Roger. *The Anatomy of a Scientific Institution: The Paris Academy of Sciences, 1666–1803.* Berkeley: University of California Press, 1971.

Haller, Albrecht von. *A Dissertation on the Sensible and Irritable Parts of Animals.* London: Printed for J. Nourse at the Lamb opposite Katherine-street in the Strand, 1755.

Hammond, William, and S. Weir Mitchell. "On the Physical and Chemical Characterization of Corroval and Vao, Two Recently Discovered Varieties of Woorara, and on a New Alkaloid Constituting Their Active Principle." *Proceedings of the Academy of Natural Sciences of Philadelphia* (1860); 4–9.

Hawgood, Barbara J. "Karl Heinrich Slotta (1895–1987) Biochemist: Snakes, Pregnancy and Coffee." *Toxicon* 39 (2001): 1277–82.

Hendon, Robert A., and Allan L. Bieber. "Presynaptic Toxins from Rattlesnake Venom." In *Rattlesnake Venoms: Their Actions and Treatment*, edited by Anthony Tu, 211–46. New York: Marcel Dekker, 1982.

Hendon, Robert A., and Heinz Fraenkel-Conrat. "Biological Roles of the Two Components of Crotoxin." *Proceedings of the National Academy of Sciences of the United States of America* 68 (1971): 1560–63.

Herschel, John. *A Preliminary Discourse of the Study of Natural Philosophy.* London: Longman, 1830.

Hoffmann, Christoph. "The Ruin of a Book: Jean Andre De Luc's *Recherches sur les modifications de l'atmosphère* (1772)." *Modern Language Notes* 118 (2003): 568–602.

Hollingsworth, J. Rogers. "Institutionalizing Excellence in Biomedical Research: The Case of the Rockefeller University." In *Creating a Tradition of Biomedical Research: Contributions to the History of the Rockefeller University*, edited by Darwin H. Stapleton, 17–64. New York: Rockefeller University Press, 2004.

Holmes, Frederic L. "Argument and Narrative in Scientific Writing." In *The Literary Structure of Scientific Argument: Historical Studies*, edited by Peter Dear, 164–81. Philadelphia: University of Pennsylvania Press, 1991.

———. "Scientific Writing and Scientific Discovery." *Isis* 78 (1987): 220–35.

Hudson, Robert G. "Mesosomes: A Study in the Nature of Experimental Reasoning." *Philosophy of Science* 66 (1999): 289–309.

———. *Seeing Things: The Philosophy of Reliable Observation*. Oxford: Oxford University Press, 2014.

Hughes, Thomas P. *American Genesis: A History of the American Genius for Invention*. New York: Penguin Books, 1989.

Hüntelmann, Axel C. "Evaluation as a Practical Technique of Administration: The Regulation and Standardization of Diphtheria Serum." In *Evaluating and Standardizing Therapeutic Agents, 1890–1950*, edited by Christoph Gradmann and Jonathan Simon, 31–51. New York: Palgrave Macmillan, 2010.

———. "Pharmaceutical Markets in the German Empire. Profits between Risk, Altruism and Regulation." *Historical Social Research* 36 (2011): 182–201.

Hunter, Michael, ed. *Robert Boyle Reconsidered*. Cambridge, UK: Cambridge University Press, 1994.

Jardine, Nicholas. "The Laboratory Revolution in Medicine as Rhetorical and Aesthetic Accomplishment." In *The Laboratory Revolution in Medicine*, edited by Andrew Cunningham and Perry Williams, 304–23. Cambridge, UK: Cambridge University Press, 1992.

Kalb, Stefanie. *Wilhelm Neumann (1898–1965): Leben und Werk unter besonderer Berücksichtigung seiner Rolle in der Kampfstoff-Forschung*. Univ. diss., University of Würzburg, 2005.

Kay, Lily. "Laboratory Technology and Biological Knowledge: The Tiselius Electrophoresis Apparatus, 1930–1945." *History and Philosophy of the Life Sciences* 10 (1988): 51–72.

———. *The Molecular Vision of Life. Caltech, The Rockefeller Fouindation, and the Rise of the New Biology*. Oxford: Oxford University Press, 1993.

Keating, Peter, and Alberto Cambrosio. *Cancer on Trial*. Chicago: University of Chicago Press, 2012.

Kellaway, C. H. "Animal Poisons." *Annual Review of Biochemistry* 8 (1939): 541–56.

Kempner, W., and E. Schepilewsky. "Ueber antitoxische Substanzen gegenüber dem Botulismusgift." *Zeitschrift für Hygiene und Infektionskrankheiten* 27 (1898): 213–22.

Kita, Atsushi. *Dr. Noguchi's Journey: A Life of Medical Search and Discovery*. Tokyo: Kodansha International, 2003.

Klauber, Laurence. "Foreword." *Toxicon* 1 (1962): 1–3.

Klobusitzky, Dionysius von, and P. König. "Biochemische Studien über die Gifte der Schlangengattung Bothrops. VI. Mitteilung: Kurzer Bericht über verschiedene, in den Jahren 1936–37 gewonnene Versuchsergebnisse. " *Naunyn-Schmiedebergs Archiv für experimentelle Pathologie und Pharmakologie* (1939): 271–75.

Knoefel, Peter K. *Felice Fontana: Life and Works*. Trento, Italy: Società di Studi Trentini di Scienze Storiche, 1984.

———. *Francesco Redi on Vipers*. Leiden, Netherlands: Brill, 1988.

Knorr-Cetina, Karin. *The Manufacture of Knowledge*. Oxford: Pergamon Press, 1981.

Kohler, Robert E. "The Enzyme Theory and the Origin of Biochemistry." *Isis* 64 (1973): 181–96.

Kuhn, Thomas S. *The Structure of Scientific Revolutions*. 2nd ed. Chicago: University of Chicago Press, 1970.

Kyes, Preston. "Ueber die Wirkungsweise des Cobragiftes." In *Gesammelte Arbeiten zur Immunitätsforschung*, edited by Paul Ehrlich, 413–41. Berlin: August Hirschwald, 1904.

Lampland, Martha, and Susan Leigh Star. *Standards and Their Stories: How Quantifying, Classifying and Formalizing Practices Shape Everyday Life*. Ithaca, NY: Cornell University Press, 2009.

Landecker, Hannah. *Culturing Life: How Cells Became Technologies*. Cambridge, MA: Harvard University Press, 2007.

Landois, Leonard. *A Textbook of Human Physiology, Including Histology and Microscopical Anatomy*. 3rd American ed. Philadelphia, PA: P. Blakiston, Son & Company, 1889.

Latour, Bruno. *Science in Action: How to Follow Scientists and Engineers through Society*. Cambridge, MA: Harvard University Press, 1987.

Latour, Bruno, and Steve Woolgar. *Laboratory Life: The Social Construction of Scientific Facts*. 2nd ed. Beverly Hills, CA: Sage, 1986. 1979.

Laudan, Larry. *Science and Hypothesis*. Dordrecht: Reidel, 1981.

———. "Why Was the Logic of Discovery Abandoned?" In *Scientific Discovery*, edited by Thomas Nickles, 173–83. Dordrecht: Reidel, 1980.

Lawrence, Christopher. "Incommunicable Knowledge: Science, Technology and the Clinical Art in Britain 1850–1914." *Journal of Contemporary History* 20 (1985): 503–20.

Leigh, Robert Adam. *"On Theriac to Piso," Attributed to Galen: A Critical Edition with Translation and Commentary*. PhD diss., University of Exeter, 2013.

Lenoir, Timothy. "A Magic Bullet: Research for Profit and the Growth of Knowledge in Germany around 1900." *Minerva* 26 (1988): 66–88.

Leonelli, Sabina. "Special Issue on Data-Driven Research in the Biological and

Biomedical Sciences." *Studies in History and Philosophy of Biological and Biomedical Sciences* 43 (2012).

Lesch, John E. *Science and Medicine in France: The Emergence of Experimental Physiology, 1790–1855*. Cambridge, MA: Harvard University Press, 1984.

———. "Conceptual Change in an Empirical Science: The Discovery of the First Alkaloids." *Historical Studies in the Physical Sciences* 11 (1981): 305–28.

Levy, Joel. *Poison: A Social History*. Stroud: The History Press, 2011.

Li, Choh Hao, and Heinz Fraenkel-Conrat. "Electrophoresis of Crotoxin." *Journal of the American Chemical Society* 64 (1942): 1586–88.

Liebenau, Jonathan. "Paul Ehrlich as a Commercial Scientist and Research Administrator." *Medical History* 34 (1990): 65–78.

Lind, James. *Treatise on the Scurvy*. 2nd ed. London: 1757.

Loeb, Jacques. "On Some Facts and Principles of Physiological Morphology." In *Biological Lectures Delivered at the Marine Biological Laboratory of Wood's Holl in the Summer Session of 1893*, 37–62. Boston: Ginn & Company, 1894.

———. *Studies in General Physiology*. Chicago: University of Chicago Press, 1905.

———. *Untersuchungen zur physiologischen Morphologie der Thiere*. Würzburg: Hertz, 1891.

Louis, Pierre Charles Alexandre. *Researches on the Effects of Bloodletting in Some Inflammatory Diseases, and on the Influence of Tartarized Antimony and Vesication in Pneumonitis*. Boston: Hilliard, Gray, and Company, 1836.

Machamer, Peter, Lindley Darden, and Carl F. Craver. "Thinking about Mechanisms." *Philosophy of Science* 67 (2000): 1–25.

Maehle, Andreas-Holger. *Drugs on Trial: Experimental Pharmacology and Therapeutic Innovation in the Eighteenth Century*. Amsterdam: Rodopi, 1999.

———. *Kritik und Verteidigung des Tierversuchs*. Stuttgart: Franz Steiner Verlag, 1990.

Maerker, Anna. *Model Experts: Wax Anatomies and Enlightenment in Florence and Vienna, 1775–1815*. Manchester: Manchester University Press, 2011.

Magendie, François. *Lectures on the Blood and on the Changes Which It Undergoes during Disease*. Philadelphia, PA: Haswell, Barrington, and Haswell, 1839.

Maienschein, Jane. "Arguments for Experimentation in Biology." *PSA: Proceedings of the 1986 Biennial Meeting of the Philosophy of Science Association* 2 (1987): 180–95.

———. *Defining Biology*. Cambridge, MA: Harvard University Press, 1986.

———. "The Origins of Entwicklungsmechanik." In *A Conceptual History of Modern Embryology*, edited by Scott Gilbert, 43–61. Baltimore: Johns Hopkins University Press, 1994.

———. "Physiology, Biology, and the Advent of Physiological Morphology." In *Physiology in the American Context, 1850–1940*, edited by Gerald Geison, 177–94. Bethesda, MD: American Physiological Society, 1987.

Mantell, Gideon Algernon. "On the Structure of the Jaws and Teeth of the Iguanodon." *Philosophical Transactions of the Royal Society of London* 138 (1848): 183–202.

Marks, Harry M. *The Progress of Experiment: Science and Therapeutic Reform in the United States, 1900–1990.* Cambridge, UK: Cambridge University Press, 1997.

Martin, H. Newell. "The Study and Teaching of Biology. An Introductory Lecture Delivered at Johns Hopkins University, October 23, 1876." *Popular Science Monthly* 10 (1877): 298–309.

Matthews, Rosser. *Quantification and the Quest for Medical Certainty.* Princeton, NJ: Princeton University Press, 1995.

Matty, Matthew. *Authentic Memoirs of the Life of Richard Mead, M.D.* London: Printed for J. Whiston, and B. White in Fleet-Street, 1755.

Maxwell, J. Clerk. "On Governors." *Proceedings of the Royal Society of London* 16 (1867): 270–83.

Mayo, Deborah G. *Error and the Growth of Experimental Knowledge.* Chicago: University of Chicago Press, 1996.

Mayo, Deborah G., and Aris Spanos. *Error and Inference: Recent Exchanges on Experimental Reasoning, Reliability, and the Objectivity and Rationality of Science.* Cambridge, UK: Cambridge University Press, 2010.

McCalla, Albert. "President's Address: The Verification of Microscopic Observation." *Proceedings of the American Society of Microscopists* 5 (1883): 1–19.

McCartney, Eugene S. "A Control Experiment in Antiquity." *The Classical Weekly* 36 (1942): 5–6.

McPherson, Alexander. "A Brief History of Protein Crystal Growth." *Journal of Crystal Growth* 110 (1991): 1–10.

McVaugh, Michael. "The 'Experience-Based Medicine' of the Thirteenth Century." In *Evidence and Interpretation in Studies on Early Science and Medicine*, edited by Edith Sylla and William R. Newman, 105–30. Leiden: Brill, 2009.

Mead, Richard. *A Mechanical Account of Poisons.* 4th ed. London: J, Brindley, 1745.

———. *A Mechanical Account of Poisons.* London: 1702.

———. *A Short Discourse Concerning Pestilential Contagion.* 8th ed. London: 1722.

———. *A Short Discourse Concerning Pestilential Contagion.* 9th ed. London: 1744.

Meade, Richard H. *In the Sunshine of Life; a Biography of Dr. Richard Mead, 1673–1754.* Philadelphia, PA: Dorrance, 1974.

Mellish, Maud. *The Writing of Medical Papers.* Philadelphia, PA: W. B. Saunders, 1922.

Ménez, André. *The Subtle Beast: Snakes, from Myth to Medicine.* London: Taylor and Francis, 2003.

Middleton, W. E. Knowles. *The Experimenters: A Study of the Accademia Del Cimento.* Baltimore: Johns Hopkins University Press, 1971.

Mill, John Stuart. *A System of Logic*. Vol. VII of *Collected Works*. Indianapolis, IN: Liberty Fund, 2006.

Mitchell, S. Weir. "Experimental Contributions to the Toxicology of Rattle-Snake Venom." *The New York Medical Journal* (1868): 289–322.

———. "Memoir of John Call Dalton, 1825–1889." *Biographical Memoirs of the National Academy of Sciences* 3 (1890): 178–85.

———. *On the Treatment of Rattlesnake Bites, with Experimental Criticisms Upon the Various Remedies Now in Use*. Philadelphia, PA: Lippincott & Co., 1861.

———. "The Poison of the Rattlesnake." *The Atlantic Monthly* (1868): 452–61.

———. *Researches on the Venom of the Rattlesnake: With an Investigation of the Anatomy and Physiology of the Organs Concerned*. Philadelphia, PA: Smithsonian Institution, 1860.

Mitchell, S. Weir, and Edward T. Reichert. *Researches upon the Venoms of Poisonous Serpents*. Washington, DC: Smithsonian Institution, 1886.

Mitchell, S. Weir, and Alonzo H. Stewart. "A Contribution to the Study of the Action of the Venom of the Crotalus Adamanteus Upon the Blood." *Transactions of the College of Physicians of Philadelphia* 19 (1897): 105–10.

Myers, Greg. *Writing Biology: Texts in the Social Construction of Scientific Knowledge*. Madison: University of Wisconsin Press, 1990.

Neumann, Wilhelm Paul. "Chromatographische Trennung von 'Crotoxin' in zwei verschiedene Wirkstoffe." *Die Naturwissenschaften* (1955): 370–76.

Neumann, Wilhelm Paul, and Ernst Habermann. "Ueber Crotactin, das Haupttoxin des Giftes der Brasilianischen Klapperschlange (Crotalus terrificus terrificus)." *Biochemische Zeitschrift* 327 (1955): 170–85.

Newman, William R. *Atoms and Alchemy: Chymistry and the Experimental Origins of the Scientific Revolution*. Chicago: University of Chicago Press, 2006.

Newton, Isaac. "A Letter of Mr. Isaac Newton, Professor of the Mathematicks in the University of Cambridge; Containing His New Theory About Light and Colors." *Philosophical Transactions of the Royal Society of London* 6 (1671): 3075–87.

Nickles, Thomas. "Beyond Divorce: Current Status of the Discovery Debate." *Philosophy of Science* 52 (1985): 177–206.

Nockels Fabbri, Christiane. "Treating Medieval Plague: The Wonderful Virtues of Theriac." *Early Science and Medicine* 12 (2007): 247–83.

Noguchi, Hideyo. *Snake Venoms: An Investigation of Venomous Snakes with Special Reference to the Phenomena of Their Venoms*. Washington, DC: The Carnegie Institution of Washington, 1909.

Numbers, Ronald L. "Science and Medicine." In *Wrestling with Nature: From Omens to Science*, edited by Peter Harrison, Ronald L. Numbers, and Michael H. Shank, 201–24. Chicago: University of Chicago Press, 2011.

Nuttall, George Henry Falkiner. "Experimente über die bacterienfeindlichen Einflüsse des thierischen Körpers." *Zeitschrift für Hygiene* 4 (1888): 353–93.

Nutton, Vivian. "The Reception of Fracastoro's Theory of Contagion: The Seed That Fell among Thorns?" *Osiris* 6 (1990): 196–234.

———. "The Seeds of Disease: An Explanation of Contagion and Infection from the Greeks to the Renaissance." *Medical History* 27 (1983): 1–34.

Olesko, Kathryn. *Physics as a Calling: Discipline and Practice in the Königsberg Seminar for Physics.* Ithaca, NY: Cornell University Press, 1991.

Olesko, Kathryn, and Frederic L. Holmes. "The Images of Precision: Helmholtz and the Graphical Method in Physiology." In *The Values of Precision*, edited by Norton Wise, 198–221. Princeton, NJ: Princeton University Press, 1995.

O'Malley, Maureen. "Exploratory Experimentation and Scientific Practice: Metagenomics and the Proteorhodopsin Case." *History and Philosophy of the Life Sciences* 29 (2007): 337–60.

Orfila, Matteu. *A General System of Toxicology, or, a Treatise on Poisons, Found in the Mineral, Vegetable, and Animal Kingdoms, Considered in Their Relations with Physiology, Pathology, and Medical Jurisprudence.* Philadelphia, PA: M. Carey & Sons, 1817.

"Osservationi Intorno Alle Vipere, fatte da Franceso Redi, in Firenze." *Journal des Sçavants* 2 (1666): 9–12.

Otis, Laura. *Membranes: Metaphors of Invasion in Nineteenth-Century Literature, Science and Politics.* Baltimore: Johns Hopkins University Press, 1999.

Pagel, Walter. "Van Helmont's Concept of Disease—to Be or Not to Be? The Influence of Paracelsus." *Bulletin of the History of Medicine* 46 (1972): 419–54.

Parke, Emily C. "Flies from Meat and Wasps from Trees: Reevaluating Francesco Redi's Spontaneous Generation Experiments." *Studies in History and Philosophy of Biological and Biomedical Sciences* 45 (2014): 34–42.

Pauly, Philip J. *Biologists and the Promise of American Life: From Meriwether Lewis to Alfred Kinsey.* Princeton, NJ: Princeton University Press, 2000.

———. *Controlling Life: Jacques Loeb & the Engineering Ideal in Biology.* Oxford: Oxford University Press, 1987.

Pera, Marcello. *The Ambiguous Frog: The Galvani–Volta Controversy on Animal Electricity.* Princeton, NJ: Princeton University Press, 1992.

Pera, Marcello, and William R. Shea, eds. *Persuading Science: The Art of Scientific Rhetoric.* Canton, MA: Science History Publications, 1991.

Phisalix, Marie. "Moyse Charas et les Vipères au Jardin du Roy." *Archives du Muséum d'Histoire Naturelle* 12 (1935): 469–72.

Platt, Thomas. "An Extract of a Letter Written to the Publisher by Mr. Thomas Platt, from Florence, August 6, 1672. Concerning Some Experiments, There Made Upon Vipers, since Mons. Charas His Reply to the Letter Written by Signor Francesco Redi to Monsieur Bourdelot and Monsieur Morus." *Philosophical Transactions of the Royal Society of London* 7 (1672): 5060–66.

Plesset, Isabel R. *Noguchi and His Patrons.* Cranbury, NJ: Associated University Presses, 1980.

Porter, Theodore M. *The Rise of Statistical Thinking, 1820–1900*. Princeton, NJ: Princeton University Press, 1986.

———. *Trust in Numbers: The Pursuit of Objectivity in Science and Public Life*. Princeton, NJ: Princeton University Press, 1995.

Post Office Dept. City of Washington. *Letter of the Postmaster General and Opinion of the Attorney General in Reference to the Power of the Circuit Court for the District of Columbia to Control Executive Officers of the United States in the Performance of Their Official Duties, Also, an Exposition of the Reasons of the Postmaster General for Refusing to Execute a Part of the Award of the Solicitor in Favor of Messrs. Stockton, Stokes, and Others United States*. Washington, DC: Blair & Rives, 1837.

Prelli, Lawrence. *A Rhetoric of Science: Inventing Scientific Discourse*. Columbia: University of South Carolina Press, 1989.

"Preparation of Copy for the *Botanical Gazette*." *Botanical Gazette* 61 (1916): 337–40.

Principe, Larry. "Virtuous Romance and Romantic Virtuoso: The Shaping of Robert Boyle's Literary Style." *Journal of the History of Ideas* 56 (1995): 377–97.

Prüll, Cay-Rüdiger. "Paul Ehrlich's Standardization of Serum: *Wertbestimmung* and Its Meaning for Twentieth-Century Biomedicine." In *Evaluating and Standardizing Therapeutic Agents, 1890–1950*, edited by Christoph Gradmann and Jonathan Simon, 13–30. New York: Palgrave Macmillan, 2010.

Putnam, James. "On the Reliability of Marey's Tambour in Experiments Requiring Accurate Notations of Time." *Journal of Physiology* 2 (1879): 209–13.

Rader, Karen. *Making Mice: Standardizing Animals for American Biomedical Research*. Princeton, NJ: Princeton University Press, 2004.

Ragland, Evan. "Chymistry and Taste in the Seventeenth Century: Franciscus Dele Boë Sylvius as a Chymical Physician between Galenism and Cartesianism." *Ambix* 59 (2012): 1–21.

Ransom, W. H. "Observations on the Ovum of Osseous Fishes." *Philosophical Transactions of the Royal Society of London* 157 (1867), 431–501.

Rasmussen, Nicolas. "Facts, Artifacts, and Mesosomes: Practicing Epistemology with the Electron Microscope." *Studies in History and Philosophy of Science* 24, no. 2 (1993): 227–65.

———. *Picture Control: The Electron Microscope and the Transformation of Biology in America, 1940–1960*. Stanford, CA: Stanford University Press, 1997.

Redi, Francesco. *Experiments on the Generation of Insects*. Chicago: Open Court, 1909. 1688.

———. "Lettera Sopra Alcune Opposizioni Fatte Alle Sue Osservazioni Intorno Alle Vipere." In *Opere Di Francesco Redi*, irreg. Naples: A spese di R. Gessari, nella stamperia di A. Carfora, 1740–41.

———. "Osservazioni Intorno Alle Vipere." In *Opere Di Francesco Redi*, irreg. Naples: A spese di R. Gessari, nella stamperia di A. Carfora, 1740–41.

Reichert, Edward T. *The Differentiation and Specificity of Corresponding Proteins and Other Vital Substances in Relation to Biological Classification and Organic Evolution: The Crystallography of Hemoglobins.* Washington, DC: Carnegie Institution of Washington, 1909.

Reinhardt, Carsten. *Shifting and Rearranging: Physical Methods and the Transformation of Modern Chemistry.* Sagamore Beach, MA: Science History Publications, 2006.

"Report of the Lancet Special Commission on the Relative Strengths of Diphtheria Antitoxic Serums." *Lancet* (1896): 182–95.

"Review: An Accompt of Some Books." *Philosophical Transactions of the Royal Society of London* 4 (1669): 1086–97.

"Review: An Account of Some Books." *Philosophical Transactions of the Royal Society of London* 7 (1672): 4071–78.

Rheinberger, Hans-Jörg. *Toward a History of Epistemic Things.* Stanford, CA: Stanford University Press, 1997.

Roe, Shirley A. "Anatomia Animata: The Newtonian Physiology of Albrecht Von Haller." In *Transformation and Tradition in the Sciences: Essays in Honor of I. Bernard Cohen*, edited by Everett Mendelsohn, 273–300. Cambridge, UK: Cambridge University Press, 1984.

Rosenberg, Philip. "Pharmacology of Phospholipase A2 from Snake Venoms." In *Snake Venoms*, edited by Chen-Yuan Lee, 403–47. Berlin: Springer, 1979.

Rosenfeld, Louis. "A Golden Age of Clinical Chemistry: 1948–1960." *Clinical Chemistry* 46 (2000): 1705–14.

Ross, Karen Deane. "Making Medicine Scientific: Simon Flexner and Experimental Medicine at the Rockefeller Institute for Medical Research, 1901–1945." PhD diss., University of Minnesota, 2006.

Rous, Peyton. "Simon Flexner. 1863–1947." *Obituary Notices of Fellows of the Royal Society.* (1947).

Roux, Wilhelm. "Einleitung." *Archiv für Entwickelungsmechanik* 1 (1894): 1–42.

———. "The Problems, Methods, and Scope of Developmental Mechanics." In *Defining Biology*, edited by Jane Maienschein, 106–48. Cambridge, MA: Harvard University Press, 1986.

Rübsamen, Klaus, Henning Breithaupt, and Ernst Habermann. "Biochemistry and Pharmacology of the Crotoxin Complex. 1. Subfractionation and Recombination of the Crotoxin Complex." *Naunyn-Schmiedebergs Archiv der Pharmakologie* 270 (1971): 274–88.

Russell, Findlay E. "History of the International Society on Toxinology and Toxicon. I. The Formative Years, 1954–1965." *Toxicon* 25 (1987): 3–21.

———. *Snake Venom Poisoning.* Philadelphia, PA: J. B. Lippincott Company, 1980.

Ryan, Kathy. "APS at 125: A Look Back at the Founding of the American Physiological Society." *Advances in Physiological Education* 37 (2013): 10–14.

Sampaio, Sandra C., Stephen Hyslop, Marcos R.M. Fontes, Julia Prado-Franceschi, Vanessa O. Zambelli, Angelo J. Magro, Patricia Brigatte, Vanessa P. Gutierrez, and Yara Cury. "Crotoxin: Novel Activities for a Classic B-Neurotoxin." *Toxicon* 55 (2010): 1045–60.

Sargent, Rose-Mary. *The Diffident Naturalist: Robert Boyle and the Philosophy of Experiment*. Chicago: University of Chicago Press, 1995.

Schaffner, Kenneth. *Discovery and Explanation in Biology and Medicine*. Chicago: University of Chicago Press, 1993.

———. "Clinical Trials: The Validation of Theory and Therapy." In *Physics, Philosophy, and Psychoanalysis: Essays in Honor of Adolf Grünbaum*, edited by R. S. Cohen and Larry Laudan, 191–208. Boston: Reidel, 1983.

Schickore, Jutta. "'. . . as Many and as Diverse Methods as Possible Ought to Be Employed . . .'—Methodological Reflections in *General Cytology* in Historical Perspective." In *Visions of Cell Biology Reflecting on Cowdry's General Cytology*, edited by Karl Matlin, Jane Maienschein, and Manfred Laubichler. Chicago: University of Chicago Press, forthcoming.

———. "'Exploratory Experimentation' as a Probe into the Relation between Historiography and Philosophy of Science." *Studies in History and Philosophy of Science* 55 (2016): 20–26.

———. "Scientists' Methods Accounts: S. Weir Mitchell's Research on the Venom of Poisonous Snakes." In *Integrating History and Philosophy of Science: Problems and Prospects*, edited by Tad Schmaltz and Seymour Mauskopf (Dordrecht: Springer, 2011), 141–61.

———. "The Significance of Re-Doing Experiments: A Contribution to Historically Informed Methodology." *Erkenntnis* 75 (2011): 325–47.

———. "Test Objects." *History of Science* 47 (2009): 117–45.

———. "Trying Again and Again: Multiple Repetitions in Early Modern Reports of Experiments on Snake Bites." *Early Science and Medicine* 15 (2010): 567–617.

———. "What Does History Matter to Philosophy of Science? The Concept of Replication and the Methodology of Experiments." *Journal of the Philosophy of History* 5 (2012): 513–32.

Senebier, Jean. *L'art d'observer*. Vol. 2. Genève, Switzerland: Cl. Philiber & Bart. Chirol, 1775.

Severino, Marco Aurelio. *Vipera Pythia. Id Est De Viperae Natura, Veneno, Medicina Demonstrationes Et Experimenta Nova*. Padua, Italy, 1651.

Sewall, Henry. "Experiments on the Preventive Inoculation of Rattlesnake Venom." *Journal of Physiology* 8 (1887): 203–10.

Shapin, Steven. "Pump and Circumstance: Robert Boyle's Literary Technology." *Social Studies of Science* 14 (1984): 481–520.

———. *The Scientific Life: A Moral History of a Late-Modern Vocation*. Chicago: University of Chicago Press, 2008.

Shapin, Steven, and Simon Schaffer. *Leviathan and the Air-Pump: Hobbes, Boyle, and the Experimental Life.* Princeton, NJ: Princeton University Press, 1985.

Silverstein, Arthur M. "The Heuristic Value of Experimental Systems: The Case of Immune Hemolysis." *Journal of the History of Biology* 27 (1994): 437–47.

———. *A History of Immunology.* San Diego, CA: Academic Press, 1989.

Slotta, Karl Heinrich. "Further Experiments on Crotoxin." In *Venoms: Papers Presented at the First International Conference on Venoms December 27–30, 1954 at the Annual Meeting of the American Association of the Advancement of Science, Berkeley, California,* edited by E. E. Buckley and N. Porges, 253–58. Washington, DC: American Association of the Advancement of Science, 1956.

———. "Zur Chemie der Schlangengifte." *Experientia* 9 (1953): 81–88.

Slotta, Karl Heinrich, and Walter Forster. "Schlangengifte, IV. Mitteilung: Quantitative Bestimmung der schwefelhaltigen Bausteine." *Berichte der deutschen chemischen Gesellschaft* 71 (1938): 1082–88.

Slotta, Karl Heinrich, and Heinz Fraenkel-Conrat. "Crotoxin." *Nature* 144 (1939): 290–91.

———. "Schlangengifte, II. Mitteilung: Ueber die Bindungsart des Schwefels." *Berichte der deutschen chemischen Gesellschaft* 71 (1938): 264–71.

———. "Schlangengifte, III. Mitteilung: Reinigung und Krystallisation des Klapperschlangen-Giftes." *Berichte der deutschen chemischen Gesellschaft* 71 (1938): 1076–81.

———. "Two Active Proteins from Rattlesnake Venom." *Nature* 142 (1938): 213.

Smith, George E. "The Methodology of the *Principia.*" In *The Cambridge Companion to Newton,* edited by George E. Smith and I. Bernard Cohen, 138–73. Cambridge, UK: Cambridge University Press, 2002.

Smith, Hamilton. *Cotton and the Only Practical Methods Presented to Its Producers of Advancing and Controlling Its Price: Article IV, in July No. 1849, of De Bow's Commercial Review.* Louisville: Prentice and Weissinger, 1850.

Sprat, Thomas. *The History of the Royal-Society of London, for the Improving of Natural Knowledge.* London: Printed by T. R. for J. Martyn at the Bell without Temple-bar, and J. Allestry at the Rose and Crown in Duck-lane, Printers to the Royal Society, 1667.

Staden, Heinrich von. *Herophilus: The Art of Medicine in Early Alexandria.* Cambridge, UK: Cambridge University Press, 1989.

Stapleton, Darwin H., ed. *Creating a Tradition of Biomedical Research: Contributions to the History of the Rockefeller University.* New York: Rockefeller University Press, 2004.

Steinke, Hubert. *Irritating Experiments: Haller's Concept and the European Controversy on Irritability and Sensibility, 1750–90.* Amsterdam: Rodopi, 2005.

Steinle, Friedrich. "Entering New Fields: Exploratory Uses of Experimentation." *Philosophy of Science Supplement* 64 (1997): S65–S74.

———. "Experiments in History and Philosophy of Science." *Perspectives on Science* 10 (2002): 408–32.

———. *Explorative Experimente: Ampere, Faraday und die Ursprünge der Elektrodynamik.* Stuttgart: Franz Steiner Verlag, 2005.

———. "Newton, Newtonianism, and the Roles of Experiment." In *The Reception of Isaac Newton in Europe*, edited by Helmut Pulte and Scott Mandelbrote. Continuum, forthcoming.

Stigler, Stephen. "Gergonne's 1815 Paper on the Design and Analysis of Polynomial Regression Experiments." *Historia Mathematica* 1 (1974): 431–47.

Sumner, James B., and Karl Myrbäck. "Introduction." In *The Enzymes: Chemistry and Mechanism of Action*, edited by James B. Sumner and Karl Myrbäck, 1–27. New York: Academic Press, 1950.

Svedberg, The. "The Ultra-Centrifuge and the Study of High-Molecular Compounds." *Nature* 139 (1937): 1051–62.

Taton, René. "Bourdelot, Pierre Michon." In *Dictionary of Scientific Biography*, edited by Charles C. Gillispie. New York: Charles Scribner's Sons.

Teich, Mikuláš. "Ferment or Enzyme, What's in a Name?" *History and Philosophy of the Life Sciences* 3 (1981): 193–215.

Tribby, Jay. "Cooking (with) Clio and Cleo: Eloquence and Experiment in Seventeenth-Century Florence." *Journal of the History of Ideas* 52 (1991): 417–39.

Trizio, Emiliano. "Achieving Robustness to Confirm Controversial Hypotheses: A Case Study in Cell Biology." In *Characterizing the Robustness of Science*, edited by Léna Soler, Emiliano Trizio, Thomas Nickles, and William Wimsatt, 105–20. Boston: Springer, 2012.

Tröhler, Ulrich. " 'To Improve the Evidence of Medicine': Arithmetic Observation in Clinical Medicine in the Eighteenth and Early Nineteenth Centuries." *History and Philosophy of the Life Sciences* 10 (1988): 31–40.

Trumpler, Maria. "Verification and Variation: Patterns of Experimentation in Investigations of Galvanism in Germany, 1790–1800." *Philosophy of Science* 64 (1997): S75–S84.

Tu, Anthony. "The Mechanism of Snake Venom Actions—Rattlesnakes and Other Crotalids." In *Neuropoisons: Their Pathophysiological Actions*, edited by Lance L. Simpson, 87–109. New York: Plenum Press, 1971.

———, ed. *Rattlesnake Venoms, Their Actions and Treatment.* New York: M. Dekker, 1982.

Tyson, Edward. "Vipera Caudi-Sona Americana, or the Anatomy of a Rattle-Snake, Dissected at the Repository of the Royal Society in January 1682/3." *Philosophical Transactions of the Royal Society of London* 13 (1683): 25–54.

Ward, James. *The True Policy of Organising a System of Railways for India: A Letter to the Right Hon. The President of the Board of Control.* London: Smith, Elder, and Co., 1847.

Warner, John Harley. *Against the Spirit of System: The French Impulse in Nineteenth-Century American Medicine.* Princeton, NJ: Princeton University Press, 1998.

———. "The Fall and Rise of Professional Mystery: Epistemology, Authority, and the Emergence of Laboratory Medicine in Nineteenth-Century America." In *The Laboratory Revolution in Medicine*, edited by Andrew Cunningham and Perry Williams, 110–41. Cambridge, UK: Cambridge University Press, 1992.

———. "Ideals of Science and Their Discontents in Late Nineteenth-Century American Medicine." *Isis* 82 (1991): 454–78.

———. *The Therapeutic Perspective: Medical Practice, Knowledge, and Identity in America, 1820–1885*. Cambridge, MA: Harvard University Press, 1986.

Wassermann, A., and Kanehiro Takaki. "Ueber tetanusantitoxische Eigenschaften des normalen Centralnervensystems." *Berliner Klinische Wochenschrift* (1898): 5–6.

Watson, G. *Theriac and Mithridatium: A Study of Therapeutics*. London: Wellcome Historical Medical Library, 1966.

Welch, William H. "Adaptation in Pathological Processes." *The American Journal of the Medical Sciences* 113 (1897): 631–55.

———. "The Huxley Lecture on Recent Studies of Immunity, with Special Reference to Their Bearing on Pathology." *British Medical Journal* 2 (1902): 1105–14.

———. "On Some of the Humane Aspects of Medical Science." In *Papers and Addresses—Medical Education—History and Miscellaneous—Vivisection*, 3–8. Baltimore: 1892.

———. "Pathology in Its Relation to General Biology." *Johns Hopkins Hospital Bulletin* 1 (1890): 25–27.

———. "Some of the Advantages of the Union of Medical School and University." In *Papers and Addresses—Medical Education—History and Miscellaneous—Vivisection*, 26–40. Baltimore: Johns Hopkins University Press, 1892.

Whewell, William. *The Philosophy of the Inductive Sciences*. Vol. I of *The Historical and Philosophical Works of William Whewell*, 2nd rev. ed., edited by Gerd Buchdahl and Larry Laudan. London: Frank Cass, 1967. 1847.

———. *The Philosophy of the Inductive Sciences*. Vol. II of *The Historical and Philosophical Works of William Whewell*, 2nd rev. ed., edited by Gerd Buchdahl and Larry Laudan. London: Frank Cass, 1967. 1847.

Whitman, Charles O. "A Biological Farm. For the Experimental Investigation of Heredity, Variation and Evolution and for the Study of Life-Histories, Habits, Instincts and Intelligence." *Biological Bulletin* 3 (1902): 214–24.

Whitney, William Dwight. *The Century Dictionary and Cyclopedia, a Work of Universal Reference in All Departments of Knowledge with a New Atlas of the World*. Vol. II. New York: Century Company, 1897.

Wild, Markus. "Marin Cureau De La Chambre on the Natural Cognition of the Vegetative Soul: An Early Modern Theory of Instinct." *Vivarium* 46 (2008): 443–61.

"William Keith Brooks: A Sketch of His Life by Some of His Former Pupils and Associates." *Journal of Experimental Zoology* 9 (1910): 1–52.

Wood, Horatio C. *A Treatise on Therapeutics, Comprising Materia Medica and Toxicology*. Philadelphia, PA: Lippincott & Co., 1874.

Woodward, James. "Some Varieties of Robustness." *Journal of Economic Methodology* 13 (2006): 219–40.

Worboys, Michael. "Practice and the Science of Medicine in the Nineteenth Century." *Isis* 102 (2011): 109–15.

———. "Was There a Bacteriological Revolution in Late Nineteenth-Century Medicine?" *Studies in History and Philosophy of Biology and Biomedical Sciences* 38 (2007): 20–42.

Yeo, Richard, and John A. Schuster, eds. *The Politics and Rhetorics of Scientific Method.* Dordrecht: Reidel, 1986.

Zeller, E. Albert. "Enzymes as Essential Components of Bacterial and Animal Toxins." In *The Enzymes: Chemistry and Mechanism of Action*, edited by James B. Sumner and Karl Myrbäck, 986–1013. New York: Academic Press, 1951.

Zuckerman, Arnold. *Dr. Richard Mead (1673–1754): A Biographical Study.* PhD diss., University of Illinois, 1965.

———. "Plague and Contagionism in Eighteenth-Century England: The Role of Richard Mead." *Bulletin of the History of Medicine* 78 (2004): 273–308.

Index